DIGITAL SIGNAL PROCESSING USING MATLAB FOR STUDENTS AND RESEARCHERS

DIGITAL SIGNAL PROCESSING USING MATLAB FOR STUDENTS AND RESEARCHERS

JOHN W. LEIS
University of Southern Queensland

A JOHN WILEY & SONS, INC., PUBLICATION

Published by John Wiley & Sons, Inc., Hoboken, New Jersey.
Published simultaneously in Canada.

Library of Congress Cataloging-in-Publication Data:

Leis, John W. (John William), 1966-
Digital Signal Processsing Using MATLAB for Students and Researchers / John W. Leis.
p. cm
Includes bibliographical references and index.
ISBN 978-0-470-88091-3
1. Signal processing–Digital techniques. 2. Signal processing–Mathematics–Data processing. 3. MATLAB. I. Title.
TK5102.9.L4525 2011
621.382′2–dc22

2010048285

10 9 8 7 6 5 4 3 2

To Debbie, Amy, and Kate

CONTENTS

CHAPTER [illegible]

[illegible]

PREFACE

I was once asked what signal processing is. The questioner thought it had something to do with traffic lights. It became clear to me at that moment that although the theory and practice of signal processing in an engineering context has made possible the massive advances of recent times in everything from consumer electronics to healthcare, the area is poorly understood by those not familiar with digital signal processing (DSP). Unfortunately, such lack of understanding sometimes extends to those embarking on higher education courses in engineering, computer science, and allied fields, and I believe it is our responsibility not simply to try to cover every possible theoretical aspect, but to endeavor to open the student's eyes to the possible applications of signal processing, particularly in a multidisciplinary context.

With that in mind, this book sets out to provide the necessary theoretical and practical underpinnings of signal processing, but in a way that can be readily understood by the newcomer to the field. The assumed audience is the practicing engineer, the engineering undergraduate or graduate student, or the researcher in an allied field who can make use of signal processing in a research context. The examples given to introduce the topics have been chosen to clearly introduce the motivation behind the topic and where it might be applied. Necessarily, a great deal of detail has to be sacrificed in order to meet the expectations of the audience. This is not to say that the theory or implementation has been trivialized. Far from it; the treatment given extends from the theoretical underpinnings of key algorithms and techniques to computational and numerical aspects.

The text may be used in a one-term or longer course in signal processing, and the assumptions regarding background knowledge have been kept to a minimum. Shorter courses may not be able to cover all that is presented, and an instructor may have to sacrifice some breadth in order to ensure adequate depth of coverage of important topics. The sections on fast convolution and filtering, and medical image processing, may be omitted in that case. Likewise, recursive filter design via analog prototyping may be omitted or left to a second course if time does not permit coverage.

A basic understanding of algebra, polynomials, calculus, matrices, and vectors would provide a solid background to studying the material, and a first course in linear systems theory is an advantage but is not essential. In addition to the aforementioned mathematical background, a good understanding of computational principles and coding, and a working knowledge of a structured programming language is desirable, as is prior study of numerical mathematics. Above all, these should not be considered as a list of essential prerequisites; the reader who is lacking in some of these areas should not be deterred.

It is hoped that the problems at the end of each chapter, in conjunction with the various case studies, will give rise to a sufficiently rich learning environment, and appropriately challenging term projects may be developed with those problems as starting points.

John W. Leis

CHAPTER 1

WHAT IS SIGNAL PROCESSING?

1.1 CHAPTER OBJECTIVES

On completion of this chapter, the reader should

1. be able to explain the broad concept of digital signal processing (DSP);
2. know some of the key terms associated with DSP; and
3. be familiar with the conventions used in the book, both mathematical and for code examples.

1.2 INTRODUCTION

Signals are time-varying quantities which carry information. They may be, for example, audio signals (speech, music), images or video signals, sonar signals or ultrasound, biological signals such as the electrical pulses from the heart, communications signals, or many other types. With the emergence of high-speed, low-cost computing hardware, we now have the opportunity to analyze and process signals via computer algorithms.

The basic idea is straightforward: Rather than design complex circuits to process signals, the signal is first converted into a sequence of numbers and processed via software. By its very nature, software is more easily extensible and more versatile as compared with hard-wired circuits, which are difficult to change. Furthermore, using software, we can build in more "intelligence" into the operation of our designs and thus develop more human-usable devices.

A vitally important concept to master at the outset is that of an *algorithm*: the logical sequence of steps which must be followed in order to generate a useful result. Although this definition is applicable to general-purpose information processing, the key difference is in the nature of the data which are processed. In signal processing, the data sequence represents information which is not inherently digital and is usually imprecise.

Digital Signal Processing Using MATLAB for Students and Researchers, First Edition. John W. Leis.

For example, the algorithm for calculating the account balance in a person's bank account after a transaction deals directly with numbers; the algorithm for determining whether a sample fingerprint matches that of a particular person must cope with the imperfect and partially specified nature of the input data. It follows that the designer of the processing algorithm must understand the nature of the underlying sampled data sequence or *signal.*

Furthermore, many signal processing systems are what is termed ***real time***; that is, the result of the processing must be available within certain time constraints for it to be of use. If the result is not available in time, it may be of no use. For example, in developing a system which records the heart signal and looks for abnormalities, we may have a time frame of the order of seconds in which to react to any change in the signal pattern and to sound an alert.

The order of steps in the algorithm, and any parameters applicable to each step, must be decided upon by the designer. This may be done via theoretical analysis, experimentation using typical signal data or, more often, a combination of the two. Furthermore, the processing time of the algorithm must often be taken into account: A speech recognition system which requires several minutes (or even seconds) to convert a simple spoken text into words may not find much practical application (even though it may be useful for theoretical studies).

Signal processing technology relies on several fields, but the key ones are

Analog electronics to capture the real-world quantity and to preprocess it into a form suitable for further digital computer manipulation

Digital representations of the real world, which requires discrete sampling since the values of the real-world signal are sampled at predefined, discrete intervals, and furthermore can only take on predefined, discrete values

Mathematical analysis in order to find ways of analyzing and understanding complex and time-varying signals; the mathematics helps define the processing algorithms required.

Software algorithms in order to implement the equations described by the mathematics on a computer system

Some examples of real-world signal processing problems are presented later in this chapter.

1.3 BOOK OBJECTIVES

This book adopts a "hands-on" approach, with the following key objectives:

1. to introduce the field of signal processing in a clear and lucid style that emphasizes principles and applications in order to foster an intuitive approach to learning and
2. to present signal processing techniques in the context of "learning outcomes" based on self-paced experimentation in order to foster a self-directed investigative approach.

It is hoped that by adopting this "learn-by-doing" approach, the newcomer to the field will first develop an intuitive understanding of the theory and concepts, from which the analytical details presented will flow logically. The mathematical and algorithmic details are presented in conjunction with the development of each broad topic area, using real-world examples wherever possible. Obviously, in some cases, this requires a little simplification of all the details and subtleties of the problem at hand.

By the end of the book, the reader will

1. have an appreciation of where the various algorithmic techniques or "building blocks" may be used to address practical signal processing problems,
2. have sufficient insight to be able to develop signal processing algorithms for specific problems, and
3. be well enough equipped to be able to appreciate new techniques as they are developed in the research literature.

To gain maximum benefit from the presentation, it is recommended that the examples using MATLAB® be studied by the reader as they are presented. MATLAB is a registered trademark of The MathWorks, Inc. For MATLAB product information, please contact

The MathWorks, Inc.
3 Apple Hill Drive
Natick, MA, 01760-2098
Tel: 508-647-7000
Fax: 508-647-7101
E-mail: info@mathworks.com

MATLAB is discussed further in Chapter 2. The reader is encouraged to experiment by changing some of the parameters in the code given to see their effect. All examples in the book will run under the academic version of MATLAB, without additional "toolboxes."

1.4 DSP AND ITS APPLICATIONS

Application areas of signal processing have grown dramatically in importance in recent times, in parallel with the growth of powerful and low-cost processing circuits, and the reduction in price of computer memory. This has led, in turn, to many new applications, including multimedia delivery and handheld communication devices, with the convergence of computer and telecommunications technologies. However, many important applications of DSP may not be as immediately obvious:

Speech recognition provides a more natural interface to computer systems and information retrieval systems (such as telephone voice-response systems).

Image recognition involves recognizing patterns in images, such as character recognition in scanned text or recognizing faces for security systems, and handwriting recognition.

Image enhancement is the improvement of the quality of digital images, for example, when degraded by noise on a communications channel or after suffering degradation over time on older recording media.

Audio enhancement and noise reduction is the improvement of audio quality, particularly in "acoustically difficult" environments such as vehicles. In cars and planes, for example, this is a desirable objective in order to improve passenger comfort and to enhance safety.

Digital music in the entertainment industry uses special effects and enhancements—for example, adding three-dimensional sound "presence" and simulating reverberation from the surroundings.

Communications and data transmission relies heavily on signal processing. Error control, synchronization of data, and maximization of the data throughput are prime examples.

Biomedical applications such as patient monitoring are indispensable in modern medical practice. Medical image processing and storage continues to attract much research attention.

Radar, sonar, and military applications involve detection of targets, location of objects, and calculation of trajectories. Civilian applications of the Global Positioning System (GPS) are an example of complex signal processing algorithms which have been optimized to operate on handheld devices.

Note that these are all the subject of ongoing research, and many unsolved problems remain. These are also very difficult problems—consider, for example, speech recognition and the many subtle differences between speakers. Exact algorithmic comparison using computers relies on precise matching—never mind the differences between people: Our own speech patterns are not entirely repeatable from one day to the next!

1.5 APPLICATION CASE STUDIES USING DSP

Some application examples of DSP techniques are given in this section. It is certainly not possible to cover all possible aspects of DSP in the short space available nor to cover them in depth. Rather, the aim is to obtain a deeper insight into some problems which can be solved using DSP.

The following sections give two brief case studies of one-dimensional signal processing applications, where we have one sampled parameter varying with time, followed by two studies into two-dimensional signal processing applications.

1.5.1 Extracting Biomedical Signals

A great many applications of DSP exist in the medical field. Various measurement modalities—ultrasound, image, X-ray, and many others—are able to yield important

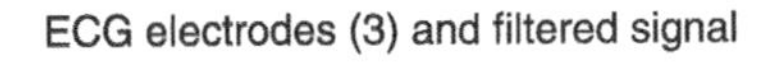

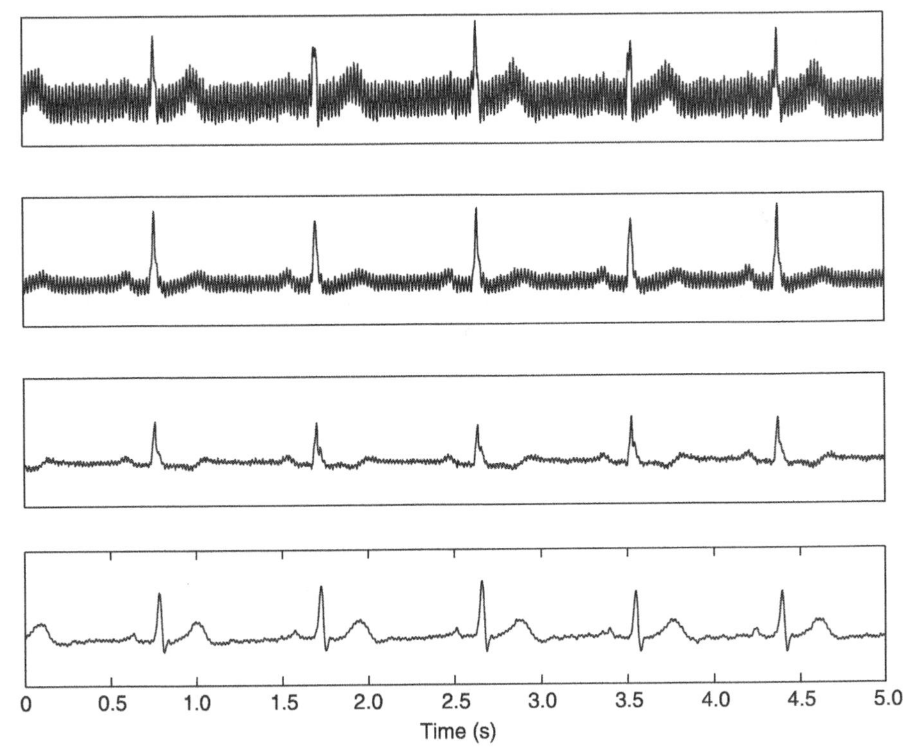

FIGURE 1.1 Three-lead electrocardiograph (ECG) signals. Note the significant amount of interference (noise) in the raw signals (top three traces).

diagnostic information to the clinician. In essence, the role of signal processing is to enhance the available measurements so as yield insight into the underlying signal properties. In some cases, the measurements may be compared to databases of existing signals so as to aid diagnosis.

Figure 1.1 shows the measurements taken in order to provide an electrocardiogram (ECG) signal, which is derived from a human heartbeat. The earliest ECG experiments were performed around a century ago, but the widespread application and use of the ECG was limited by the very small signal levels encountered. In the case illustrated in Figure 1.1, a so-called three-lead ECG uses external connections on the skin, to the wrists, and ankles. It may be somewhat surprising that such external connections can yield information about the internal body signals controlling the heart muscles, and indeed the signals measured are quite small (of the order of microvolts to millivolts). The large amount of amplification necessary means that noise measurement is also amplified. In addition, both the sampling leads and the body itself act as antennas, receiving primarily a small signal at the frequency of the main electricity supply. This, of course, is unwanted.

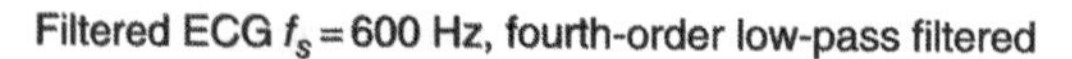

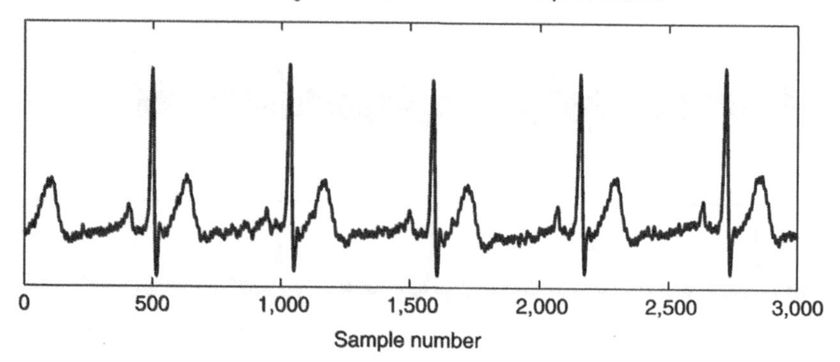

FIGURE 1.2 A filtered electrocardiograph (ECG) signal. Note that the horizontal axis is now shown as the sample number; hence, the sample period must be known in order to translate the signal into time. The sample period is the reciprocal of the sample frequency f_s (in this case, 600 samples per second).

The unprocessed lead traces of Figure 1.1 clearly show the result of such interfering signals. Figure 1.2 shows a representative waveform after processing. The signal is sampled at 600 samples per second, and a digital filter has been applied to help reduce the unwanted interference components. This type of *filtering* operation is one of the fundamental DSP operations. How to parameterize the digital filter so that it removes as much as possible the unwanted interfering signal(s) and retains as much of the desired signal without alteration is an important part of the DSP filter design process. Algorithms for digital (or discrete-time, quantized-amplitude) filters and the design approaches applicable in various circumstances are discussed in Chapters 8 and 9.

A great deal of additional information may be gleaned from the ECG signal. The most obvious is the heart rate itself, and visual inspection of the filtered signal may be all that is necessary. However, in some circumstances, automatic monitoring without human intervention is desirable. The problem is not necessarily straightforward since, as well as the aforementioned interference, we have to cope with inherent physiological variability. The class of *correlation* algorithms is applicable in this situation; correlation is covered in Chapter 6.

1.5.2 Audio and Acoustics

Audio processing in general was one of the first application areas of DSP—and continues to be—as new applications emerge. In this case study, some parameters relating to room acoustics are derived from signal measurements. The experimental setup comprises a speaker which can synthesize various sounds, together with a microphone placed at a variable distance from the speaker, all in a room with unknown acoustic properties.

Consider first the case of generating a pure tone or sinusoid. We need to generate the samples corresponding to the mathematical sine function and do so at the

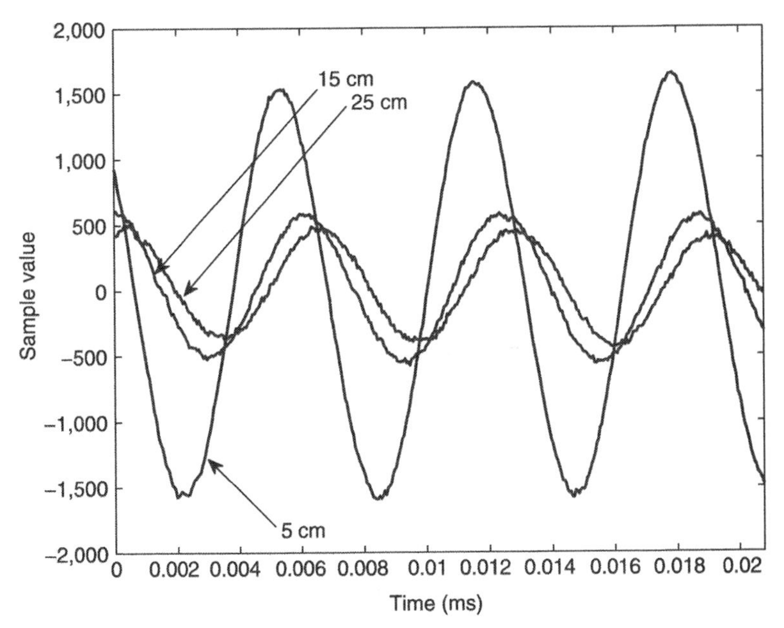

FIGURE 1.3 The response of the speaker–microphone system to a pure tone, with the microphone placed at various distances away from the source. The relative delay may be used to estimate distance. Note that the signals, particularly those at larger distances, have a larger component of noise contamination.

required sampling rate (in this example, 96,000 samples per second). Figure 1.3 shows the resulting measurements, with the microphone placed at various distances from the speaker. What is immediately clear is that the amplitude of the received signal decreases with the distance, as would be expected. Furthermore, the relative delay or *phase* of the tone changes according to the distance. This change could be used to estimate the distance of the microphone from the speaker. One complication is that the signal clearly contains some additional noise, which must be reduced in order to make a more accurate measurement.

The estimates derived from Figure 1.3 yield a figure of 12-cm movement for the case where the microphone was moved from 5 to 15 cm, and an estimate of 18 cm for the 5- to 25-cm microphone movement. What factors influence the accuracy of the result? Clearly, we must make an assumption for the speed of sound, but in addition, we need to determine the relative phase of the sinusoids as accurately as possible. Correlation algorithms (mentioned earlier) are of some help here. It also helps to maximize the sampling rate since a higher rate of sampling means that the sound travels a shorter distance between samples.

Other information about the speaker–microphone–room coupling can be gleaned by using a different test signal, which is not difficult to do using DSP techniques. Figure 1.4 shows the result of using random noise for the output signal (the random noise is also termed "white noise" by analogy with white light, which comprises all wavelengths). This type of signal is a broad-spectrum one, with no one

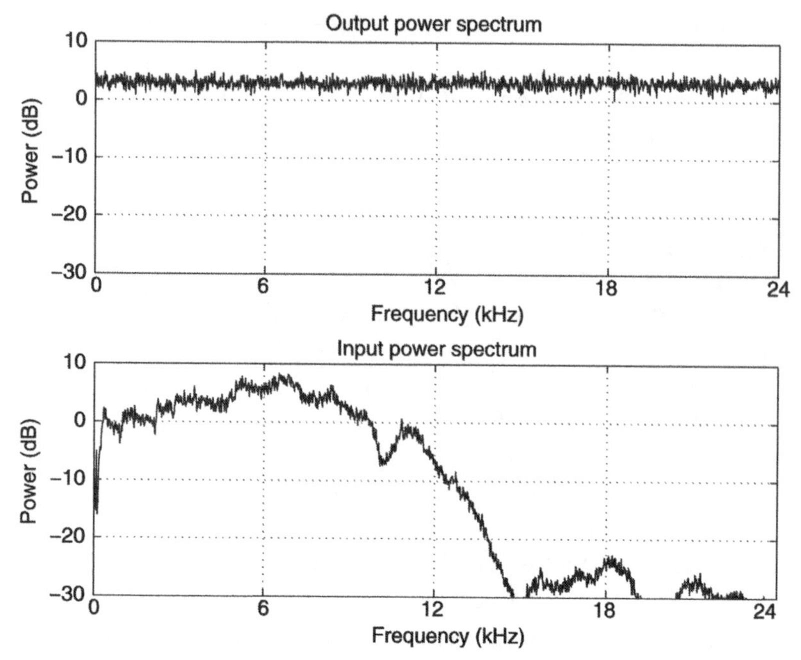

FIGURE 1.4 The estimated frequency response of the audio system. This particular system clearly shows a marked reduction in the transmission of higher-frequency audio components.

pure tone dominating over any other. The figure shows the energy content over the frequency range of the output signal and the estimated energy content over the same range of frequencies as received by the microphone. This is a derived result in that all that is available directly are the sample values at each sampling instant, and it is necessary to derive the corresponding frequency components indirectly. The algorithms for performing this analysis are introduced in Chapter 7. From Figure 1.4 it is evident that the speaker–microphone channel exhibits what is termed a "bandpass" characteristic—lower frequencies are reduced in amplitude (attenuated), and higher frequencies are also reduced substantially. This information may be used to compensate for the shortcomings of a particular set of conditions—for example, boosting frequencies which have been reduced so as to compensate for the acoustics and speaker setup.

Figure 1.5 shows a different set of results as calculated from the broad-spectrum experiment. Here, we perform a system identification algorithm using the correlation function (as discussed in Chapter 6) at each of the microphone displacements. Each response has been normalized to a ±1 amplitude level for easier comparison. It is clear that the displacement of each waveform corresponds to the time delay, as discussed. However, the shape of the graphs is approximately the same. This characteristic response shape is termed the *impulse response*. This is a key concept in signal processing—the impulse response is that response produced by a system (in this case, the electroacoustic system) as a result of a single pulse. In most

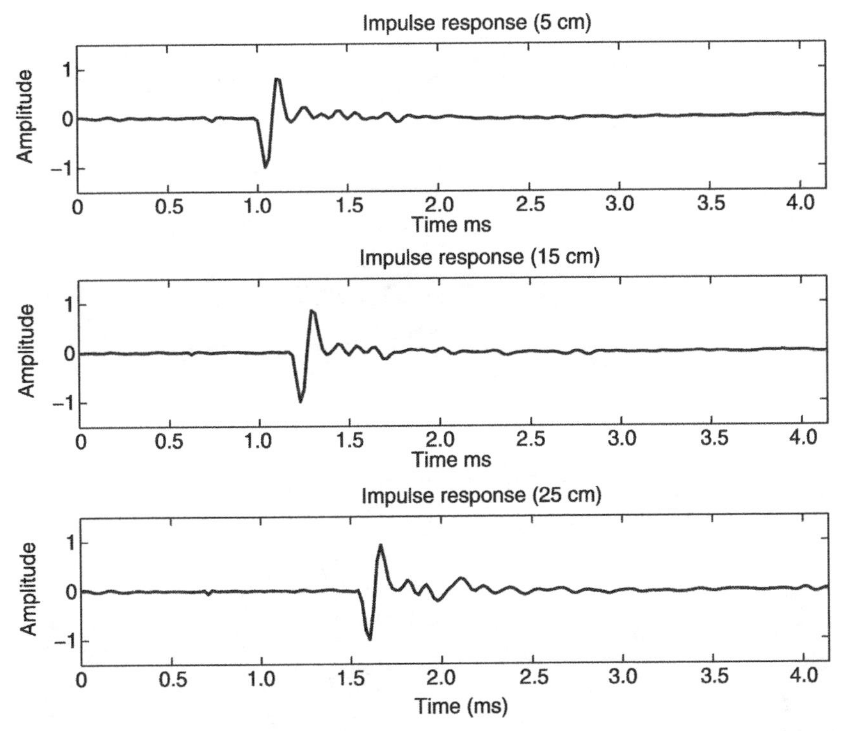

FIGURE 1.5 The computed impulse response of the audio system, with the microphone placed at varying displacements. As well as the delay due to acoustic propagation, it is possible to derive the estimated impulse response, which is characteristic of the system.

cases, a single pulse is not a feasible test signal, and thus the impulse response cannot be measured directly. Rather, we use methods such as that described using white noise, to estimate the impulse response of the system. Chapter 6 discusses correlation and system identification, and Chapter 7 further examines the concept of frequency response.

1.5.3 Image Processing

The processing of digital pictures, in particular, and digital images from sensor arrays, in general, is an important aspect of DSP. Because of the larger processing and memory requirements inherent in two-dimensional pictures, this area was not as highly developed initially. Today, however, two- and even three-dimensional signals are routinely processed; one may even consider some implementations to be four dimensional, with the fourth dimension being time t, along with spatial dimensions x, y, and z.

Figure 1.6 illustrates the problem of determining whether a particular image taken with a digital camera is in focus. A number of images are taken with the lens at various positions relative to the image sensor. From the figure, it is clear which of the four images presented is closest to being in focus. So we ask whether it is

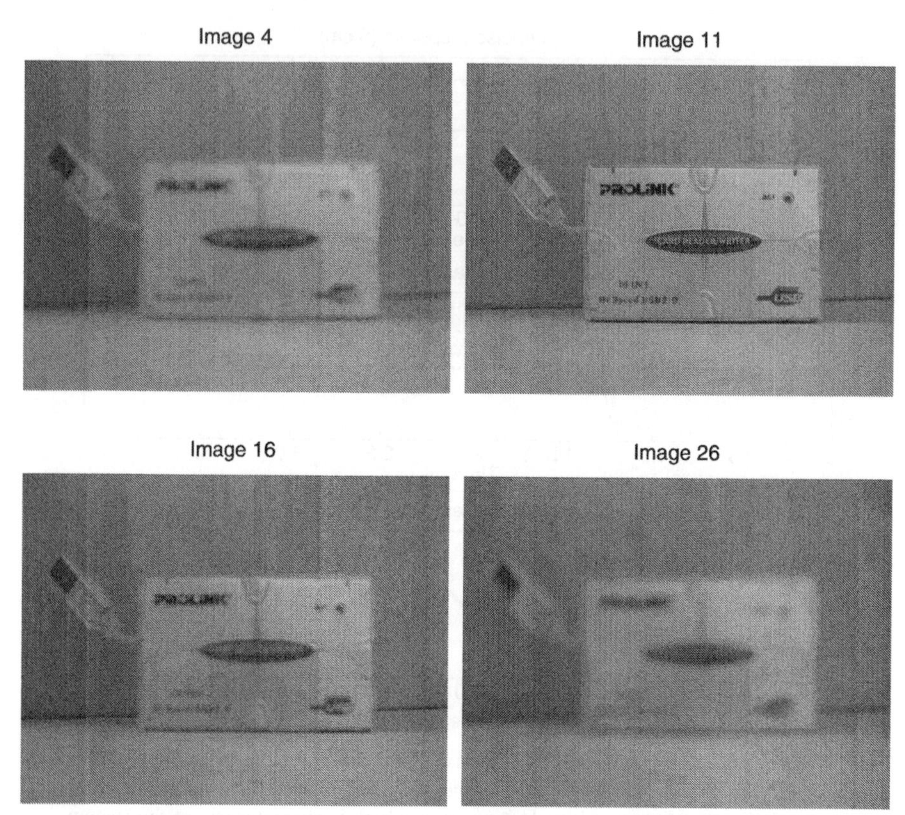

FIGURE 1.6 Images for the focusing experiment. The key question is to be able to determine which of the set is closest to being in focus.

possible to develop an algorithm for determining this automatically. Figure 1.7 shows the results of testing across a set of 40 images. We need to derive one parameter which represents the concept of "focus" as we would interpret it. In this case, the algorithms developed consist of filtering, which is akin to determining the sharpness of the edges present in the image, followed by a detection algorithm. Figure 1.7 shows the residual signal energy present in the filtered signal for each image, with the camera's autofocus image shown for comparison. Clearly, the peak of the calculated parameter corresponds to the relative degree of focus of the image. Chapter 3 introduces the digital processing of images; Chapters 8 and 9 consider digital filters in more detail.

1.5.4 Biomedical Visualization

Finally, we consider another biomedical application in the visual domain, but one that is not "imaging" in the conventional sense of taking a picture. In this case, we investigate the area of computerized tomography (CT), which provides clinicians

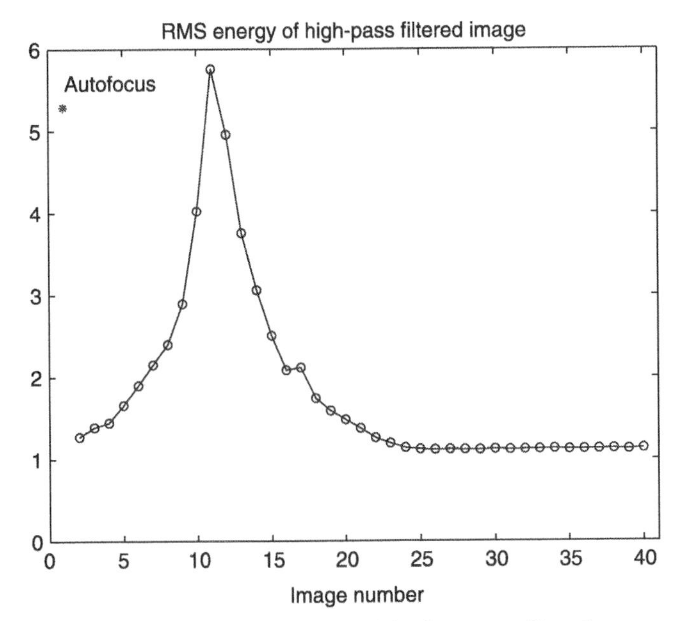

FIGURE 1.7 The relative value of the frequency-filtered root mean square energy derived for each image. This is seen to correspond to the relative focus of each image. In this case, we need to apply several DSP algorithms to the image data to synthesize one parameter which is representative of the quantity we desire, the degree of focus. The asterisk (*) indicates the energy in the camera's auto-focused image.

with an unprecedented view inside the human body from external noninvasive scanning measurements.

The fundamental idea of tomography is to take many projections through an object and to construct a visualization of the internals of the object using post-processing algorithms. Figure 1.8 shows a schematic of the setup of this arrangement. The source plane consists of X-ray or other electromagnetic radiations appropriate to the situation. A series of line projections are taken through the object, and the measurements at the other side are collated as illustrated by the graph in the figure. This gives only one plane through the object—a cross-sectional view only. What we desire is a two-dimensional view of the contents of the object. In a mathematical sense, this means a value of $f(x, y)$ at every (x, y) point in a plane. A single cross-sectional slice does not give us an internal view, only the cross-sectional projection.

The key to the visualization is to take multiple cross sections, as illustrated in Figure 1.9. This shows two cross sections of the object, each at an angle θ from the x axis. The projection of a particular point, as illustrated in the figure, yields the equivalent density at the point (x, y) inside the object, shown as the dot where the two ray traces intersect. The figure shows only two projections; in reality, we need to take multiple projections around the object in order to resolve all ambiguities in the value of the intensity $f(x, y)$.

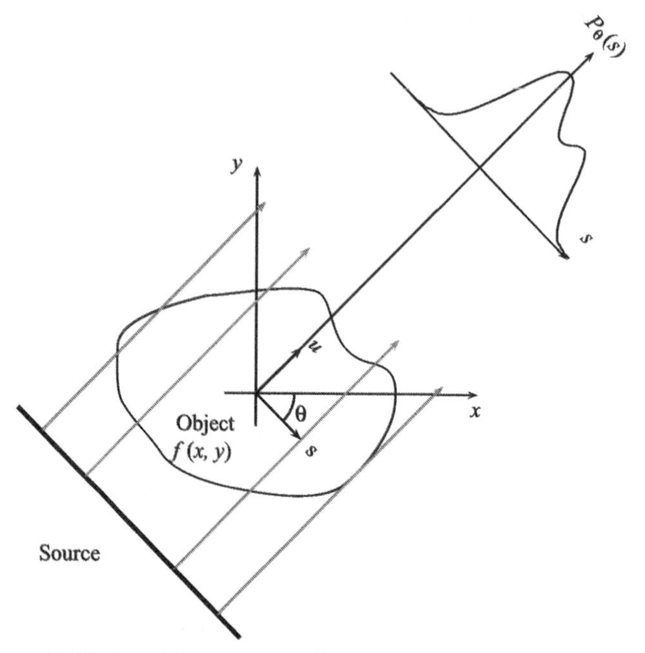

FIGURE 1.8 The projection of a source through an object, which can be interpreted as the line integral or the Radon transform from source to detector through the object.

If we take sufficient projections around the object, its internals can be visualized. Figure 1.10 illustrates this process. On the left, we see the so-called Shepp–Logan phantom, which is an "artificial" human head containing ellipses representing tissue of various densities (Shepp and Logan 1974). The image on the right shows the reconstruction of the phantom head, using a back projection algorithm which accumulates all the projections $P_\theta(s)$ for many angles, θ, so as to approximate the internal density at a point $f(x, y)$. Figure 1.10 shows a deliberately lower-resolution reconstruction with a limited number of angular measurements so as to illustrate the potential shortcomings of the process. Chapter 6 investigates the signal processing required in more detail.

1.6 OVERVIEW OF LEARNING OBJECTIVES

This text has been designed to follow a structured learning path, with examples using MATLAB being an integral part. The chapters are organized as follows:

Chapter 2 covers the basics of the MATLAB programming language and environment. The use of MATLAB is central to the development of the learn-by-doing approach, and the treatment given in this chapter will serve as a grounding for subsequent chapters.

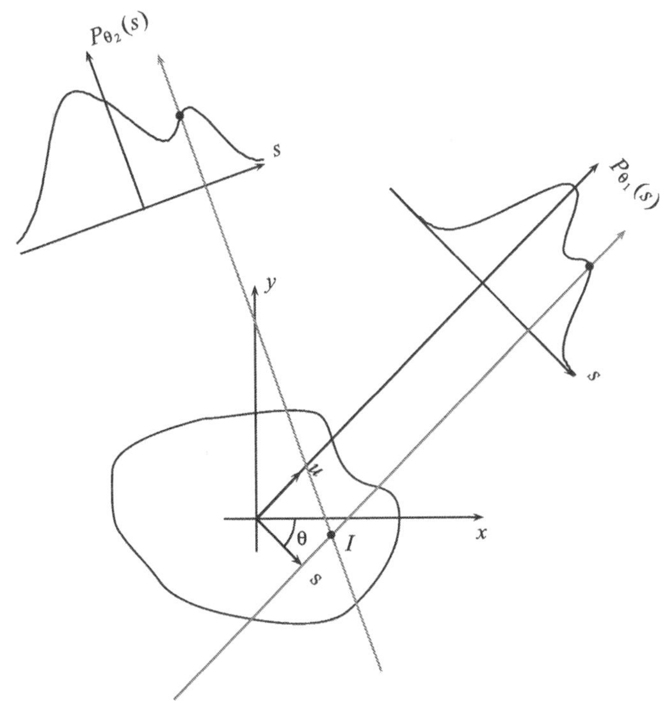

FIGURE 1.9 Interpretation of the back projection algorithm for reconstructing the internal density of an object, commonly known as a CT scan. By taking multiple external measurements, a cross section of the human body can be formed. From this cross section, it is possible to estimate the internal density at the point *I*, where the lines intersect. To obtain sufficient clarity of representation, a large number of such points must be produced, and this can be very computationally intensive.

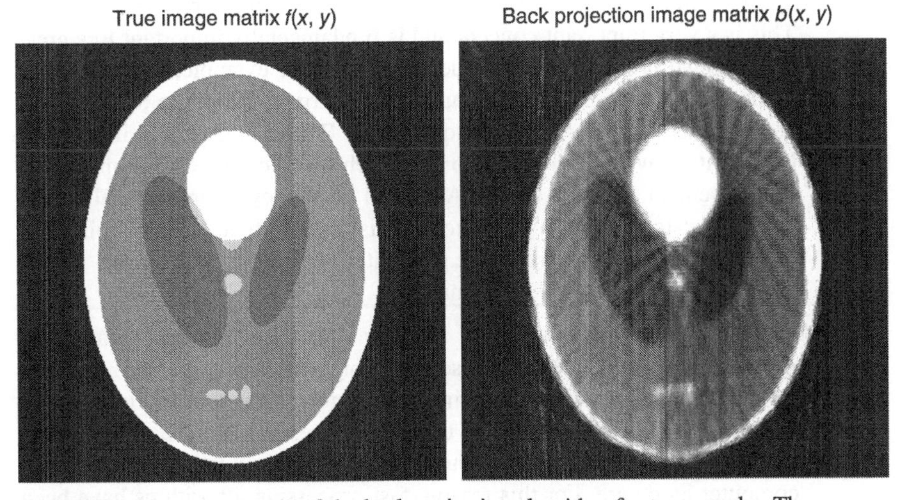

FIGURE 1.10 An example of the back projection algorithm for tomography. The Shepp–Logan head phantom image is shown on the left, with a low-resolution reconstruction shown on the right. The lower resolution allows the scan lines to be seen, as illustrated in the previous figures.

Chapter 3 looks at how signals in the real world are acquired for processing by computer. Computer arithmetic is introduced since it is fundamental to any signal processing algorithms where the processing is performed digitally. How signals in the real world are acquired is discussed, and from all these aspects, we can gauge the processing accuracy, speed, and memory space required for any given application.

Chapter 4 looks at random signals. The "noise" which affects signals is often random, and thus it is important to understand the basics of how random signals are characterized. The examples in the chapter look at noise in both audio signals and images.

Chapter 5 introduces the representation of known signals via mathematical equations. This is important, as many signals used in communications systems, for example, must be generated in this way. This is followed by a look at how systems alter signals as they pass through, which is the fundamental mode of operation of many signal processing systems.

Chapter 6 looks at how we can sample signals in the time domain and the spatial domain. One important principle in this regard is called "correlation," which is essentially comparing two signals to determine their degree of similarity. In the real world, any two given signals won't ever be identical in a sample-for-sample sense. Correlation allows us to tell if two signals are "somewhat similar." The basic concept of correlation is then extended into noise filtering, where a signal is contaminated by noise which we can estimate. Finally, spatial signal processing is examined, with a particular emphasis on the tomography problem as described in the example above.

Chapter 7 looks at signals from another perspective: their frequency content. This is a very important concept and is fundamentally important to a great many signal processing techniques. Consider, for example, two keys on a piano: They sound different because of the particular mixture of frequency components present. The technique called Fourier analysis is introduced here, as it allows the determination of which specific frequency components are present in a given signal. A related frequency approach, called the cosine transform, is then introduced. This finds a great deal of application in digital media transmission (e.g., digital television, compressed music, and the JPEG digital image format).

Chapter 8 examines how we can filter, or alter subject to specifications, a given signal to suit some purpose. This might be, for example, removing some interfering noise or enhancing one or more of the frequencies present. These algorithms, although very useful in processing signals, often require a large number of computations and hence can be slow to run on affordable computer hardware. For this reason, fast processing algorithms have been developed to obtain the same result with fewer computations. These are also covered in this chapter.

Chapter 9 introduces a different type of discrete-time filter, the so-called recursive filter, which is used for more efficient processing in some circumstances. It is useful to have a good theoretical understanding of the possible approaches to digital filtering, and this chapter complements the previous chapter in this regard.

1.7 CONVENTIONS USED IN THIS BOOK

Because signal processing is based on algorithms, and algorithms are based on mathematics, a large proportion of time must be spent explaining algorithms and their associated mathematics. A basic grounding in linear algebra and calculus is necessary for some sections, as is an understanding of complex numbers. Because we often have to explain the use of blocks of samples and how they are stored and processed, the concepts of vectors and matrices are essential. To avoid confusion between scalar values and vectors, vectors and matrices, and constants and variables, the conventions used throughout the book are shown in Table 1.1.

MATLAB code in a script file is shown as follows. It may be typed into the MATLAB command window directly or entered into a file.

TABLE 1.1 Mathematical Notation Conventions Used in the Book

x	Scalar variables (lowercase)
N	Integer constants (uppercase)
n	Integer variable over a range, for example, $\sum_{n=0}^{N-1} x(n)$
j	Unit complex number, $\sqrt{-1} = 1e^{j\pi/2}$
$\mathbf{x}$	Column vector (bold font, lowercase)
$\mathbf{x}_k$	*k*th column vector (in a matrix)
$\mathbf{A}$	Matrix (bold font, uppercase)
a_{ij}	Element (i, j) of matrix **A**
$\mathbf{A}^{-1}$	Inverse of matrix **A**
$\mathbf{A}^{+}$	Pseudoinverse of matrix **A**
$\mathbf{p}^*$	Complex conjugate of **p**
$\mathbf{R}_{xy}$	Correlation matrix of **x** and **y**
$E\{\cdot\}$	Expectation operator, average of a sequence
$\mathbb{R}^N$	*N*-dimensional parameter space
$\lvert x \rvert$	Absolute value (magnitude) of x
$\lVert \mathbf{x} \rVert$	Norm (length) of vector **x**
$\hat{x}$	Estimate of x
δx	Change in x

```
d = zeros(2 * N + 1, 1);
d(2 * N + 1) = 1;
d(1) = (1/j) ^ (2 * N);
```

Where MATLAB script is entered interactively and the result is shown, it is presented as follows:

```
m = rand(2, 2)
m =
        0.9501    0.6068
        0.2311    0.4860
inv(m)
ans =
        1.5117   -1.8876
       -0.7190    2.9555
```

Where a MATLAB variable or function is described in-text, it is shown as follows: `fft()`.

Each chapter concludes with a set of problems. Some of these focus on a mathematical solution only; some require the further development of MATLAB code described in the chapter; and some require both mathematical analysis and algorithm development.

1.8 CHAPTER SUMMARY

The following are the key elements covered in this chapter:

- The role of DSP and some application areas
- The learning objectives for subsequent chapters
- The notational conventions used throughout the book

REVIEW QUESTIONS

1.1. Two of the case studies cited in this chapter have been associated with Nobel Prizes in the past. Which two? How has DSP enabled the original discoveries to fulfill their promise?

1.2. Describe any applications of DSP in the room in which you are sitting.

1.3. From the ECG in Figure 1.2, estimate the heart rate. Write down the steps you took to do this. What practical problems may arise in implementing your algorithm? For example, a step such as "find the largest peak" could be ambiguous if there are several peaks.

1.4. Suppose the head cross-sectional slice as depicted in Figure 1.10 is required to have a resolution of 2,048 points horizontally and 1,024 points vertically. How many data points are there in total? If the cross section has to be produced within 30 seconds, roughly how long would be available for the computation of each point?

1.5. Consider the sound delay measurement system described in this chapter. Suppose the samples are taken at a rate of 10,000 samples per second. Look up an estimate of the speed of sound. How far would a sound wave travel in the time taken between samples? How could we increase the accuracy of the distance measurements? What other variable aspects should be considered?

CHAPTER 2

MATLAB FOR SIGNAL PROCESSING

2.1 CHAPTER OBJECTIVES

On completion of this chapter, the reader should be able to:

1. use the MATLAB interactive window command line, and enter MATLAB scripts.
2. use MATLAB function files.
3. be able to display image data from a matrix.
4. be able to play back audio data from a vector.
5. understand the concepts of signal data file storage and data formats.

2.2 INTRODUCTION

MATLAB® is used extensively throughout this text to illustrate practical signal processing concepts. This tutorial chapter introduces some of the features which are useful in this regard. Note that the introductory nature of the tutorial means that it is by no means exhaustive in its examination of MATLAB's capabilities.

2.3 WHAT IS MATLAB?

MATLAB is a commercial software product available from The Mathworks.[1] It consists of a main "engine" to perform computations, and several (optional) extended-function libraries (called "toolboxes") for special-purpose applications. The main MATLAB engine has many robust mathematical functions built-in. It is an *interpreted* language—meaning that each line of source is parsed (analyzed and converted) in text form, without being converted into a native machine code executable. This means that it may be slower to execute the given code in many circumstances than equivalent code written for a specific processor.

[1]Website: http://www.mathworks.com/.

Digital Signal Processing Using MATLAB for Students and Researchers, First Edition. John W. Leis.

It is worth noting that DSP systems are often developed in MATLAB, and implemented on the target hardware using the C language. This is because MATLAB provides an easier environment in which to design and test algorithms, and the so-called "target" devices (the end-product) often have limited memory, display, or other constraints. The C language (and by extension, C++) is able to produce assembly language code for target processors, which is optimized for that particular device's instruction set. The MATLAB syntax is, in many ways, somewhat similar to the C programming language.

MATLAB may be used interactively or in a "batch" mode for long computational scripts. Simply typing "help" at the command prompt provides assistance with various features. MATLAB has a rich set of constructs for plotting scientific graphs from raw or computed data, suitable for inclusion in reports and other documents. One particular advantage of MATLAB is that it is available on several different computer operating system platforms. This text makes no assumption about the particular version which is available, and does not assume the presence of any toolboxes.

2.4 GETTING STARTED

When MATLAB is started, the command window will appear. MATLAB may be used in one of two ways: by entering interactive commands, or to process a set of functions in a script file. "Interactive mode" means typing commands at the prompt in the MATLAB window. Command script files, or "m-files," may be called from the command window prompt or from other m-files. The convention used in this text is that commands that are entered into the MATLAB command window or placed in a script file are shown in a frame. If entered in a script file, the file becomes the command or function name, and must have a `.m` extension (file name suffix).

Assistance for commands may be obtained by typing **`help`** on the command line, followed by the specific keyword. For example, in using the **`disp`** command to display text, typing **`help disp`** shows the syntax of the command, together with a brief explanation of its operation. Two useful commands are **`clear all`** and **`close all`**, which clear all current variables, and close all display windows, respectively.

2.5 EVERYTHING IS A MATRIX

MATLAB may be used from the interactive window, and does not require variables to be declared, as in many procedural languages. Simply typing the name of the variable prints its value. One very important distinction is that *all variables are matrices*—a scalar (single value) is just the special case of a 1×1 matrix. A vector is of course a matrix with one dimension equal to unity. By convention (as in mathematics), vectors are assumed to be "column vectors"—that is, of dimension $N \times 1$ (N rows in one column). To picture this, we might have:

$$\begin{pmatrix} 9 & 3 \\ 5 & 4 \\ 8 & 3 \end{pmatrix} \qquad \begin{pmatrix} 1 \\ 2 \\ 3 \end{pmatrix}.$$

A 3×2 Matrix A 3×1 Column Vector

2.6 INTERACTIVE USE

Suppose we have a matrix **M** defined as

$$\mathbf{M} = \begin{pmatrix} \pi & e^{\pi} \\ 0 & -1 \end{pmatrix}.$$

This matrix may be entered into MATLAB using

```
M = [ pi exp(pi); 0 -1 ]
M =
    3.1416    23.1407
    0         -1.0000
```

Note how the semicolon is used to indicate "the next row," and that MATLAB echoes the variable after input. A semicolon *at the end of the input line* suppresses this echo. If we want the transpose of **M**, simply type `M'`. To transform a matrix to a vector "column-wise," we type `M(:)` as shown below.

```
M = [ pi exp(pi); 0 -1 ];
M'
ans =
    3.1416          0
    23.1407   -1.0000

M(:)
ans =
    3.1416
    0
    23.1407
    -1.0000
```

If we have a vector **v** defined as

$$\mathbf{v} = \begin{pmatrix} 3 \\ 4 \end{pmatrix},$$

we could find the matrix–vector product **Mv** as follows:

```
v = [3 ; 4]
v =
    3
    4
M*v
ans =
    101.9875
    -4.0000
```

Now suppose we want the vector dot-product (also called the "inner product"),

$$\mathbf{p} = \mathbf{v} \cdot \mathbf{v}.$$

To evaluate this, first type

```
p = v.*v
p =
    9
    16
```

Then entering

```
sum(p)
ans =
    25
```

will compute the column sum of the vector. Now suppose we wish to find

$$\mathbf{q} = \mathbf{v}\mathbf{v}^T.$$

Note that $\mathbf{v}$ is a column vector (dimension 2×1), and thus $\mathbf{q}$ will be a 2×2 matrix:

```
size(v)
ans =
    2    1
q = v*v'
q =
    9    12
    12   16
```

In order to find

$$\mathbf{r} = \mathbf{v}^T \mathbf{v},$$

we would use

```
size(v)
ans =
     2     1
r = v'*v
r =
    25
```

This is of course the vector dot product, computed in a different way.

Matrix–vector dimensionality can be the source of much confusion when using MATLAB—care must be taken to ensure the mathematical correctness of what we are asking it to calculate:

```
v
v =
    3
    4
v*v
    ??? Error using ==> *
    Inner matrix dimensions must agree.
```

The fact that this example produces an error may be checked as above, using **size**(v). Of course, a matrix may in fact not contain any elements. If this has to be checked, the operation **isempty**(a) will determine if matrix a is empty. MATLAB will print empty matrices as [].

2.7 TESTING AND LOOPING

The basic structure of if-then-else tests is explained below. The code may be entered into an m-file by typing `edit` at the command prompt, entering the text, and saving the file. If the file were saved as, say, `examp.m`, then it would be invoked by typing `examp` in the command window.

```
a = 4;
b = 6;
if( a == 4 )
    disp('equality');
end
```

Note that testing for equality (the **if**() test) requires the use of *two* equals signs "==." A single "=" implies assignment, as shown for variables a and b. Adding the following text to the above file shows the use of two (or more) conditions. The first test is to determine whether a is equal to 4, *and* b is *not equal* to 4. The second test is true if a is greater than 0, *or* b is less than or equal to 10.

```
if( (a == 4) && (b ~= 4) )
     disp('AND-NOT condition');
end

if( (a > 0) || (b <= 10) )
     disp('OR condition');
end

if( a > 0 )
     disp('case one');
else
     disp('case two');
end
```

In some cases, multiple conditions have to be tested. That is, a value could be one of several possible candidates, or it could be a none of those. We could use multiple if..else..if..else statements, but that becomes unwieldy for more than a few cases. It is better to use the `switch-case` construct, which conveniently gives us an "otherwise" clause as follows.

```
a = 2;
b = 3;
switch (a*b)
    case 4
         disp('answer is 4');
    case a*b
         disp([num2str(a*b) 'of course']);
    otherwise
         disp('not found');
end
```

The main constructs for iteration (looping) in MATLAB are the **`for`** and **`while`** statements. Note the use of the step increment of 0.1 in the following. The variable x takes on values 0, 0.1, 0.2,... 10 on each iteration:

```
for x = 0:0.1:10
     fprintf(1, 'x is %d\n', x);
end
```

The **`while`** loop form may be used as shown in the following example. It is typically used where the terminating condition for the iteration is calculated within the loop, rather than iterating over a fixed range of values as with the **`for`** loop.

```
x = 1;

while( x < 10 )
    x = x*1.1;
end
```

Care should be exercised when testing for *exact* equality, and where very large or very small numbers might be encountered. MATLAB has the built-in variables **realmin** and **realmax** for the smallest and largest positive numbers, and **eps** for floating-point relative accuracy. While the largest number MATLAB can represent is quite large, and the smallest number is quite small, certain situations may cause problems in signal processing code, particularly where iterative calculations are involved. For example, the following loop *should* never terminate, because it calculates the sequence 1, 0.5, 0.25, ... In theory, this sequence never reaches zero. However, entering and running the loop below shows that the loop does indeed terminate eventually:

Listing 2.1 Unexpected Loop Termination

```
r = 1;
while( r > 0 )
    r = r/2;
    fprintf(1, 'r=%d\n', r);
end
```

2.8 FUNCTIONS AND VARIABLES

As mentioned, MATLAB code is placed in "m-files," which are just plain text files with extension (suffix) `.m`. It is possible to edit these files using the MATLAB built-in editor (or in fact any plain-text editor). In order to be able to find the place (folder/directory) where the m-files have been placed, they must be in the current MATLAB directory or in the MATLAB search path. Type **pwd** to see the current directory, and **cd** to change to another directory—for example, typing **cd** `c:\matlab\work` will change the current working directory to disk drive `c:` in directory `\matlab\work`. Type **path** to see the current search path.

The commands **clear all** and **close all** are often used when beginning a MATLAB session. They are used to clear all variables and close all figure windows, respectively.

MATLAB will use a file called `startup.m` in the current directory when it starts up, to set some initial conditions. This file is optional, but can be useful in larger projects. A simple `startup.m` file is shown below:

```
% add extra paths - user library files
addpath('c:\matlab\lib');

% start in this directory
cd c:\matlab\work
```

The comment character % makes the remainder of the line a comment (which is ignored by MATLAB). The **addpath** command adds a new directory path to the existing search path (where MATLAB searches for files). **path** may be used to show the current search path.

As with any programming script, MATLAB allows the decomposition of problems into smaller sub-problems via "functions." Functions normally exist in separate m-files, and are given an identical file name to the function name itself, but with .m appended. Arguments *passed to* the function are placed in braces after the function name. Arguments *returned from* the function are placed on the left-hand side in square brackets.

Suppose we wish to calculate the infinite series for e^x. This is defined for a scalar value x as

$$\begin{aligned} f(x) &= 1 + x + \frac{x^2}{2!} + \frac{x^3}{3!} + \cdots \\ &= \sum_{n=0}^{N} \frac{x^n}{n!}. \end{aligned}$$

This definition may also be used for a matrix-valued $\mathbf{x}$. The declaration in the file eseries.m to compute this above would look like:

```
function [y, epsilon] = eseries(x, N)
```

The values x and N are passed to the m-file function eseries(). The returned values are y and epsilon. Usually, the returned values are in order of importance from left to right, because in some situations, some of the return arguments are not required. In the present example, y is the series approximation (which depends on the number of iterations, N), and epsilon (ε) is the series approximation error for N terms, defined as

$$\varepsilon = e^x - f(x).$$

We may not always wish to use the value of ε. The MATLAB code for the function is shown in Listing 2.2. Note that this function is intended for demonstration purposes only: MATLAB has a built-in function expm() for this computation. Our eseries function is used in a MATLAB session as follows:

```
x = 2.4;
N = 10;
[y, epsilon] = eseries(x, N);
y
y =
   11.0210
epsilon
epsilon =
        0.0022
```

Comments (starting with percent signs %) at the beginning of the file up to the first blank line may be used to provide information to the user on the calculation performed by the function and the expected arguments. Typing **help** eseries will present this information to the user. Thus, it should normally contain a brief description of the m-file, its inputs and outputs, and any usage limitations. It's usually a good idea to develop a template with sufficient information for future use, and to use it consistently. For example:

```
% eseries.m - Exponent of a scalar/matrix using N terms of a series
% function [y, epsilon] = eseries(x, N)
%
% Calculate the exponent of a scalar
% or a matrix using a series approximation
% e^x = 1 + x + (x^2)/2! + (x^3)/3!
%
% Inputs:
%    x = Exponent is calculated on this number.
%        May be a matrix.
%    N = The number of terms to use
%        in the expansion.
% Outputs:
%    y = the resulting expansion.
%    epsilon = the error
%            = true value - estimated value
%
% Author:
% Date:
% Revision:

function [y, epsilon] = eseries(x, N)

% Remainder of function follows ...
```

Listing 2.2 An Example MATLAB Function File. Note the **function** keyword. The file name must match the function name, with a **.m** extension.

```
% eseries.m - Exponent of a scalar/matrix using N terms of a series
% function [y, epsilon] = eseries(x, N)
% compute exponential of a matrix
% input:
%    x = matrix to find exponent of
%    N = number of terms in approximation
% output:
%    y       = matrix result
%    epsilon = error compared to true value

function [y, epsilon] = eseries(x, N)

% input argument checking
if( nargin ~= 2 )
    disp('eseries: requires');
    disp('x = value for exponentiation');
    disp('N = number of terms to use.');

    error('Invalid number of input arguments.');
end

[m, n] = size(x);

% accumulates the result
res = eye(m, n);

% numerator in each term of the expansion
% eye function gives an identity matrix
num = eye(m, n);

% denominator in each term of the expansion
% this is a scalar quantity
den = 1;

% first term is unity (or an identity matrix)
% and is already calculated on initialization
for term = 2:N

    % update numerator & denominator
    num = num*x;
    den = den*(term - 1);
```

```
    % update the running sum
    res = res + (num/den);
end

% returned values
y = res;
epsilon = exp(x) - y;
```

A related command is **lookfor**, which is useful if you know some keywords but not the function name. This command looks for the specified keyword in the first comment line. So **lookfor** `exponent` would return the comment line from the above (as well as other functions containing this keyword).

It's also wise to perform some argument checking: the variable **nargin** contains the number of input arguments to the function:

```
% input argument checking
if( nargin ~= 2 )
    disp('eseries: requires: ');
    disp('x = value for exponentiation');
    disp('N = number of terms to use.');
    error('Invalid number of arguments.');
end
```

More sophisticated display of variables may be performed using the **fprintf**`()` function. For example,

```
for k = 1:20
    fprintf(1, 'k=%d, ans=%.2f \n', k, k*pi);
end
```

The built-in function **fprintf**`()`, discussed in the next section, is used for writing to files. Here, the variable 1 refers to the file number (1 being the standard output, or the MATLAB command window). The format specifier `%d` is used for integers, and `%.2f` displays a floating-point number with two digits after the decimal place. The current display line may be terminated and a new one started using the control character `\n` within a format string. Of course, there are many other formatting specifications, such as `%s` for strings—type **help fprintf** for more information.

MATLAB has a large number of built-in functions. For example, to sort an array of numbers, the **sort**`()` function sorts the numeric values in ascending order, and optionally returns the index of the sorted values in the original array:

```
x = [4 3 8 2];
[array index] = sort(x)
array =
   2  3  4  8
index =
   4  2  1  3
```

Note that MATLAB array indexing starts at 1 and not 0. This can be a source of confusion, especially since many algorithms are developed using zero as a base. Indexes `x(1)` through to `x(4)` in the above are valid, whereas index `x(0)` produces the error message `Index into matrix is negative or zero.`

2.9 PLOTTING AND GRAPHING

In addition to computation, MATLAB includes extensive facilities for the display of data in the form of various graphical plots. Here's a simple example:

```
x = 0:pi/100:2*pi;

wave1 = sin(x*10);
wave2 = sin(x*10) + cos(x*14);

plot(wave1);
hold on;
plot(wave2);
```

This generates a vector of x-axis data `x` from 0 to 2π in steps of $\pi/100$. The two waveforms `wave1` and `wave2` are then plotted. The **`hold`** on stops MATLAB from replacing the first plot with the second. At this point, our plot looks as shown in Figure 2.1.

For two or more plots superimposed, we would want to have a legend to distinguish them, plus some sensible labels on the x- and y-axes, together with an overall title:

```
close all
plot(x, wave1, 'b-', x, wave2, 'b:');
legend('wave 1', 'wave 2');
title('Some Example Data Waveforms');
xlabel('Angle');
ylabel('Amplitude');
```

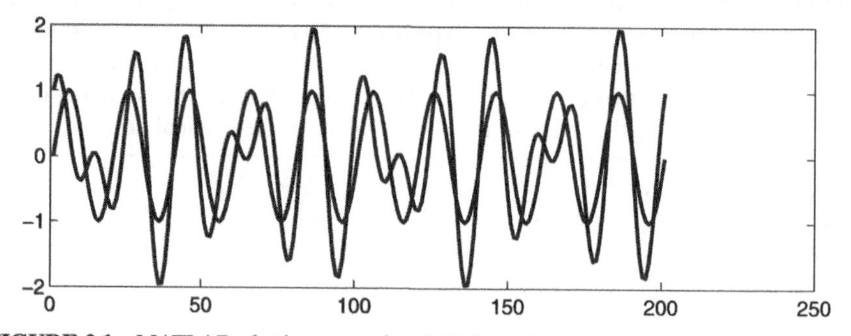

FIGURE 2.1 MATLAB plotting example—initial version.

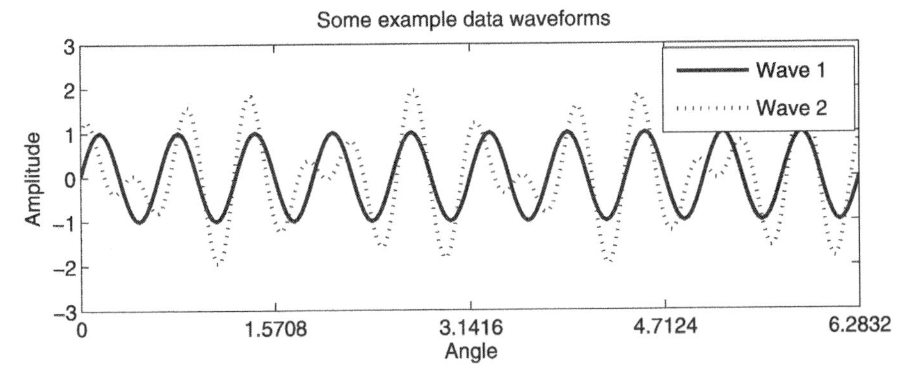

FIGURE 2.2 MATLAB plotting example (continued).

We would like the extent of the x-axis to encompass the entire plot and not to have the legend box partially obscure the plot. The MATLAB function **gca** (get current axes) is useful for that:

```
set(gca, 'xlim', [0 2*pi]);
set(gca, 'xtick', [0 0.25 0.5 0.75 1]*2*pi);
set(gca, 'ylim', [-3 3]);
```

Note that the axis limits could also be set using `axis([0 2*pi -3 3]);`. The graph now looks as shown in Figure 2.2. There are a number of other plotting functions that are useful, including `mesh()` and `surf()` for three-dimensional plots, and `contour()` and `subplot()` for multiple small plots in a group.

2.10 LOADING AND SAVING DATA

It is often necessary to load some data from a file, to save calculated data, or just to save the current variables from an interactive session for later resumption. The simplest case is to type `save myfile`, which will save the current variables to a file `myfile.mat`. Typing `load myfile` will retrieve them. However, what about data files which may be supplied via some other means?

A distinction must be made between *binary* and *text* (or "ASCII") formatted files. Text files contain plain, readable text that can be viewed with any text editor. A simple text file, such as `mydata.dat` containing plain text on each line, might be as follows:

```
line one
line two
```

This data file may be read using

```
FileId = fopen('mydata.dat', 'r');
s = fgetl(FileId)
s =
   line one

s = fgetl(FileId)
s =
   line two

s = fgetl(FileId)
s =
   -1

fclose(FileId);
```

Note how the value –1 is returned if we attempt to read past the end of the file. To read a *formatted* text data file, we must open the file, read the data, and close the file. Suppose now we have a data file `myfile.dat` containing numeric data:

```
23.4
45.6
76.8
```

We can use the **`fscanf()`** function, in conjunction with the formatting specifier `%f`, to interpret the text as a floating-point number:

```
FileId = fopen('mydata.dat', 'r');

DataPoints = fscanf(FileId, '%f');
fclose(FileId);

DataPoints
DataPoints =
    23.4000
    45.6000
    76.8000
```

It's always a good idea to check the value returned from **`fopen()`** to ensure that the file exists. In an m-file, we could use:

```
FileName = 'mydata.dat';
FileId = fopen(FileName, 'r');

if( FileId < 0 )
    fprintf(1, '%s does not exist\n', FileName);
    error('open file');
end
```

To write (output) to text files, we can use the **fprintf()** function and open the file in write (w) mode:

```
M = rand(4, 2)
M =
    0.7012    0.0475
    0.9103    0.7361
    0.7622    0.3282
    0.2625    0.6326
FileId = fopen('mydata.dat', 'w');
fprintf(FileId, '%.2f\n', M);
fclose(FileId);
```

This gives an output file, `mydata.dat`, containing only one column (one value per line) as follows:

```
0.70
0.91
0.76
0.26
0.05
0.74
0.33
0.63
```

It appears that MATLAB has not written the data in the "expected" format. The reason is that MATLAB takes variables *column-wise* to fulfill the output request. Thus the matrix M has been taken down the first column, then down the second. To get a data file of the same "shape," it is necessary to add two format fields, and transpose the matrix so that M^T is now a 2×4 matrix. The construct:

```
fprintf(FileId, '%.2f %.2f\n', M');
```

will give an output file `mydata.dat`:

```
0.70    0.05
0.91    0.74
0.76    0.33
0.26    0.63
```

In addition to text files as discussed above, one may encounter *binary* data files. These are *not* viewable using text editors—they consist of "raw" 8 or 16-bit quantities (usually), which are machine readable and not human readable. The exact representation depends on the central processing unit (CPU) being used, and there are different conventions in common use depending on the manufacturer. Integer representations may be converted between formats, but the situation for floating-point numbers is much more problematic.

For this reason, the Institution of Electrical and Electronic Engineers (IEEE) floating-point format is commonly used (IEEE floating-point numbers are discussed in detail in Section 3.4). If the need to read or write binary files arises, the appropriate source code methods will often be supplied with the data. If not, they must be written using a combination of **fread**() and **fwrite**().

Consider the following, which opens a file for reading in binary mode using the IEEE floating-point format:

```
FileId = fopen(FileName, 'rb', 'ieee-le');
if( FileId < 0 )
    fprintf(1, 'Cannot access %s\n', FileName);
    error('myfunction()');
end
```

The `ieee-le` indicates the use of the IEEE floating-point format, using "little-endian" byte ordering. This means that the least-significant byte is read first. Note the format specifier `'rb'` (read-only, in binary mode) as opposed to the text files discussed previously, which were opened using `'r'` mode (text is the default). To read the file, instead of the **fscanf**() as before with appropriate formatting conversion specifiers like `%s` for string, `%d` for numeric and so forth, we are restricted to intrinsic machine data types such as `'uchar'` for an unsigned character (8 bit), `'ulong'` for an unsigned long (32 bit) and so forth. To read two characters followed by one 32-bit integer, we could use:

```
ID       = fread(FileId, 2, 'uchar');
FileSize = fread(FileId, 1, 'ulong');
```

If the signal data file is stored in 8- or 16-bit binary format, then it may be necessary to use the above functions to read in the raw data. In that case, it is usually necessary to convert to floating point format using the **double** conversion function. In the following, note how the number is truncated when converted to `uint8` format (unsigned 8-bit integer), and precision is lost when converted to `uint16` format (unsigned 16-bit integer).

```
a = 321.456;
i8 = uint8 (a)
i8 =
    255
i16 = uint16 (a)
i16 =
    321
```

```
d = double(i16)
d =
   321
Whos
Name     Size      Bytes    Class
A        1 × 1     8        double array
D        1 × 1     8        double array
i16      1 × 1     2        uint16 array
i8       1 × 1     1        uint8 array
```

In a similar manner, the **fwrite**() function allows binary output to files, using the 'wb' format specifier for the **fopen**() call. Note that there is considerable scope for generating code which is machine dependent—that is, code scripts which will work on one platform (such as Windows) and not on another (such as Unix), and vice-versa. This is of course undesirable, and should be avoided whenever possible.

Most binary files will contain some information at the start of the file pertaining to the characteristics of the data, in addition to the data itself. This *header* information is somewhat specific to the type of data file. For example, an image file will require information about the width of the picture, the height of the picture, the number of colors present, and so forth.

A great many standards for data representation exist in practice. For example, a common format found on Windows is the "bitmap" file format with extension .bmp, whereas sound (audio, or "wave") files often use extension .wav. Unfortunately, however, there is a plethora of different file formats, even for (ostensibly) the same type of data. Other file formats in common use include JPEG for image data (.jpg extension), and MP3 for audio (.mp3 extension).

Examples of MATLAB built-in functions which read binary data include **imread**() to read image data, and **wavread**() to read audio (wave) files. These will be discussed in Section 2.14.

2.11 MULTIDIMENSIONAL ARRAYS

In addition to scalar quantities, vectors (one-dimensional arrays), and matrices (two-dimensional arrays), MATLAB allows for multidimensional arrays. The **cat** command is used to concatenate matrices along a given dimension to form multidimensional arrays.

Multidimensional arrays are useful in certain types of signal processing algorithms. Consider a video or television picture: it is comprised of successive still images, with each individual image in effect comprised of a matrix of pixels. What is needed is a structure to hold a series of matrices. This would, in effect, require three-dimensional indexing.

The **cat**() operator may be understood with reference to Figure 2.3. The matrices may be joined together either by abutting them horizontally, vertically, or one behind the other.

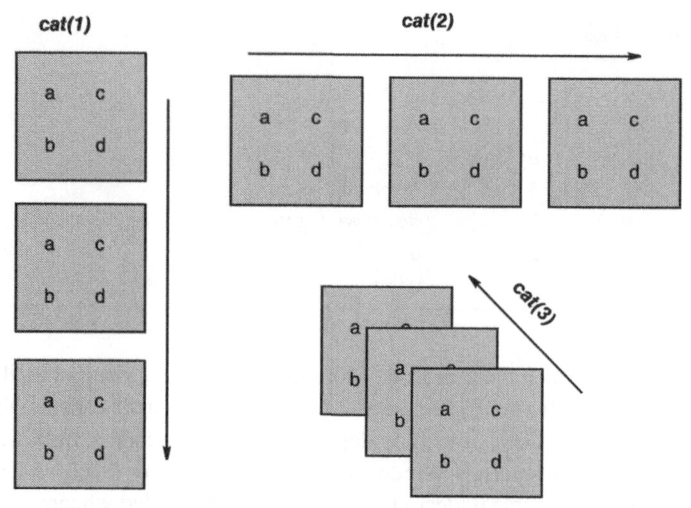

FIGURE 2.3 The MATLAB **cat**() function for multidimensional arrays. Each matrix is shown as a square, and the possible ways three matrices may be aligned and joined ("concatenated") is shown. This may be imagined as stacking the matrices vertically, horizontally, or depth-wise.

Such a data structure may be useful in processing video sequences: imagine each frame of video as a still picture. Each frame may be stored in a matrix, with the sequence of matrices comprising the video sequence stored as shown in the **cat**(3) diagram in Figure 2.3. The video image is formed by replaying the still images at the appropriate rate. Some MATLAB numerical examples of manipulating multidimensional arrays are shown below.

```
a = [0 1; 2 3];
b = [4 5 ; 6 7];
c = [8 9 ; 10 11];

md1 = cat(1, a, b, c)
md1 =
     0     1
     2     3
     4     5
     6     7
     8     9
    10    11
size(md1)
ans =
     6 2
ndims(md1)
```

```
ans =
    2

md2 = cat(2, a, b, c)
md2 =
    0   1   4   5    8    9
    2   3   6   7   10   11
size(md2)
ans =
    2   6
ndims(md2)
ans =
    2

md3 = cat(3, a, b, c)
md3(:, :, 1) =
    0   1
    2   3
md3(:, :, 2) =
    4   5
    6   7
md3(:, :, 3) =
     8    9
    10   11
size(md3)
ans =
    2   2   3
ndims(md3)
ans =
    3
```

2.12 BITWISE OPERATORS

As well as operations on integers and floating-point numbers, operations on one or more binary digits (bits) are sometimes necessary. These "bitwise" operations involve selecting out certain bits contained within an integer, inverting certain bits (converting binary 1s to 0s and vice-versa), and moving the position of bits within an integer. Bitwise operators are used in signal processing for encoding binary information for communications and storage. After all, any storage or transmission is ultimately just a sequence of bits, grouped into bytes, characters, words, and so forth. MATLAB provides many functions which operate on individual bits within an integer, some of which are listed in Table 2.1. Section 3.4 deals with binary numbers in much greater detail.

TABLE 2.1 Some MATLAB Functions that Operate at the Bit Level

`bitand`	Logical AND operator
`bitor`	Logical OR operator
`bitxor`	Logical XOR operator
`bitget`	Return a bit's value
`bitset`	Set a bit's value
`bitcmp`	Complement bits
`bitshift`	Shift a word left or right

The example below illustrates the use of the **`bitand()`** function. The use of the other bitwise functions is similar. The format specifier `%x` is used to specify hexadecimal (base-16) format display.

```
a = 83;      % 0101 0011 = 83H
mask = 15;  % 0000 1111 = 0FH
r = bitand(a, mask);
fprintf(1, 'AND(%02x, % 02x) = %02x\n', a, mask, r);
AND(53, 0 f) = 03
```

2.13 VECTORIZING CODE

Generally speaking, it is desirable to try to *vectorize* MATLAB code for efficiency, rather than use for/end and while/end loop constructs. The term "vectorize" comes from the implication that the operation is being performed on a sequence (or vector) of numbers all at once. For example, adding the constant 4 to a vector of 200 numbers can be optimized by loading the constant 4 and the loop counter (which counts up to 200) into registers of the CPU. Fortunately MATLAB hides most of this complexity from the programmer, but nevertheless must be given some "hints" to help it perform these types of optimizations.

The reason for the increased efficiency of code vectorization is due to several factors. Of course, a significant factor is the way MATLAB is internally optimized to deal with matrices. Other reasons include the interpretive nature of MATLAB, the way MATLAB allocates variables dynamically, and the presence of large amounts of high-speed cache memory on contemporary computer systems. Since variables are not explicitly declared, the interpreter does not know the memory size needed in advance.

The obvious way to encode the zeroing of a 1,000-element array is:

```
for ArrayIndex = 1:1000
    SampleValues(ArrayIndex, 1) = 0;
end
```

It is possible to "vectorize" the loop into a single operation:

```
SampleValues = zeros(1000, 1);
```

A range of values may be initialized as follows (here, it is integers 1, 2, ... , 999, 1,000):

```
SampleIndex = 1:1000;
```

The related **linspace()** function can be used to initialize an array of known size over a range of values from minimum to maximum. Even if it is not possible to do away with **for** loops, pre-allocating memory space saves execution time. This is typically done using an allocation of a zero matrix like that above. A little thought in using the most efficient coding method can result in considerable saving in computation time. Consider the following code:

```
% using loops
clear all

N = 1e4;
tstart = cputime;
for k = 1:N
    x(k) = k;
    y(k) = N-k + 1;
end

for k = 1:N
    z(k) = x(k) + y(k);
end
tstop = cputime;
fprintf(1, 'Time using loops = %.3f\n', tstop - tstart);
```

Simply by pre-allocating the required space for vectors x, y, and z as follows, it is possible to speed up the execution by a great deal:

```
% preallocated using loops
clear all
N = 1e6;
tstart = cputime;
x = zeros(N, 1);
y = zeros(N, 1);
z = zeros(N, 1);
for k = 1:N
    x(k) = k;
```

```
    y(k) = N-k + 1;
end

for k = 1:N
    z(k) = x(k) + y(k);
end
tstop = cputime;
fprintf(1, 'Using preallocation = %.3f\n', tstop - tstart);
```

Finally, it is often possible to take advantage of the fact that MATLAB treats all variables as matrices. The following code performs the same computation as the two previous examples, although significantly faster:

```
% using vectorized code
clear all

N = 1e6;
tstart = cputime;
k = 1:N;
x = k;
y = N-k + 1;
z = x + y;
tstop = cputime;
fprintf(1, 'Using vectorization = %.3f\n', tstop - tstart);
```

2.14 USING MATLAB FOR PROCESSING SIGNALS

We are now in a position to use MATLAB to process some signals. Once a signal is sampled (in digital form), it can be manipulated using mathematical methods. To illustrate, suppose we generate a set of N random samples in MATLAB as follows:

```
N = 1,000;
x = rand(N, 1);
```

Now we can plot the random vector x and play it through the speakers at the sample rate of 8,000 samples per second:

```
plot(x);
sound(x, 8,000);
```

Next, consider simple image manipulation in MATLAB. Create an $N \times M$ image of random pixels as follows. Note that an image is usually thought of as being *width* × *height*, whereas a matrix is specified as *rows* × *columns*.

```
M = 64;
N = 128;

x = rand(M, N);
x = floor(x*256);

image(x);
set(gca, 'PlotBoxAspectRatio', [N M 1]);
axis('off');
box('on');
```

The size of the image on screen corresponds to the size of the matrix, with the image pixels scaled to integer values in the range 0 to 255. The reason for this will be explained further in Section 3.7. The command to set the aspect ratio ensures that the pixels are square, rather than distorted to satisfy the size constraints of the available display area.

To extend this example further, we will make the image into a grayscale one, where 0 represents black, and 255 represents white. For convenience, the axis markers are turned off using **axis**('off'). The resulting image is shown in Figure 2.4.

```
image(x);
colormap(gray[256]);
axis('off');
box('on');
```

FIGURE 2.4 A random grayscale image of size 128 × 64 (that is, 64 rows and 128 columns of data).

Finally, it is time to look at some "real" signals: first, an image data file. In the following, the bitmap image file `ImageFile.bmp` is loaded using **`imread()`**, and an inverted or negative image is displayed in a figure window. The exact details of the various functions required will be deferred until Section 3.7, when the nature of the image signal itself is discussed.

```
[img cmap] = imread('ImageFile.bmp');
img = double(img);

invimg = 255 - img;
image(invimg);
colormap(cmap);

axis('off');
set(gca, 'DataAspectRatio', [1 1 1]);
```

Note that the bitmap (bmp) format in the above example stores the color palette alongside the image. If we use the above code directly on a JPEG file (.jpg extension rather than .bmp), two changes will occur. First, the image matrix returned will be a three-dimensional array, with the third dimension being 3. These are for the red, green, and blue color components. Second, even if the image is a grayscale one, the colormap will not usually be present. This is easily remedied if we assume that the image has 256 levels, and we create a synthetic colormap:

```
[img cmap] = imread('ImageFile.jpg');
img = double(img);

if( isempty(cmap) )
    % for JPEG, make 256 level colormap
    cmap = [0:255; 0:255; 0:255]'/255;
end
```

For audio (sound) signals, the **`wavread()`** function is able to load an audio file—in this case, the file `AudioFile.wav`. In the following code, the sound is played through the speakers at the correct rate, and at a slower rate (thus altering the pitch). Again, the exact details of the various functions required will be deferred until subsequent chapters, when the nature of the audio signal itself is discussed.

```
[snd fs] = wavread('AudioFile.wav');
sound(snd, fs);

tplay = length(snd)/fs;
pause(tplay + 1);
sound(snd, fs/2);
```

2.15 CHAPTER SUMMARY

The following are the key elements covered in this chapter:

- a review of MATLAB constructs and usage.
- the role of MATLAB in digital signal processing.
- reading and writing data files, audio, and image files.

PROBLEMS

2.1. Enter and test the example in Listing 2.1.

(a) Replace the loop condition with `while(r > eps)`. Is this a more appropriate test for terminating the loop? Why?

(b) What are the values for **realmin**, **realmax**, and **eps** in MATLAB?

2.2. Generate and play back a sinusoidal signal using

```
t = 0: 1/22,050: 2;
y = sin(2*pi*400*t);
sound(y, 22050);
```

(a) What is the significance of 22,050?

(b) How long will this sound be, and how many samples will there be?

(c) Repeat the sound playback using a sample rate of 8,000 Hz (samples/second) and then at 44,100 Hz. What do you notice? Can this be explained? Repeat for 44,100 Hz.

2.3. Repeat the steps used to generate Figure 2.4, substituting `x = floor(x*32);`

(a) What is the pixel range?

(b) What is the grayscale color range?

(c) Now invert the image before display, by inserting after the `rand()` function the line `x = 1 - x;` Verify that that the pixels have been inverted in magnitude.

2.4. Section 2.14 showed how to create a grayscale image, with one byte per pixel, using a colormap. To create a full color image, we need 3 bytes for red, green, and blue color components.

(a) Create a random array of size 20×10 for the red component. Repeat for blue and green.

(b) Combine these three separate arrays into one $20 \times 10 \times 3$ array as shown in Section 2.11.

(c) Display this multidimensional array using the `image()` function.

(d) Set the pixel in the upper-left corner of the array to purely red (that is, $R = 1, G = 0, B = 0$). Redisplay the image array and verify that the pixel is shown as red only.

(e) Repeat for green only and red only.

(f) Try differing color combinations. What do you need to create a yellow pixel? Black? White?

2.5. The following outline shows how to use MATLAB to sample audio input.

```
fs = 8,000;
tmax = 4;
nbits = 16;
nchan = 1;
Recorder = audiorecorder(fs, nbits, nchan);
record(Recorder);
fprintf(1, 'Recording ... \n');
pause(tmax);
stop(Recorder);
% convert to floating-point vector and play back
yi = getaudiodata(Recorder, 'int16');
y = double(yi);
y = y/max(abs[y]);
plot(y);
sound(y, fs);
```

(a) Sample your own voice, using the MATLAB code above.

(b) Determine the size of the sample vector. Does this correspond to the sample rate you selected when saving the file?

(c) Plot the entire waveform (as above), and portions of it using vector subscripting. Note how the 16-bit audio samples are converted to floating-point values for calculation.

2.6. A certain DSP problem requires a buffer of 32,768 samples. Explain why the following initialization is wrong: `x = zeros(32768);`

2.7. What does the `any()` command do? Which one of these loop constructs is probably what was intended? Why?

```
while(any[h > 0])
   h = h - 2*pi;
end
while(any(h) > 0)
   h = h - 2*pi;
end
```

2.8. We will later use the "sinc" function, defined as sin $(\pi t/T)/(\pi t/T)$, where t is a vector of time instants and T is a constant. Explain why the following is incorrect, and how it may be corrected. What is the purpose of the `eps` value?

```
T = 0.1;
t = -0.5: 0.001: 0.5;
x = pi*t/T;
y = sin(x + eps)/(x + eps);
plot (t, y);
```

2.9. Enter the example multidimensional arrays as shown in the code example in Section 2.11. By writing out the matrices and their elements, verify that the `cat()` command operates as you expect.

CHAPTER 3

SAMPLED SIGNALS AND DIGITAL PROCESSING

3.1 CHAPTER OBJECTIVES

On completion of this chapter, the reader should be able to

1. explain the concepts of *sampling* and *reconstruction*;
2. convert decimal numbers into *unsigned binary*, 2's complement *signed* or *floating-point* formats, and convert binary numbers back to decimal;
3. explain *quantization* and signal-to-noise ratio (SNR);
4. be able to use *difference equations* and manipulate them for a given task;
5. be able to iterate difference equations and turn this into code for a digital signal processing *(DSP) implementation*; and
6. explain ideal *reconstruction* and derive the polynomial *interpolation* for a signal.

3.2 INTRODUCTION

Signals invariably come from the real world. To process signals via computer, it is necessary to have some systematic means to acquire the signal so as to capture all necessary information about the signal. This signal acquisition must take into account not only the physical characteristics of the signal but also the necessity of using a numerical representation of the signal at discrete instants in time.

3.3 PROCESSING SIGNALS USING COMPUTER ALGORITHMS

Most signals are analog in nature. Sound, for example, is a pressure variation over time. Images are composed of various light wavelengths at various intensities, over a two-dimensional plane. Many other signal types exist, such as radio frequency waves, bioelectric potentials, and so forth. The key element is to first convert the

Digital Signal Processing Using MATLAB for Students and Researchers, First Edition. John W. Leis.

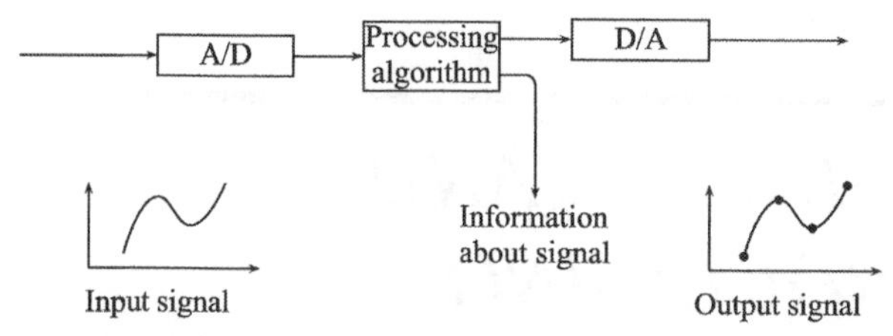

FIGURE 3.1 Simplified view of the process of discrete-time sampling. A/D represents the analog-to-digital conversion, and D/A is the digital-to-analog conversion. The purpose of sampling and analyzing the signal may be either to find some information about the signal or to output the signal in some altered or enhanced form.

signal to an electrical voltage followed by discrete sampling to convert the signal into a sequence of numbers over time. Once the signal has been captured in a digital form, it opens the door to a wide variety of signal processing techniques.

As shown in Figure 3.1, the conversion from an analog form into a digital form is accomplished using an analog-to-digital (A/D) converter (also called an ADC). This is followed by some form of algorithm (or several algorithms used in conjunction) to process the discrete sequence of numbers that represent the signal. At this point, we may be interested in some information about the signal—for example, if the signal were a heart rhythm, we may be interested in detecting the onset of an abnormal pattern and sounding an alarm.

In other situations, the output may be fed to the converse of the A/D converter, the digital-to-analog (D/A) converter (also called a DAC), in order to produce an output signal. The output may be, for example, music that has been filtered to enhance certain frequencies or an image which has had noise removed. In communications systems, the output may be a sequence of binary digits representing the samples, termed a "bitstream." This is then suitable for transmission or storage—for example, encoded audio on a CD or DVD, or perhaps audio streamed or downloaded over the Internet.

In Figure 3.1, the output is represented as dots at the discrete sampling instants. The A/D conversion and subsequent computer processing takes a certain finite amount of time. In theory, the faster the processing, the better the representation of the signal. However, this has several implications, particularly in terms of the cost of equipment (faster equipment is invariably more expensive). In addition, the amount of information processed need only be enough to faithfully represent the signal for the user—this consideration is particularly important in digital communications and storage. Naturally, the more data that are required to represent a signal, the longer it takes to transmit and the less that can be stored on a given medium.

There are several implications of discrete-time sampling followed by digital processing. The questions which immediately arise are

1. What is an appropriate sampling rate, and does it depend on the signal being sampled?

2. Since digital computers use the binary system for representing numbers, what is a suitable number of bits to use for each sample? And does that also depend on the signal being sampled?

Although the exact requirements will take some time to spell out more precisely, some "order-of-magnitude" figures may be useful at this point in time:

Digital audio uses sampling rates of the order of 40kHz, at 16 bits per sample (and of course two channels, left and right, if stereo reproduction is required). For lower quality (such as telephone speech), an 8-kHz sample rate is often employed.

Digital photographs and computer images require the order of 8 bits per sample for each primary color (red, green, and blue). An image may be of the order of 2,000 × 1,000 samples (width × height).

3.4 DIGITAL REPRESENTATION OF NUMBERS

Since we are dealing with representing signals using numbers, it is a worthwhile aim to have a working knowledge of how numbers are represented in the binary number system. This helps us in two regards. First, when an analog signal is converted into a digital representation (or vice versa), we need to be conversant with the internal number format; second, when we implement equations to manipulate those sampled signals, we need to perform arithmetic operations. The latter can be problematic and may be the source of unexpected or erroneous results. We need to be conversant with issues such as rounding and truncation of numerical results, as well as with less obvious issues with precision. Finally, calculation time is often an important issue in real-time DSP, and some appreciation of the computational complexity of the various arithmetic operations helps improve our designs. In the following sections, the intention is not to give an exhaustive treatment since there is ample material in this area to fill entire textbooks. Rather, the salient features are addressed as they apply to DSP, as a starting point for further investigation if warranted.

3.4.1 Representing Integers

The basic representational unit in the binary system is the *bit* or the binary digit. This is a base 2 number, having values of 0 or 1 only. Combining several bits into higher groupings allows us to represent a range of numerical values. Typical groupings used are a nibble (4 bits), a byte (8 bits), and a word (16, 32, or 64 bits, depending on the context). The place value system follows the same rules as the decimal system, except that the base is 2 rather than 10. The following shows an example of an 8-bit unsigned number:

2^n	2^7	2^6	2^5	2^4	2^3	2^2	2^1	2^0	
Decimal	128	64	32	16	8	4	2	1	
Example	0	0	1	0	1	0	0	1	$(1 + 8 + 32 = 41)$

This is an 8-bit quantity, so we have $N = 8$ bits. The 2^0 bit position is usually called bit 0 or the least significant bit (LSB), and the 2^{N-1} bit position is called the most significant bit (MSB). Mathematically, the bits a_i sum together to form a positive integer, A, according to

$$A = \sum_{i=0}^{N-1} 2^i a_i \quad \text{for } A \geq 0 \quad \text{and} \quad a_i = 0 \text{ or } 1. \tag{3.1}$$

An N-bit number with all 1's is the equivalent of $2^N - 1$. Thus, an 8-bit number would have $N = 8$, and the binary value 1111 1111 is 255 in decimal $(1 + 2 + \cdots + 64 + 128 = 255)$. This is equivalent to $256 - 1 = 255$ in this case, or $2^N - 1$ in the general case for any N.

3.4.2 Signed Integers

Signed integers are usually (though not always) represented by so-called 2's complement notation. In this approach, the MSB represents the sign of the number: 0 for positive, 1 for negative. For positive numbers, we take the same approach as with unsigned numbers to determine the value. For negative numbers (MSB = 1), we must first invert all binary values (convert 1 to 0 and 0 to 1) and add one. As a simple example, consider the 8-bit number

$$1111\,1110.$$

This is a negative number, so we must follow the 2's complement procedure. Inverting all bits, we obtain

$$0000\,0001.$$

Then, adding one, we have

$$0000\,0010.$$

This is +2 in decimal, so the original (negative) number was −2.

If we were to interpret 1111 1110 as an unsigned number, it would be 254. So the 2's complement notation is equivalent to $2^N - A$, where A is the unsigned representation. In this case, $2^8 - 2 = 256 - 2 = 254$.

Whether a number is signed or unsigned is a matter of interpretation in the context in which it is used. Table 3.1 shows the signed and unsigned equivalent values of a 4-bit binary number.

The key advantage of 2's complement representation is that arithmetic operations perform as expected. For example, $-2 + 5 = 3$, $-4 + 4 = 0$, and so forth. Note that in Table 3.1, the range of unsigned numbers is $0 \rightarrow (2^N - 1)$ and that for signed numbers is $-2^{N-1} \rightarrow +(2^{N-1} - 1)$. For positive and negative numbers, the 2's complement notation may be unified to become

$$A = -2^{N-1} a_{N-1} + \sum_{i=0}^{N-2} 2^i a_i. \tag{3.2}$$

TABLE 3.1 Signed and Unsigned Interpretation of a 4-Bit Quantity

Binary	Unsigned	Signed	Binary	Unsigned	Signed
0000	0	0	1000	8	−8
0001	1	1	1001	9	−7
0010	2	2	1010	10	−6
0011	3	3	1011	11	−5
0100	4	4	1100	12	−4
0101	5	5	1101	13	−3
0110	6	6	1110	14	−2
0111	7	7	1111	15	−1

This is valid for both positive (sign bit $a_{N-1} = 0$) and negative ($a_{N-1} = 1$) numbers.

MATLAB has data types for commonly used standard integer representations. These are given mnemonics such as `uint8` for unsigned 8-bit integers, `int8` for signed 8-bit integers, and so forth. The following example shows how to determine the maximum and minimum ranges for some integer types—these should be studied in conjunction with Table 3.1. Note also how the logarithm to base 2 is used to determine the number of bits employed.

```
intmin('uint8')
ans =
    0
intmax('uint8')
ans =
    255
intmax('int8')
ans =
    127
intmin('int8')
ans =
    -128
log2(double(intmax + 1))
ans =
    31
intmax('int16')
ans =
    32767
intmin('int16')
ans =
    -32768
intmax('int32')
```

```
ans =
      2147483647
intmax('int64')
ans =
      9223372036854775807
```

3.4.3 Binary Arithmetic

We now turn to the basic arithmetic of binary numbers. The basic rules are identical to decimal arithmetic with regard to place value and carry from one place to the next. So, for adding pairs of bits, we have

$$
\begin{aligned}
0+0&=0\\
0+1&=1\\
1+0&=1\\
1+1&=10 \text{ (0 with 1 carry, decimal 2).}
\end{aligned}
$$

It will be helpful in the following examples to also note that

$$1+1+1=11 \text{ (1 with carry 1, decimal 3).}$$

Consider the binary addition of 8-bit numbers 13 + 22, as follows. We add bitwise as shown next, noting the carry of 1's. The decimal values of each binary number are shown in brackets for checking:

$$
\begin{array}{rll}
 & 0\,0\,0\,0\ \ {}^{1}1\,1\,0\,1 & (1+4+8=13) \\
+ & & \\
 & \underline{0\,0\,{}^{1}0\,{}^{1}1\ \ 0\,1\,1\,0} & (2+4+16=22) \\
 & \underline{0\,0\ \ 1\ \ 0\ \ 0\,0\,1\,1} & (1+2+32=35).
\end{array}
$$

Next, consider multiplication of binary numbers. For the specific case of multiplication by 2, a shift-left by one bit position is all that is required. For multiplication by 4, a shift of two bit positions is required, and so forth. Similarly, for division by two, a shift-right by one bit position gives the correct result (assuming we ignore any overflow bits and do not require fractions).

Generally speaking, multiplication algorithms are implemented in DSP hardware and are not usually required to be coded in software. However, understanding the underlying algorithms can yield significant insight, especially into performance-related issues. To understand the procedure, it helps to revisit how we approach multiplication in decimal arithmetic.

For example, if we had to calculate 234 × 23, it would be equivalent to adding the value 234 to itself in a loop of 23 iterations or, equivalently, adding 23 to itself over 234 iterations. Such an approach would be quite slow to execute, depending as it does on the magnitude of the numbers, and so highly undesirable. But using the concept of place value, we can employ the elementary multiplication algorithm as follows:

$$
\begin{array}{rrrrl}
 & {}^{1}2 & {}^{1}3 & 4 & \times \\
 & & 2 & 3 & \\
\hline
 & 7 & 0 & 2 & \\
+ & & & & \\
{}^{1}4 & 6 & 8 & \cdot & \\
\hline
5 & 3 & 8 & 2. &
\end{array}
$$

Here, we multiply 3 by 234 and add it to the shifted product of 2 and 234. The multiplications are by one digit, and the shifting is in effect multiplication by 10 (the base). A little thought shows that the time taken to perform this operation is independent of the magnitude of the values. It is dependent on the number of figures in each value, but this is essentially constant in any given problem, although we may prepend leading zeros (thus, for calculation to four digits, 234 becomes 0234, and 23 becomes 0023).

In the binary system, we can follow a similar procedure. Consider the following example of the binary multiplication of 13×6:

$$
\begin{array}{rl}
1101\times & (1+4+8=13) \\
0110 & (2+4=6) \\
\hline
0000 & \\
1101\cdot & \\
1101\cdot\cdot & \\
0000\cdot\cdot\cdot & \\
\hline
1001110 & (2+4+8+64=78).
\end{array}
$$

In fact, it is somewhat simpler than decimal multiplication in that the multiplication by one digit at each step is actually multiplication by zero (with a zero result always) or one (which is the original number repeated).

3.4.4 Representing Real Numbers

Consider the following hypothetical example, in which we take the square root of a number N times and then square the result N times iteratively.

```
x = 11;
N = 50;
for k = 1:N
    x = sqrt(x);
    fprintf(1, '%d %.20 f\n', k, x);
end
for k = 1:N
    x = x^2;
    fprintf(1, '%d %.20 f\n', k, x);
end
```

One would expect that the result ought to always be what we started with—in this case, 11. But running the above for a reasonable number of iterations (say, 50) yields the following:

```
1     3.31662479035539980000
2     1.82116028683787180000
3     1.34950371871954160000
...
48    1.00000000000000840000
49    1.00000000000000420000
50    1.00000000000000200000

1     1.00000000000000400000
2     1.00000000000000800000
3     1.00000000000001600000
...
48    1.75505464632777850000
49    3.08021681159672370000
50    9.48773560644308670000
```

So we do not in fact end up where we started. What we have ended up with is a result of numerical (im)precision. If we increased N above, eventually we would have a result of 1 exactly after the first loop, and it would not matter how many iterations the second loop performs; the end result would still be 1. This may seem like a contrived case, but it illustrates the fact that numerical precision cannot be taken for granted. DSP algorithms often run continually, essentially indefinitely, and any rounding errors may well accumulate for long running times and/or large data sets.

So that brings us to the question of how to represent fractions. One method using an implied or fixed binary point is shown next:

2^n	2^3	2^2	2^1	2^0	$\cdot$	2^{-1}	2^{-2}	2^{-3}	2^{-4}	
Decimal	8	4	2	1		$\frac{1}{2}$	$\frac{1}{4}$	$\frac{1}{8}$	$\frac{1}{16}$	
Example:	0	0	1	0	$\cdot$	1	0	0	1	$\left(2+\frac{1}{2}+\frac{1}{16}=2.5625\right)$.

This *fixed-point* format satisfactorily represents a range of numbers but has some shortcomings. For one, the maximum value is limited by the number of bits to the left of the binary point. Similarly, the minimum fraction is limited by the number of bits allocated to the right (here, the minimum step size is $\frac{1}{16}$). A little thought also reveals that not all numbers can be represented exactly, and we must be satisfied with the nearest approximation. For example, 0.21 decimal could not be represented exactly. The closest we could come is 0000·0011 to give 0.1875 decimal.

What is needed is a more general system for representing very large and very small numbers as accurately as possible. Floating-point numbers give us this flexibility.

3.4.5 Floating-Point Numbers

The final extension in the representation of numbers is the *floating-point* format. In fixed-point binary, the binary point is fixed and cannot move. In floating-point representations, the binary point *can* move, giving extra range and precision than if we used the same number of bits in a fixed-point representation. There have been many floating-point schemes proposed and implemented over the years, and here we cover the basic concepts using the Institution of Electrical and Electronic Engineers (IEEE) 754 standard for floating-point numbers.[1]

The fundamental change with floating-point numbers is that the binary point can move, and this fits naturally with a normalized representation. Normalized numbers are where we have one nonzero digit to the left of the decimal point. For example, the following are normalized decimal numbers:

$$2{,}153 = 2.153 \times 10^{3}$$
$$0.0167 = 1.67 \times 10^{-2}.$$

So in general, we have

$$\text{Number} \rightarrow \pm sF \times B^{\pm E}, \tag{3.3}$$

where

s = sign, 0 or 1 (representing + or −);

B = base, implied 2 for binary (base 10 for decimal);

F = fractional part; and

E = exponent (power of the base).

The fractional part F is also called the "significand" or "mantissa." The exponent E could be negative to represent small numbers. In binary, we have

$$\text{Number} \rightarrow \pm 1.ffff \ldots \times 2^{\pm E}. \tag{3.4}$$

The f's represent bits (0 or 1), and there is an implied 1 to the left of the significand. In normalizing decimal numbers, this position would be a zero. But because we are using the binary system, this position could only be a 1 or a 0, and so we may as well make it a 1 to gain an extra bit's worth in our representation. The IEEE754 standard specifies a total of 32 bits for a single-precision quantity, as well as 64-bit extended precision and 80-bit double precision (4, 8, and 10 bytes, respectively). In the C language, these are called float, double, and long double, which may give rise

[1] http://grouper.ieee.org/groups/754/.

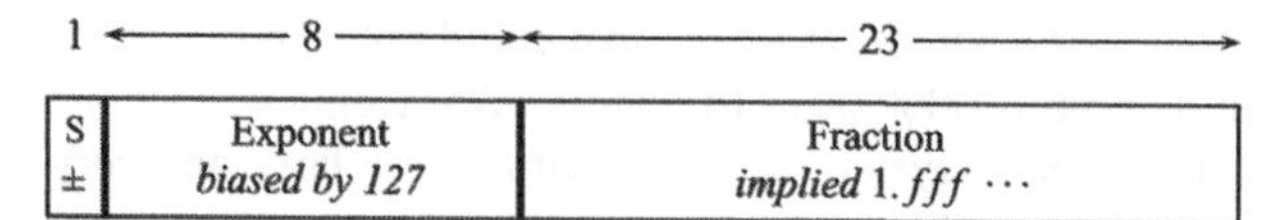

FIGURE 3.2 Floating-point format for a single-precision 32-bit format. The 64- and 80-bit formats follow a similar pattern, with larger exponent and fraction fields, to give both a larger range of values able to be represented, as well as higher precision.

to confusion in the terminology. On general-purpose CPUs, floating-point registers are usually 10 bytes, so there is no speed advantage in using lesser precision (and it may actually be slower due to the conversion required). This is summarized below for convenience.

C language data type	IEEE754 name	Size (bits)	Size (bytes)
Float	Single precision	32	4
Double	Double precision	64	8
Long double	Double extended	80	10

We now describe in more detail the 32-bit single-precision format of IEEE754. The same principles apply to 64- and 80-bit floating-point data types, but the 32-bit format is convenient to deal with for the purposes of explanation.

The bit layout is shown in Figure 3.2. In this layout, we have a single sign bit (1 = +, 0 = −), an 8-bit exponent, and a 23-bit fraction. The exponent is stored as a biased number, and so it is effectively the actual exponent plus 127. Thus, a binary range of 0–255 is actually an exponent range of −127 to +128. The 23-bit fraction part has an implied 1 before the binary point. Below are some examples to illustrate the storage format and the conversion process.

Binary to floating point is accomplished as follows. The binary representation is known, and we wish to determine the corresponding floating-point value.

Start with a given single-precision binary value:

$$01000010011000101000000000000000.$$

Referring to Figure 3.2, this is stored as

Sign	Exponent	Fraction
0	10000100	11000101000000000000000
+	132	1.76953125

This is derived as follows. The sign bit is 0; hence, the number is positive. The exponent is 10000100 binary or 132 decimal. So, the actual multiplier is $2^{132-127} = 2^5 = 32$.

The fractional part is assembled by taking the bits of the fraction and multiplying by the place value:

$$
\begin{aligned}
&1\times 2^{-1}+1\times 2^{-2}+\\
&0\times 2^{-3}+0\times 2^{-4}+0\times 2^{-5}+\\
&1\times 2^{-6}+0\times 2^{-7}+1\times 2^{-8}+\\
&0\times 2^{-9}+\cdots\\
&=\frac{1}{2}+\frac{1}{4}+\frac{1}{64}+\frac{1}{256}\\
&=0.7695312500.
\end{aligned}
$$

The number to the left of the decimal point is 1 (a 1 bit is implied), so that the number is

$$1.7695312500\times 2^{(132-127)}=56.6250.$$

Floating point to binary is accomplished as follows. We are given a floating-point value of 56.625 and wish to determine the corresponding binary value. The binary value is derived as follows. First, $2^5=32$ is the highest 2^n integer below 56, so the exponent is 5. This is stored as $127+5=132$. The fraction is then $56.625/2^5=1.76953125$. We remove the implied 1 at the start, so the value stored is 0.76953125. We now need to reduce this fraction to binary, using an iterative procedure:

0.76953125	×2	=	1.5390625	→	1
0.5390625	×2	=	1.0781250	→	1
0.0781250	×2	=	0.156250	→	0
0.156250	×2	=	0.312500	→	0
0.312500	×2	=	0.6250	→	0
0.6250	×2	=	1.250	→	1
0.250	×2	=	0.50	→	0
0.50	×2	=	1.0	→	1
0.0	×2	=	0.0	→	0
0.0	×2	=	0.0	→	0
⋮	⋮	⋮	⋮	⋮	⋮

At each stage, we multiply by 2 and take the value to the left of the decimal point. This can only be 1 or 0. We take the fractional remainder and repeat on the next line. Finally, the binary numbers are read down the right-hand column. The resulting 32-bit value is formed as follows:

Sign	Exponent	Fraction
0	10000100	11000101000000000000000
+	132	1.76953125

Finally, there are some special values defined by the standard worth mentioning. Zero is stored as all zeros, except for the sign, which can be 1 or 0. Plus/minus infinity (as might result from, say, 1/0) is stored as 1's in the exponent, with the significand all zeros. Non-a-number (NaN) is an invalid floating-point result, which sometimes arises (e.g., 0/0). It is stored as all ones in the exponent with a significand that is not all zeros.

In the light of the above explanations, it is instructive to see how MATLAB reports the results of some floating-point constants. In the following, *single* is single-precision 32-bit floating point; *double* is 64-bit format, and *eps* is the smallest increment defined for the given data type:

```
log2(realmax('single'))
ans =
    128
log2(eps('single'))
ans =
    -23
log2(realmax('double'))
ans =
    1024
```

3.4.6 The Coordinate Rotation for Digital Computers (CORDIC) Algorithm

We finish this section with a discussion on the more advanced topic of how to calculate trigonometric and other functions. This can be accomplished using a power series iteration (as will be seen in subsequent chapters); however, one particular algorithm of note has found widespread application. The CORDIC algorithm is an ingenious method of computing various arithmetic functions using only straightforward binary operations with some precomputed values.

The basis of the CORDIC algorithm is the anticlockwise rotation of a point, P, through an angle, θ, as shown in Figure 3.3. Of course, a clockwise rotation is easily accomplished by rotating through $-\theta$.

A standard derivation yields the rotation matrix for computing the new point from the old:

$$\begin{pmatrix} x_{\text{new}} \\ y_{\text{new}} \end{pmatrix} = \begin{pmatrix} \cos\theta & -\sin\theta \\ \sin\theta & \cos\theta \end{pmatrix} \begin{pmatrix} x \\ y \end{pmatrix}. \tag{3.5}$$

So in other words, a rotation of θ yields

$$\begin{aligned} x_{\text{new}} &= x\cos\theta - y\sin\theta \\ y_{\text{new}} &= x\sin\theta + y\cos\theta. \end{aligned} \tag{3.6}$$

The fundamental insight of the CORDIC family of algorithms is that the $\tan(\cdot)$ function can be expressed as a power 2^{-i}, where i is an integer. Furthermore, since

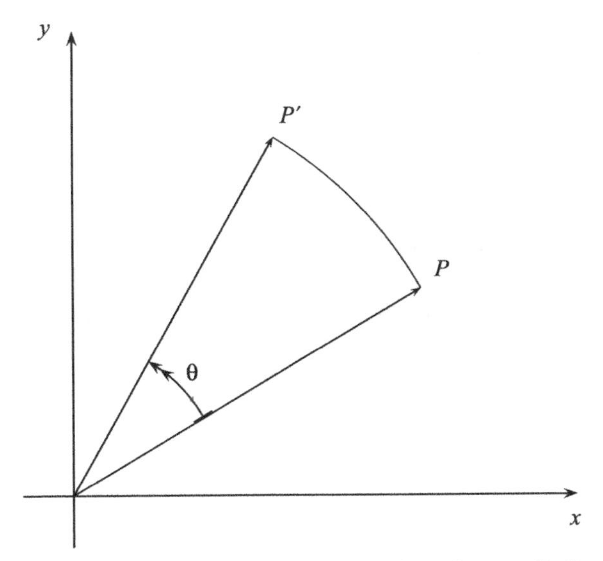

FIGURE 3.3 Rotation of a point, P, through an angle, θ, to point P'.

multiplication by a negative power of 2 (i.e., division) is a shift-right of the binary number, the operations can be accomplished using a relatively straightforward digital logic.

We first rearrange the matrix form into an iterative form, where smaller rotations of θ_i are used to successively rotate the initial vector to the final vector, through an angle, θ. The iterations are denoted by i, which corresponds to the ith bit:

$$\begin{pmatrix} x_{i+1} \\ y_{i+1} \end{pmatrix} = \begin{pmatrix} \cos\theta_i & -\sin\theta_i \\ \sin\theta_i & \cos\theta_i \end{pmatrix} \begin{pmatrix} x_i \\ y_i \end{pmatrix}. \tag{3.7}$$

We now factor out the $\cos(\cdot)$ function from the matrix:

$$\begin{pmatrix} x_{i+1} \\ y_{i+1} \end{pmatrix} = \cos\theta_i \begin{pmatrix} 1 & -\tan\theta_i \\ \tan\theta_i & 1 \end{pmatrix} \begin{pmatrix} x_i \\ y_i \end{pmatrix}. \tag{3.8}$$

To achieve the aim of a straightforward binary implementation, we let each $\tan\theta_i$ value be defined as a power of 2:

$$\begin{aligned} \tan\theta_i &= 2^{-i} \\ \therefore\quad \theta_i &= \tan^{-1} 2^{-i}. \end{aligned} \tag{3.9}$$

This has the aforementioned effect of allowing the multiplication to be performed by a binary shift. We apply the rotation successively, starting at angle θ and initial points (x, y). The successive rotations by angle θ_i mean that the multiplicative constant over N bits becomes the product of all the $\cos\theta_i$ terms:

$$K = \prod_{i=0}^{N-1} \cos\theta_i \tag{3.10}$$

$$= \prod_{i=0}^{N-1} \cos\left(\tan^{-1} 2^{-i}\right). \tag{3.11}$$

For example, letting $N = 8$ bits, $K = 0.60725911229889$. Since this constant is independent of the rotations, it can be precomputed and stored. The iteration itself becomes a process of forcing θ to zero and of selecting a direction of rotation at each iteration. We select the rotation according to whether θ is positive or negative, as shown in Equation 3.12:

$$\begin{array}{ll} \theta < 0 & \theta > 0 \\ \delta x = 2^{-i} y & \delta x = -2^{-i} y \\ \delta y = -2^{-i} x & \delta y = 2^{-i} x \\ \delta\theta = \theta_i & \delta\theta = -\theta_i. \end{array} \tag{3.12}$$

These come directly from the rotation matrix using tan $(\cdot)$ functions. Each step is now a simple addition (or subtraction), with the the $2^{-i}\ x$ and $2^{-i}\ y$ terms easily computed, since they are binary shifts. At each step, the x, y, and θ values are updated according to their respective increments:

$$\begin{aligned} x + \delta x &\rightarrow x \\ y + \delta y &\rightarrow y \\ \theta + \delta\theta &\rightarrow \theta. \end{aligned}$$

At the end of the iterations, the multiplication by K performs the necessary scaling. This constant can be precomputed to any desired precision and stored in memory.

Listing 3.1 shows a straightforward implementation of the above algorithm. Note that we have used floating-point calculations throughout, but of course the multiplication by 2^{-i} requires only a shift operation if we use binary registers.

Figure 3.4 illustrates the rotation for a certain starting point and angle. Figure 3.5 shows the value of θ over each iteration as it is forced to zero.

Since the factor K was taken out of the iteration, we must multiply by it back in at the end. For the case illustrated in Figure 3.4 with $N = 8$ bits, we initialize the iterations with $x = 1$, $y = 0$, and $\theta = \pi/3$. Note that $N = 8$ does not yield a very high level of accuracy but is useful for the purposes of example to show the errors incurred.

Equation 3.6 yields $\cos\frac{\pi}{3}$ and $\sin\frac{\pi}{3}$, and we have the following results:

Parameter	Value
N	8
K	0.60725911229889
$\cos\frac{\pi}{3}$	0.50000000000000
Kx	0.50399802184864
$\sin\frac{\pi}{3}$	0.86602540378444
Ky	0.86370480719553

Listing 3.1 CORDIC Algorithm Demonstration

```
x = 1;
y = 0;
theta = pi/3;
nbits = 8;
% determine K = cos(atan()) constant
K = 1;
for i = 0:nbits-1
    thetai = atan( 2^(-i) );
    K = K*cos(thetai);
end
xtarget = (x*cos(theta) - y*sin(theta))/K;
ytarget = (x*sin(theta) + y*cos(theta))/K;
for i = 0:nbits-1
    % atan (theta)—since it is constant for each step,
    % it would be precomputed in practice and stored in a
    % table
    p = 2^(-i);
    thetai = atan(p);
    if ( theta < 0.0 )
        dx = y*p;
        dy = -x*p;
        dtheta = thetai;
    else
        dx = -y*p;
        dy = x*p;
        dtheta = -thetai;
    end
    x = x + dx;
    y = y + dy;
    theta = theta + dtheta;
end
```

Lastly, note that CORDIC is not one algorithm but a family of algorithms for computing various trigonometrical and other functions. The above is only an introduction to the topic in order to give sufficient basis for understanding the fundamental principles.

3.4.7 Other Issues

Complex numbers are often required in signal processing algorithm implementations. These may be stored as real numbers in rectangular form $(x + \jmath y)$ or polar form $(re^{\jmath\theta})$. When assessing computational complexity, it is important to remember

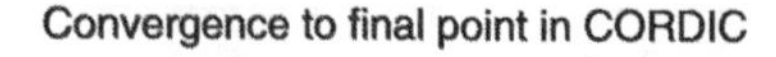

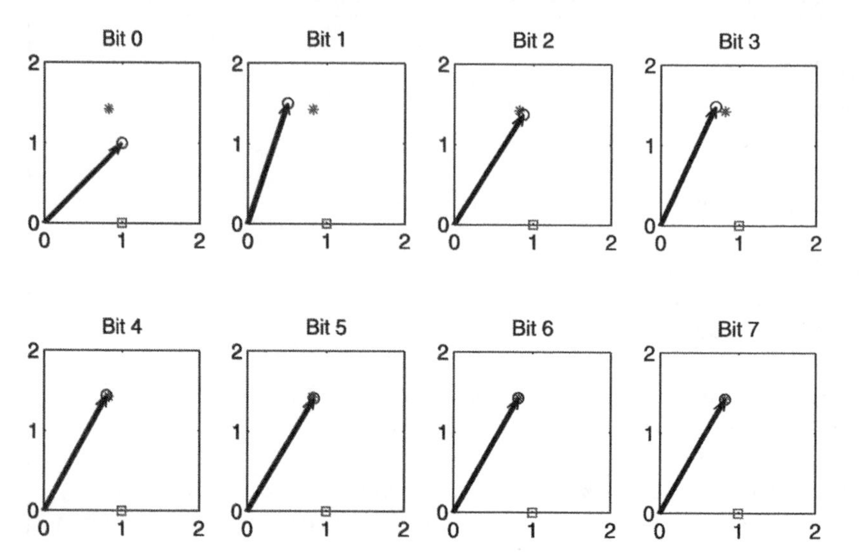

FIGURE 3.4 Convergence of the CORDIC algorithm over an 8-bit range. The box (□) shows the starting point (in this case, $x = 1$, $y = 0$). The asterisk (*) shows the target, which in this case is $((x \cos \theta - y \sin \theta)/K, (x \sin \theta + y \cos \theta)/K)$, with angle $\theta = \pi/3$. The circle and vector show the current point at each iteration.

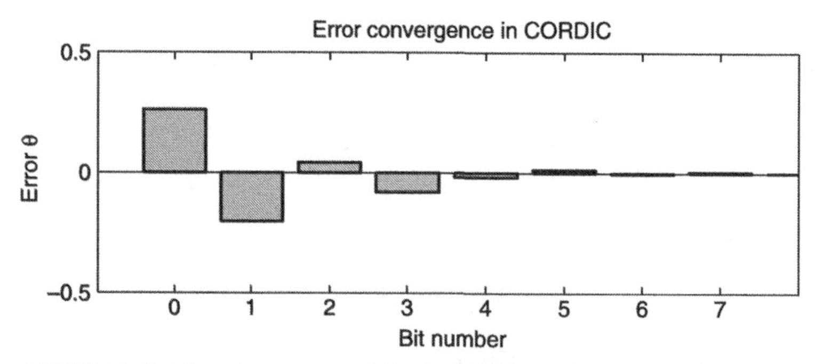

FIGURE 3.5 The convergence of the CORDIC algorithm. At each step, the angle θ is forced toward zero. Each angular update step corresponds to an angle of 2^{-i}, where i is the bit position starting at zero.

that multiplication of two complex numbers effectively requires several real-number multiplication and addition/subtraction operations as is easily seen by expanding the rectangular form:

$$(a + jb)(c + jd) = (ac - bd) + j(ad + bc). \tag{3.13}$$

In a practical sense, it is also important to remember to initialize all variables at the start of any iterative algorithms. As will be seen, we normally assume zero initial conditions, meaning we assume all past sample values are zero at the start of an iteration. Since memory may be initialized to random values, it follows that the integer or floating-point values we see, if not initialized, may start at random or unpredictable values. This can lead to puzzling and sometimes catastrophic behavior, for example, if a value happened to be initialized to the floating-point equivalent of infinity.

3.5 SAMPLING

Sampling takes place at discrete-time instants, and the signal amplitude is approximated by a binary level. Regular time sampling is examined in this section, followed by "amplitude quantization" (or simply "quantization") in subsequent sections.

Figure 3.6 shows a conceptual realization of sampling. Since the real analog signal might vary in the time it takes to sample it, the signal must be "frozen" while sampling takes place. This is shown as a switch which closes every T seconds, followed by the "hold" device. T is termed the *sample period* and is measured in seconds or, more commonly, in milliseconds or microseconds. The reciprocal, f_s, is the sample rate and is measured in samples per second or hertz.

$$\text{Sample period} = T \text{ s}$$

$$\text{Sample rate} = f_s \text{ Hz}$$

$$f_s = \frac{1}{T}.$$

The term "zero-order hold" (ZOH) is used to describe the fact that the signal is held constant during the sampling. The reconstruction after this is the "stairstep" pattern as shown in Figure 3.6. In the absence of any further information about the behavior of the signal *between* the sampling instants, this may appear to be the best we can do. But, as will be shown in subsequent sections, it can be improved upon.

The output of the A/D converter is then a sequence of binary numbers representing the amplitudes at the sampling instants. These are termed $x(n)$, where the subscript n means "the sample at time instant n." Thus, $x(n-1)$ is the measured

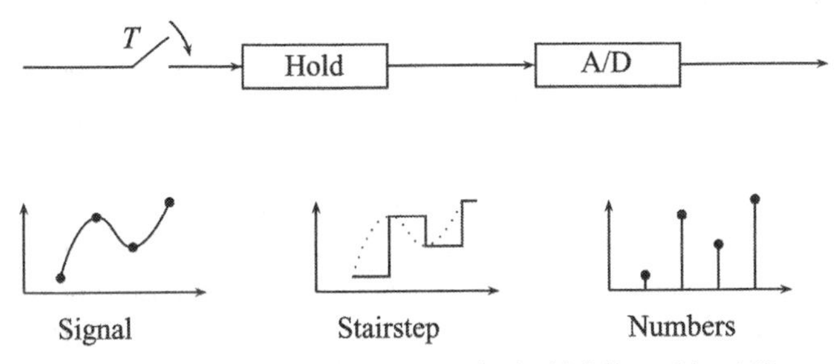

FIGURE 3.6 Sampling as a zero-order hold followed by A/D conversion. The switch closes at intervals of T in order to sample the instantaneous value of the input.

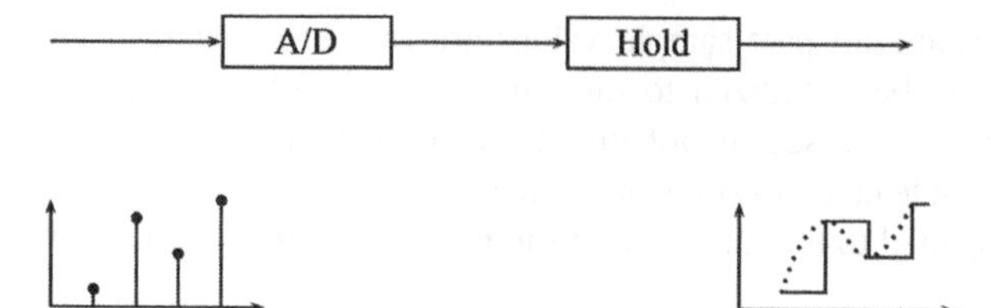

FIGURE 3.7 Reconstruction performed as digital-to-analog conversion followed by a zero-order hold.

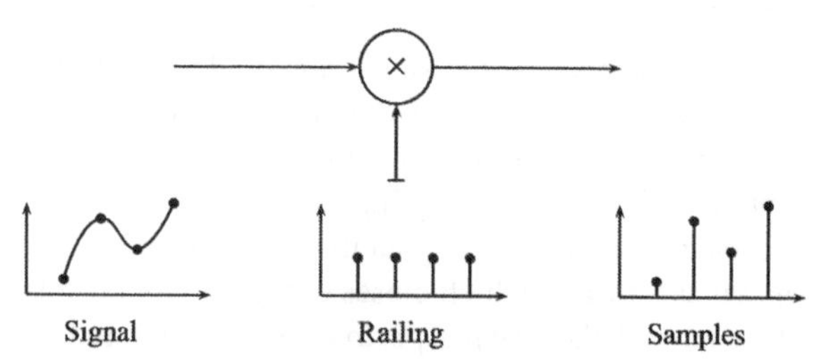

FIGURE 3.8 The process of sampling may be imagined as the continuous signal being multiplied by unit sample impulses, each spaced at the precise sampling interval.

sample value one sample instant ago, and $x(n-4)$ is the sample value $4T$ seconds ago.

If the discrete number stream were to be reversed, that is, converted back into an analog form, we would require an arrangement like that shown in Figure 3.7. The D/A converter converts the digital (binary) value into an analog voltage and holds it for T seconds. Note that the ZOH operation is in fact implicit in the D/A operation (again, because of the absence of any meaningful information about the true signal value between the sampling instants).

Mathematically, the operation of sampling may be realized as a multiplication of the continuous signal $x(t)$ by a discrete sampling or "railing" function $r(t)$, where

$$r(t)=\begin{cases}1.0 & : \quad t=nT \\ 0 & : \quad \text{otherwise.}\end{cases} \tag{3.14}$$

These sampling impulses are simply pulses of unit amplitude at exactly the sampling time instants. The sampled function is then as illustrated in Figure 3.8, and mathematically is

$$\begin{aligned}x(n)&=x(t)r(t)\\&=x(nT).\end{aligned} \tag{3.15}$$

Note that some texts use the notation $x(n)$ or $x(nT)$ rather than $x(n)$ to indicate that the signal x is sampled rather than the continuous time signal $x(t)$.

Figure 3.9 illustrates this quantization process operating on a representative signal. The "stem-and-circle" plots are used to indicate a sample. Each circle represents the amplitude of the particular sample. This type of plot is easily generated in MATLAB using the `stem()` command, as shown next for a 10-point random signal:

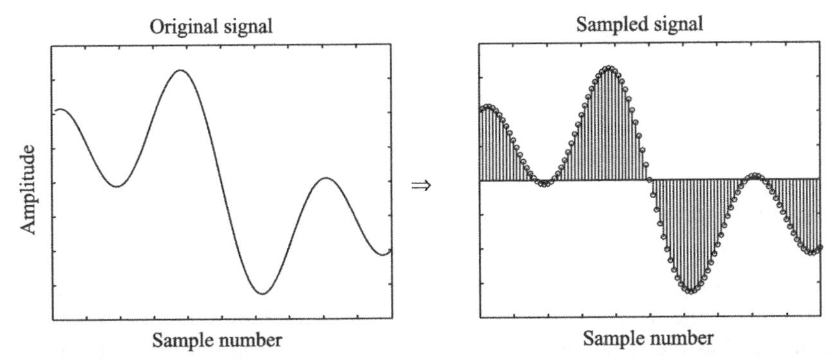

FIGURE 3.9 Converting a continuous signal to the corresponding discrete-time representation. The smoothly varying, continuous analog signal is sampled at discrete instants. These are represented by the lines at each sample instant, with the sample value represented as a circle.

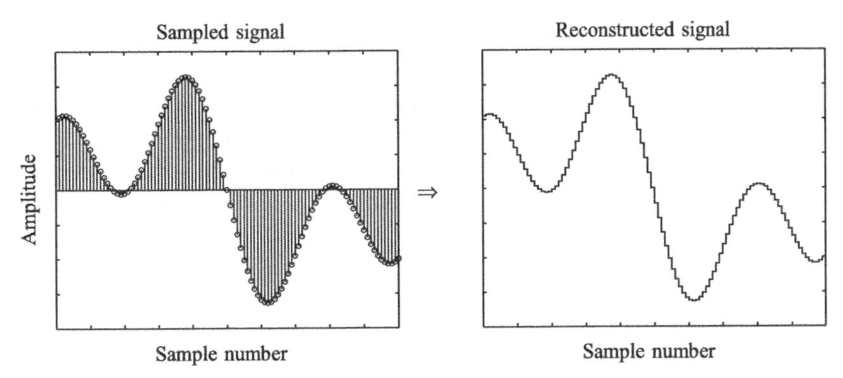

FIGURE 3.10 The process of signal reconstruction. Each discrete sample value is held until a new sample arrives, thus approximating the original continuous signal as a stairstep approximation. Further filtering smooths out the edges of the step, yielding the reconstructed signal.

```
x = randn(10,1);
stem(x);
grid on;
```

Figure 3.10 shows the process of reconstruction from the discrete samples through D/A conversion. This type of plot is easily generated in MATLAB using the **stairs**() command, as shown in the following example for a 10-point random signal:

```
x = randn(10,1);
stairs(x);
grid on
```

3.6 QUANTIZATION

Amplitude quantization is the process of representing the real analog signal by some particular level in terms of a certain N-bit binary representation. Obviously, this introduces some error into the system because only 2^N discrete levels may be represented rather than the theoretically infinite number of levels present in the analog signal. To illustrate this, consider Figure 3.11, which shows a 3-bit quantizer. The input or "decision" levels are shown on the horizontal axis as $x(t)$ since the input signal is continuous. The output or "reconstruction" levels are shown in the vertical axis as x_k for discrete integer k. Any of the three input amplitudes illustrated for $x(t)$ will be encoded as the same binary value of 010 and will produce the same output level when reconstructed at the D/A converter.

We now further explore the question of the number of bits to use in quantizing a particular signal. We do this first from an intuitive point of view and subsequently derive a theoretical result regarding the effect of the number of bits on the error introduced when a signal is quantized.

There are two issues which naturally arise from the process of quantizing a real-valued signal into a discrete representation. First, and easily overlooked, is the issue of the maximum range of the signal. Since we are using a finite number of representational levels, they must span a certain signal range. Any input signal outside that range will be truncated or "clipped." The second issue is that, necessarily, a finite number of levels reduce the representational accuracy. This is termed the "quantization noise."

Both of these factors—the maximum or dynamic range and the quantization accuracy—should be considered together. For a fixed number of bits to allocate, we

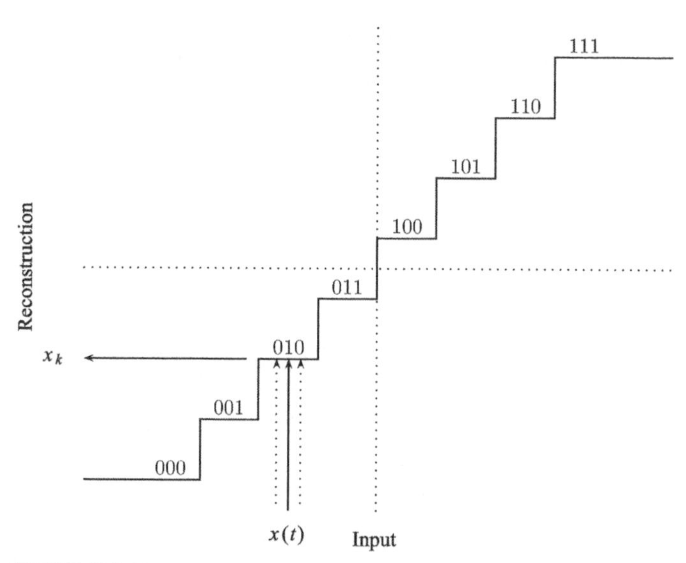

FIGURE 3.11 A scalar quantizer characteristic, showing the input level $x(t)$, the binary output for each step, and the corresponding discrete reconstruction level x_k.

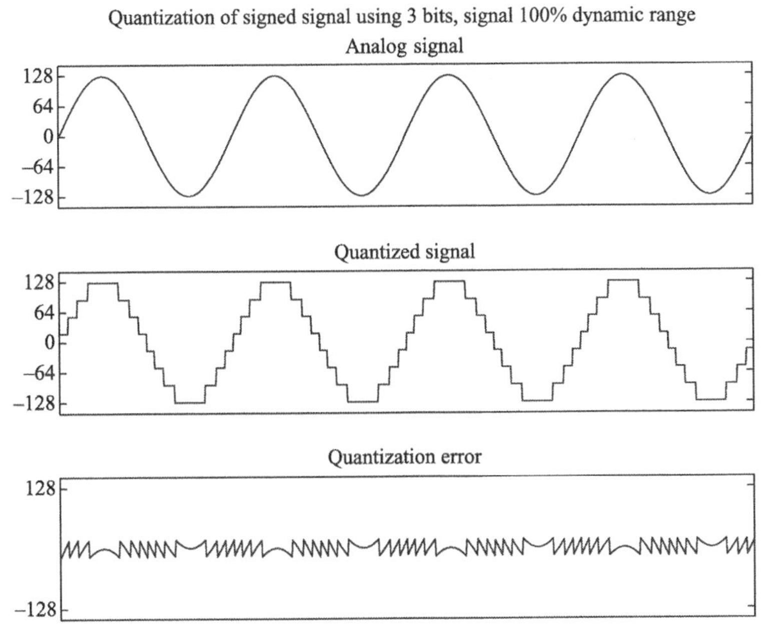

FIGURE 3.12 Quantization of signed signal using 3 bits, when the signal is 100% of the dynamic range. Signed value sampling is often used for audio (sound) sampling.

can span a certain range of signal, but that then compromises the accuracy of representation of each level. Alternatively, if we allocate the available bits in such a way that the step size of the quantizer is small, the dynamic range will be limited.

Figures 3.12–3.14 illustrate these issues. A sinusoidal signal is used as an input, which is plotted against a range of 256 levels. Figure 3.12 shows the case of a 3-bit quantizer (eight levels), where the signal span equals the quantizer span. The signal has positive and negative ranges, hence an average of zero. The lower panel shows the quantization error, which is simply the difference between the input signal and the quantized signal. Figure 3.13 shows the same experiment but with an unsigned signal source. Clearly, the same situation is occurring, and the quantizer span should be adjusted to 0–255, rather than ±128, as in the case of the signed input signal. The main point is that the quantization error is effectively the same and has an average value centered around zero.

Figure 3.14 shows the same quantization scenario as Figure 3.13 but with the signal occupying only 40% of the dynamic range of the quantizer. Clearly in this case, the quantization becomes "coarser" even though the number of levels remains the same. This is because the quantizer step size is fixed. The result is that the quantization noise becomes larger when measured as a proportion of the input signal amplitude.

It should be pointed out that a 3-bit quantization characteristic is somewhat unrealistic in practice. Such a value serves to illustrate the point graphically, however. So now let us turn to a statistical view of matters. Figure 3.15 shows the distribution

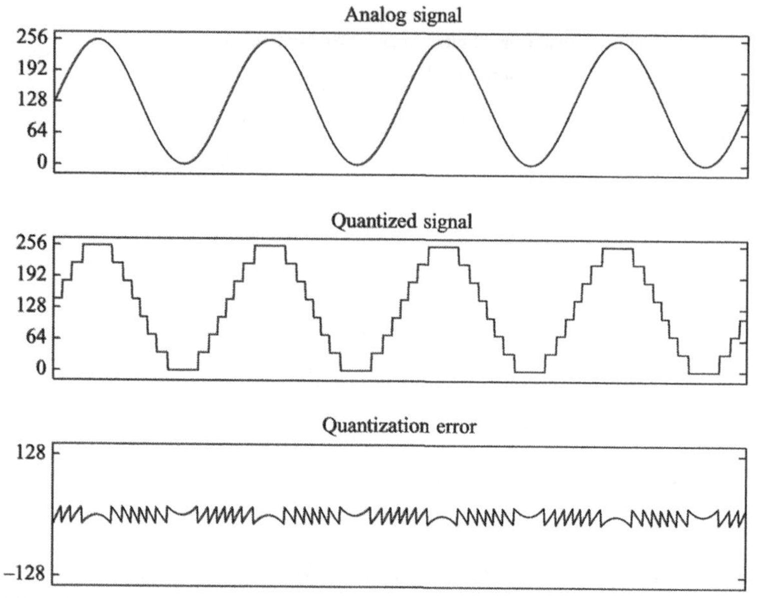

FIGURE 3.13 Quantization of an unsigned signal using 3 bits, when the signal is 100% of the dynamic range. The quantization error is the same as in the signed case (Figure 3.12). Unsigned value sampling is often used for image color components.

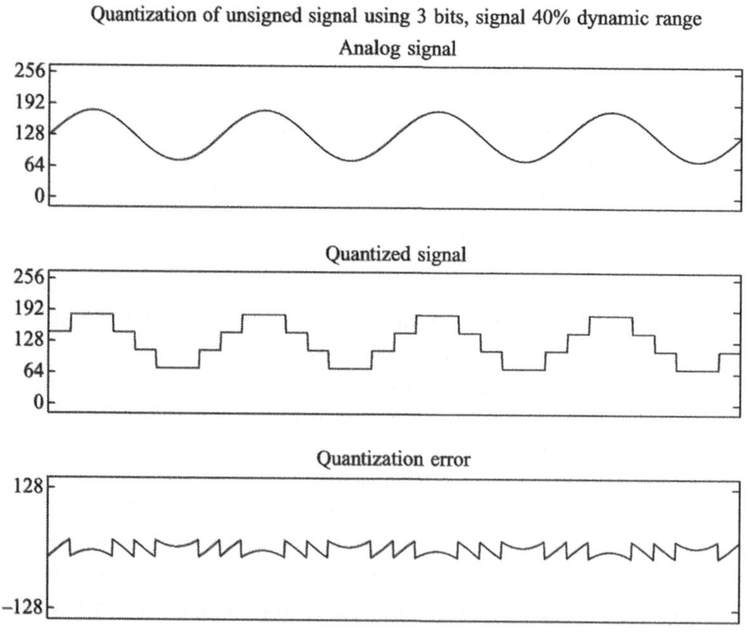

FIGURE 3.14 Quantization of an unsigned signal using 3 bits, when the signal is 40% of the dynamic range. Note how much larger the quantization noise is as a proportion of the source signal amplitude compared to when the signal occupies the full dynamic range (Figure 3.13).

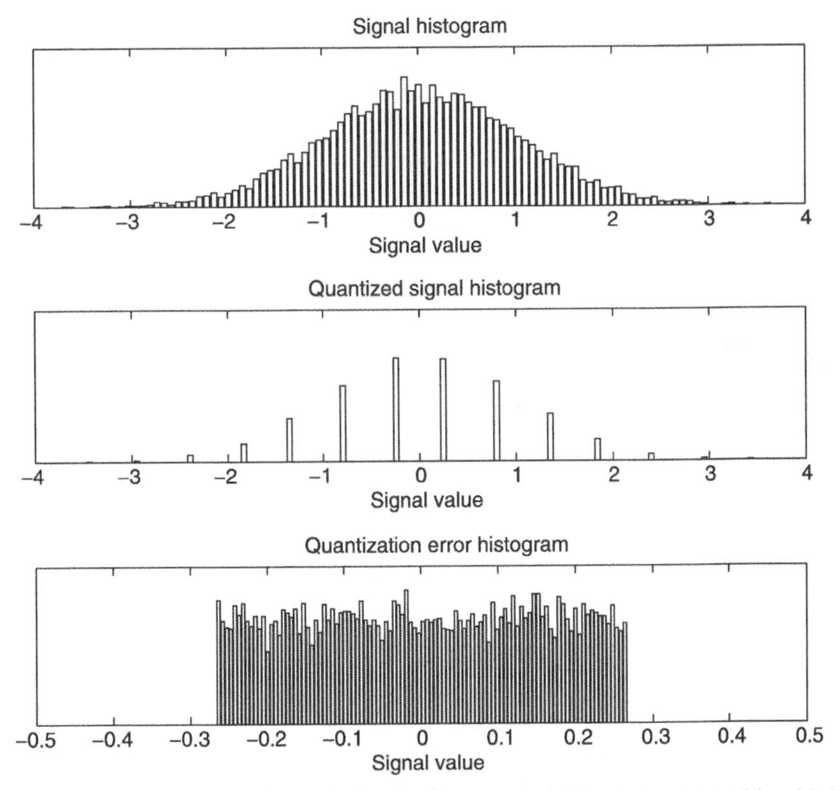

FIGURE 3.15 Quantization of a signal with a range of ±4 to $N = 4$ bits (16 levels). From top to bottom, we have the source signal probability distribution, the distribution of the quantized signal, and the distribution of the error.

of a signal with a range of ±4, its quantized version (with $N = 4$ bits, and hence 16 levels), and the distribution of the quantization error. We can see two things: first, that the operation of quantization serves to "thin out" the probability density, so that the quantized signal is representative of the original in a statistical sense; second, the quantization error distribution is seen to be plus and minus half a quantizer step value.

3.6.1 Experiments with Quantization

In order to base our understanding of the effect of quantization, we consider some experiments based on real signals. What happens when we remove the bits from a sampled signal starting at the LSB? Starting at the MSB?

Figure 3.16 shows an image with the original 8-bit resolution, and the successive removal of bits from the least significant end of the samples to obtain 4 and then 2 bits per pixel (bpp) images. The perceptual quality of such images depends on the display media and the relative size of the image, but it is quite clear that the 2-bit image is unacceptable in terms of quality. We can repeat the same experiment,

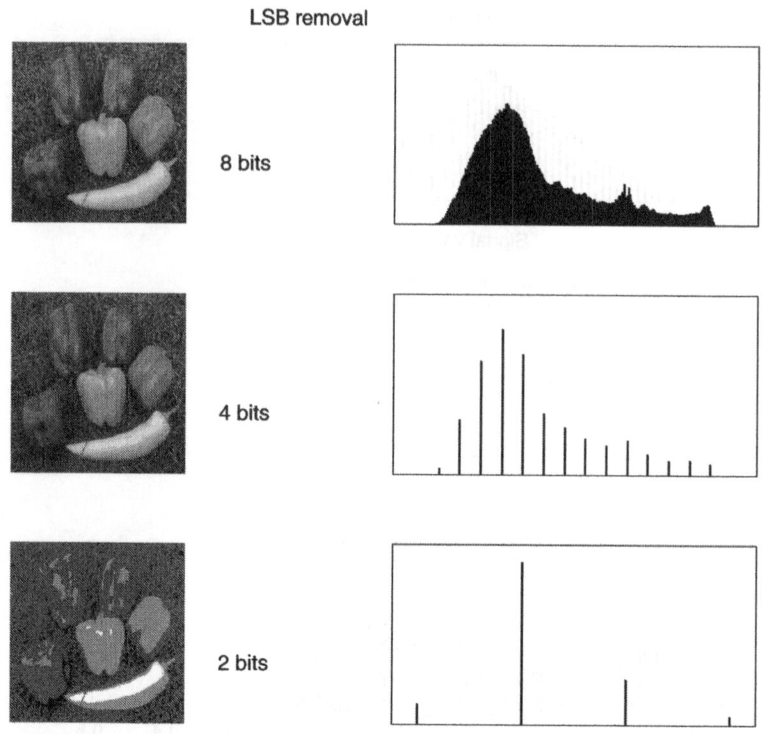

FIGURE 3.16 Image quantization: removal of the least significant bits (LSBs). The corresponding histogram for each image is shown on the right. The quantization effectively increases the quantizer step size and thus reduces the accuracy of representation of the signal. In the case of images, this manifests itself as a stepping or "contouring" of the image and is especially noticeable in regions of smoothly varying intensity.

this time removing bits from the most significant end of each sample, as shown in Figure 3.17. Comparing these, we can see that the MSBs in each sample effectively set the range of the signal, and when removed, the dynamic range is reduced. This is clearly shown in the histograms to the right of each image. However, removing the LSBs effectively increases the step size without changing the span of the representation.

We repeat the same experiments—removing bits to (re)-quantize a signal—this time with an audio signal. Figure 3.18 shows that removing the LSBs reduces the resolution by increasing the step size of the quantizer. The underlying signal is still present and indeed visible even at 8-bit resolution. Removing the MSBs, as in Figure 3.19, shows that the dynamic range of the signal is compromised. When played back, the sound of the latter is very much reduced in amplitude, whereas in the former case (removing LSBs, thus giving a coarser step), the presence of quantization noise becomes very much apparent.

What conclusion do we draw from these experiments? First, the number of bits required for a faithful representation is very much dependent on the type of

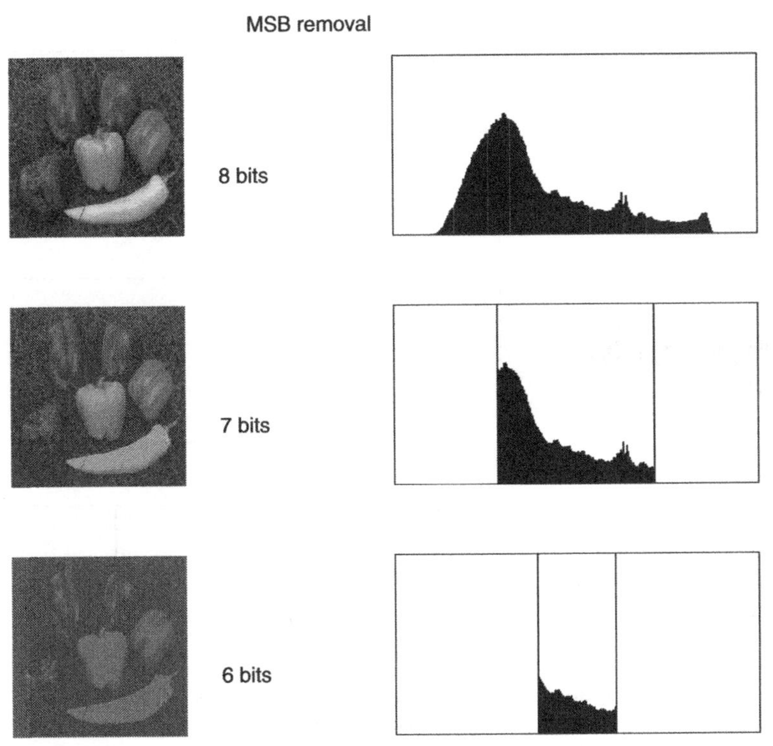

FIGURE 3.17 Image quantization: removal of the most significant bits (MSBs). The corresponding histogram for each image is shown on the right. The quantization effectively reduces the dynamic range of the signal, moving it toward black. In this example, we have compensated for the movement of the average so that the average is still midway brightness. Observation of the images shows that the range of intensities represented for 6-bit resolution is unacceptable.

signal. Here, images were represented with less than 8 bits per sample with reasonable accuracy. However, audio was severely compromised when reduced from 16 to 8 and fewer bits. The second conclusion is that rather than simply removing bits from the most or least significant end of a sample, the act of quantization really ought to be considered as a mapping (or remapping) of signal amplitudes according to a characteristic which is appropriate to the interpretation of the signal itself.

Finally, we need some basis to quantify the "noise" introduced by the act of quantization. This noise could be visual noise (as in the preceding pictures) or audio noise, but in general terms, it is simply an error which we need to minimize. The following section investigates the nature of the noise in a mathematical sense.

3.6.2 Quantization and SNR

The preceding examples have shown that we need some basis on which to allocate bits to a sample, so as to minimize the error or noise signal. In effect, we wish to

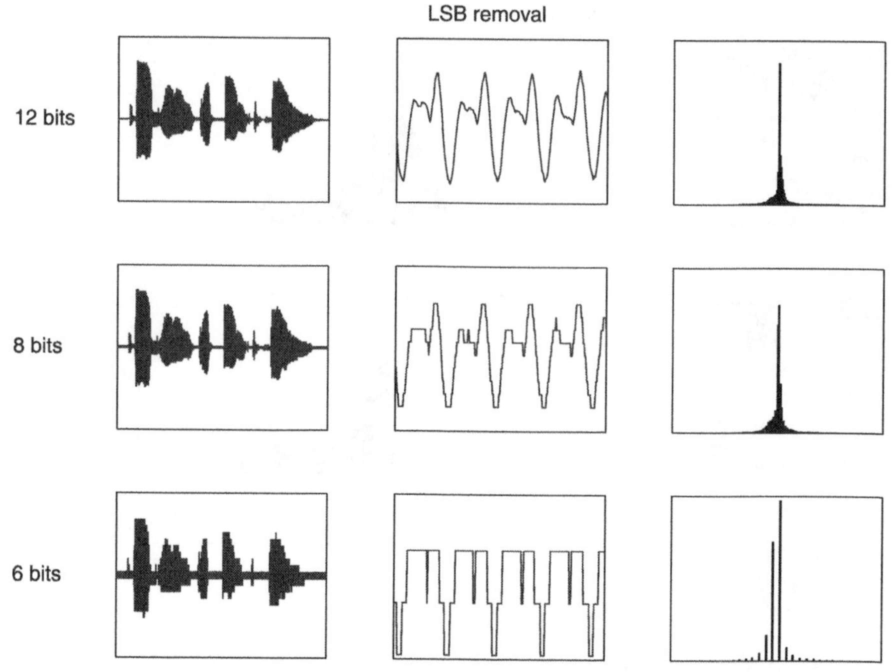

FIGURE 3.18 Sound quantization: removal of the least significant bits (LSBs). The corresponding histogram for each sound block is shown on the right. Starting at 16 bits, as we reduce the number of bits allocated, the dynamic range remains the same, but the step size becomes coarser. This manifests itself as an audible noise in the reconstructed signal, which is especially noticeable in softer sections of the audio (middle panels).

minimize the noise power. What performance metric can we devise for this, and how is such a metric related to the number of bits actually employed in quantization?

Let the signal be $x(n)$ and the quantized (or approximated) signal be denoted as $\hat{x}(n)$. The quantization error is

$$e(n) = x(n) - \hat{x}(n). \tag{3.16}$$

It does not particularly matter whether we define this as $(x(n) - \hat{x}(n))$ or $(\hat{x}(n) - x(n))$ since invariably we are not interested in whether the error is positive or negative, but rather the magnitude of the error. Furthermore, we are interested in the average error over a large number of samples. Thus, the performance measure will be related to the mean square error (MSE),

$$\mathrm{MSE} = \frac{1}{N} \sum_{N} e^2(n), \tag{3.17}$$

where the length of the block N is large enough to provide a good estimate of the average, and the error is squared. Squaring the error effectively negates the need for an absolute value. As it turns out, a squared error performance metric is not only intuitive (proportional to the signal power) but is also often mathematically easier

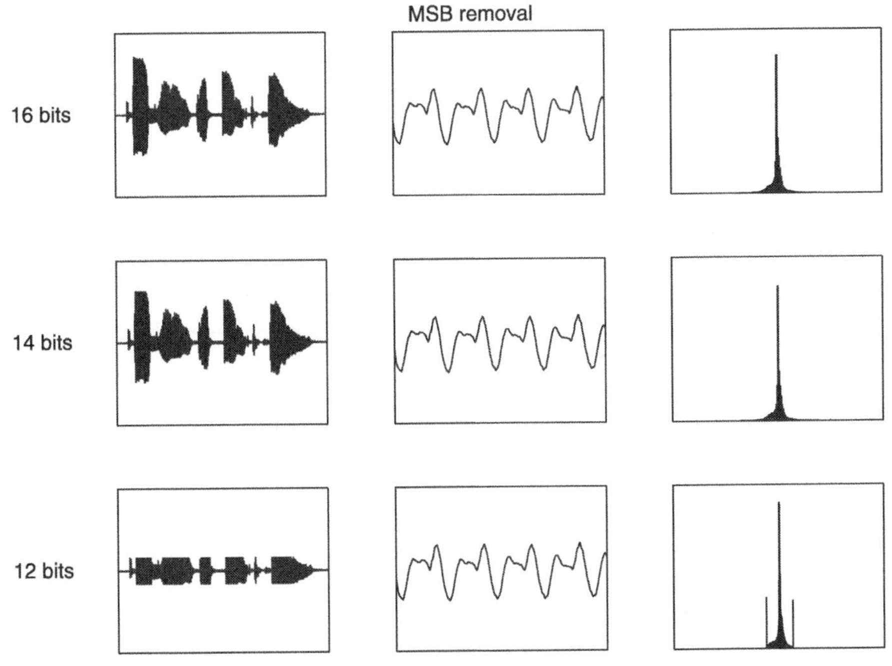

FIGURE 3.19 Sound quantization: removal of most significant bits (MSBs). The corresponding histogram for each sound block is shown on the right. The quantization effectively reduces the dynamic range of the signal. Since sounds tend to have a very wide dynamic range from the quietest sound to the loudest, we have severely compromised the performance of the system in doing this. For relatively quiet sections of the audio (as shown in the middle panels), the accuracy is still maintained, however.

to manipulate. In some problems, particularly optimal and adaptive filtering, we need to differentiate the signal, and differentiation of a squared value is easily handled by the chain rule of calculus.

The MSE as defined above is useful, but the significance of the error is clearly also dependent on the signal itself. In other words, we need to determine the energy in the error signal in relation to the signal power itself. If the error is large, and the signal is large, the effect is less significant than if the error is large and the signal is small. So a better metric of performance is the quantizing SNR, defined as the ratio of the signal power to the noise power and expressed in decibels:[2]

$$\text{SNR} = 10\log_{10}\left(\frac{\sum_n x^2(n)}{\sum_n e^2(n)}\right)\text{dB}. \tag{3.18}$$

For an equal step-size N-bit quantizer, the step size Δ is the ratio of the peak amplitude to the number of steps. Assuming a range of $\pm x_{\max}$,

[2]A decibel is defined in terms of logarithm of power: $10\log_{10}$ (power).

$$\Delta = \frac{2x_{\max}}{2^N}. \tag{3.19}$$

As Figure 3.15 indicates, it is reasonable to assume that the quantizing noise has an average of zero, and any particular value is equally likely plus or minus half a quantization step above and below the mean. This means that we have a zero mean value and a uniform distribution.

For any data set $\{x\}$, probability theory tells us that, given the probability density function $f(X)$, the mean and variance may be calculated as

$$\mu = \int_{-\infty}^{+\infty} X \, f(X) \, dX \tag{3.20}$$

$$\sigma^2 = \int_{-\infty}^{+\infty} (X - \mu)^2 \, f(X) \, dX. \tag{3.21}$$

The variance of the quantizing noise is calculated by substituting the appropriate values into the above equation for the variance. We assume that the probability density $f(X)$ may be described by a uniform density with zero mean, and this seems correct given our previous observations on the distribution of quantizing noise. The error signal must range over one step size Δ, and so the appropriate limits for integration are $\pm(\Delta/2)$. Additionally, probability theory tells us that the area under a density curve must be unity. Since the range of x is now known to be $-(\Delta/2)$ to $+(\Delta/2)$, the span of x is Δ, and thus $\Delta \cdot f(X) = 1$. It follows that $f(X) = 1/\Delta$, which is a constant. So the equation for the variance of the quantization error becomes

$$\begin{aligned}
\sigma_e^2 &= \int_{-\frac{\Delta}{2}}^{+\frac{\Delta}{2}} X^2 \frac{1}{\Delta} \, dX \\
&= \frac{1}{\Delta} \cdot \frac{1}{3} X^3 \Bigg|_{X=-\frac{\Delta}{2}}^{X=+\frac{\Delta}{2}} \\
&= \frac{1}{3\Delta} \left(\left(\frac{\Delta}{2} \right)^3 - \left(\frac{-\Delta}{2} \right)^3 \right) \\
\sigma_e^2 &= \frac{\Delta^2}{12}.
\end{aligned} \tag{3.22}$$

This gives us the quantization error variance (or equivalently, the error power) in terms of the quantizer step size. But we would like to relate this to the number of levels rather than the step size. Equation 3.19 gives the link we need since it relates the signal maximum dynamic range to the number of levels. So substituting Equation 3.19 into Equation 3.22 yields

$$\sigma_e^2 = \frac{\Delta^2}{12} \tag{3.23}$$

$$\begin{aligned} &= \frac{4x_{\max}^2}{12 \cdot 2^{2N}} \\ &= \frac{x_{\max}^2}{3 \cdot 2^{2N}}. \end{aligned} \tag{3.24}$$

Therefore, the SNR is

$$\begin{aligned} \text{SNR} &= \frac{\sigma_x^2}{\sigma_e^2} \\ &= \frac{\sigma_x^2}{\left(\frac{x_{\max}^2}{3 \cdot 2^{2N}}\right)} \\ &= 3\frac{\sigma_x^2}{x_{\max}^2} \cdot 2^{2N}. \end{aligned}$$

Such a signal-to-noise power is normally expressed in decibels, so we have

$$\begin{aligned} \text{SNR} &= 10\log_{10} 3 + 10\log_{10} 2^{2N} + 10\log_{10}\left(\frac{\sigma_x^2}{x_{\max}^2}\right) \\ &\approx 4.77 + 6.02N + 20\log_{10}\left(\frac{\sigma_x}{x_{\max}}\right) \text{dB}. \end{aligned} \tag{3.25}$$

This result depends on the ratio $x_{\max}/\sigma_x$, which is the ratio of the peak amplitude to the standard deviation.

But since σ_x^2 is effectively the signal power, we can say that σ_x is really the root mean square (RMS) of the signal, defined as

$$\text{RMS} = \sqrt{\frac{1}{N}\sum x^2(n)}. \tag{3.26}$$

So, the above result really depends on the ratio of the peak signal amplitude to the RMS power of the signal, and we need an estimation of this value. Clearly, it will depend on the type of signal, but a reasonable assumption for this "loading" factor is $x_{\max} = 4\sigma_x$. In that case,

$$\text{SNR} \approx 6N - 7.3 \text{ dB}. \tag{3.27}$$

Equation 3.27 informs us of the important fact that the SNR is proportional to the number of bits used:

$$\text{SNR} \propto N. \tag{3.28}$$

Note that this important general result does *not* state that the SNR is proportional to the number of levels, for in that case we would have 2^N in the above. Put another way, adding 1 bit to the quantizer resolution gains an additional 6 dB of signal-to-noise performance.

TABLE 3.2 Signal-to-Noise as a Result of Quantization

Bits N	Levels 2^N	Measured SNR (dB)	Theoretical SNR (dB)
2	4	1.9	4.8
3	8	9.7	10.8
4	16	16.2	16.8
5	32	22.5	22.8
6	64	28.7	28.8
7	128	34.8	34.8
8	256	40.9	40.8
9	512	46.9	46.8
10	1,024	53.0	52.8
11	2,048	59.0	58.8
12	4,096	65.0	64.8
13	8,192	71.1	70.8
14	16,384	77.1	76.8

The experimental result is obtained via simulation as described in the text using Gaussian noise with $\sigma = 1$. The theoretical result is obtained using Equation 3.25 with $\sigma_x = 1$, $x_{max} = 4$.

Note that there are several assumptions inherent in the above theory. First, σ_x / x_{max} is really the RMS to peak value (since RMS $= \sqrt{\sigma_x^2}$ when the average is zero). For a sine wave, this is $1/\sqrt{2}$. For other waveforms, it will be different. For example, a "noiselike" waveform may only have occasional peaks, so on average, the SNR will differ. The result is also only valid for a large number of quantizer steps (since we assumed that the quantization noise distribution is approximately uniform), and that relatively few samples are clipped (outside the range $\pm 4\sigma_x$).

Moreover, the ultimate definition of noise is perceptual—so we really need a subjective judgment rather than an objective measure like SNR. Finally, other elements in the signal chain also have an effect—for example, the quality of the viewing device for images and the speaker and audio amplification for sound.

So how does this theoretical result compare in practice? Table 3.2 summarizes the theoretical result as compared to an experimental result using a random signal source. As can be seen, the figures are in close agreement. The approximate increase in SNR per bit is 6 dB, as predicted. This can be seen in Figure 3.20. This is valid only for a large number of levels; hence, for 2-bit quantization, the result does not agree as well. A second assumption is that the range of the quantizer is

$$\frac{|\text{Maximum}|}{\text{RMS}} = \frac{x_{max}}{\sqrt{\sigma_x^2}} = \pm 4.$$

3.7 IMAGE DISPLAY

Quantization and representation of images deserve special mention because of the many possible ways a color image can be stored. This section explains these concepts using some examples to illustrate.

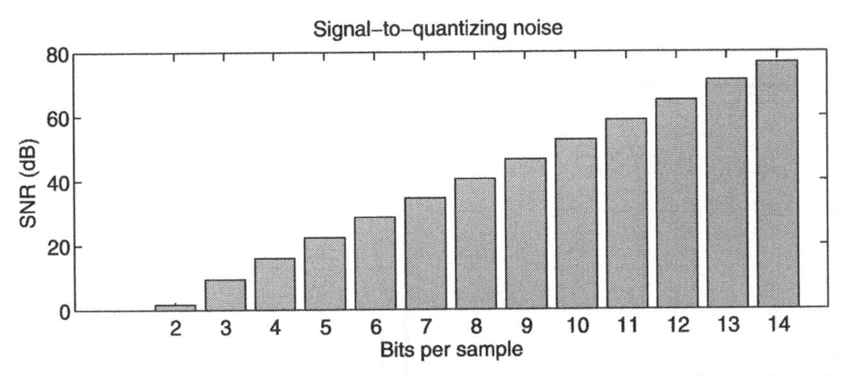

FIGURE 3.20 Quantization and measured signal to noise for a Gaussian random signal.

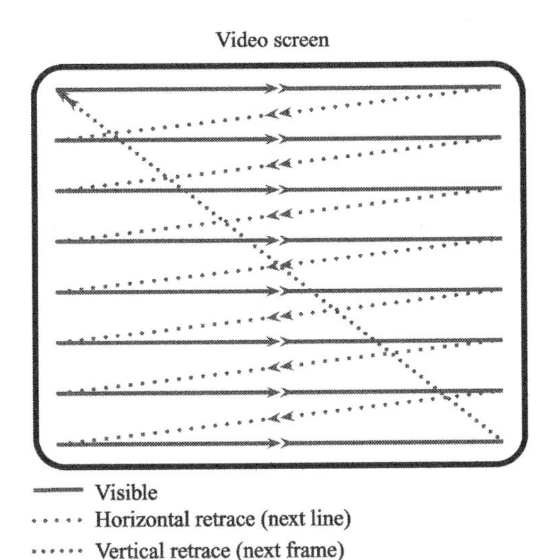

FIGURE 3.21 Images are sampled from left to right, top to bottom. The horizontal lines represent each row of pixels; the dotted lines represent the return back to the start of each scan line but are not visible. Finally, the dotted line from lower right to upper left represents the return to the start of the image. This would occur when video is being displayed, since each frame of video is, in effect, a still-image snapshot.

3.7.1 Image Storage

Images are sampled spatially, in a grid pattern. A matrix of samples is the ideal mathematical representation for these samples, termed "pixels" or "pels" (picture elements). Of course there is a temporal (time) relationship to the position of each pixel—images are normally scanned left to right, top to bottom, as depicted in Figure 3.21.

Since memory space is linear, the pixels themselves that form the image matrix $\mathbf{X}(m, n)$ are mapped in the scan order. This is depicted in Figure 3.22. Video

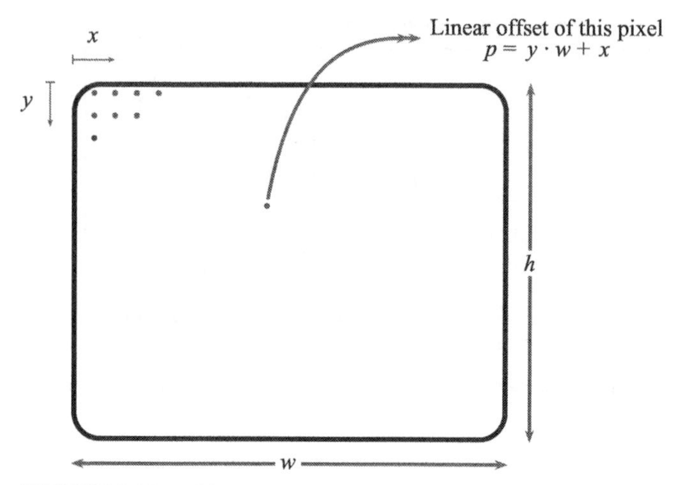

FIGURE 3.22 Video memory mapping. A two-dimensional image is mapped onto a one-dimensional memory.

sequences are effectively a concatenation of successive still-image frames, replayed at the appropriate frame rate. For the kth frame, the pixels may be denoted $\mathbf{X}_k(m, n)$. Note that this illustrates a row-wise ordering of the pixels. Alternatively, the pixels could be stored columnwise. This in fact is the representation returned if we used the "colon" operator in MATLAB. For example, if we have a hypothetical set of pixels as shown below, the colon operator extracts the values into a linear vector:

```
X = [1 2 3 ; 4 5 6 ; 7 8 9]
X =
    1   2   3
    4   5   6
    7   8   9
X(:)'
ans =
     1    4    7    2    5    8    3    6    9
```

If we desired the pixels in row order, we simply transpose the matrix:

```
Xt = X'
Xt =
    1   4   7
    2   5   8
    3   6   9
Xt(:)'
ans =
     1    2    3    4    5    6    7    8    9
```

Note that Figure 3.22 indicates the memory offset as $p = wy + x$, where p is the linear offset; w is the width of the image; and (x, y) are the pixel indexes (column, row). This assumes the (x, y) offsets are zero based. If using MATLAB indexes, it is necessary to subtract one from the x and y values before being used in this way. Finally, the conventional indexing of matrices is (row, column), whereas image coordinates are usually defined as (x, y), which is the reverse order.

3.7.2 RGB Color Representation

Color representation is done using the three primary colors—red, green, and blue. Normally, 8 bits are reserved for each color separately. This is sometimes termed RGB or *full-color* representation, and 2^{24} colors are possible in theory.

To see how this works in practice, consider reading in a real image (in this case, a JPEG image file).

```
FileName = 'MyImage.Jpg';
[ImageMat cmap] = imread(FileName);
[ImageHeight ImageWidth NumColorPlanes] = size(ImageMat);
image(ImageMat)
set(gca, 'DataAspectRatio', [1 1 1]);
axis('off');
```

If this is an RGB image, the number of color planes stored in `NumColorPlanes` will equal 3. In the following, we loop over all image pixels horizontally and over all rows vertically. For each pixel value, we extract the red, green, and blue color values. So, the red value is stored first and extracted using `redval = double (ImageMat(y, x, 1))`. The next numerical value is the green value of that pixel and finally the blue (indexes 2 and 3, respectively).

```
NewImageMat = zeros(ImageHeight, ImageWidth, NumColorPlanes);
for y = 1:ImageHeight
    for x = 1:ImageWidth
        % get original pixel RGB value
        redval = double(ImageMat(y, x, 1));
        greenval = double(ImageMat(y, x, 2));
        blueval = double(ImageMat(y, x, 3));
        % alter one or more color planes
        %redval = 0;
        greenval = 0;
        %blueval = 0;
        % save pixel RGB value
        NewImageMat(y, x, 1) = redval;
```

```
        NewImageMat(y, x, 2) = greenval;
        NewImageMat(y, x, 3) = blueval;
        end
end
% convert to 8-bit per RGB component
NewImageMatRGB8 = uint8(NewImageMat);
image(NewImageMatRGB8);
set(gca, 'DataAspectRatio', [1 1 1 ]);
axis('off');
```

In this example, we have (arbitrarily) set the green component of each pixel to zero. Thus, the resulting image has no green components at all. Finally, note that the displayed image is converted to a `uint8` value to yield 1 byte for each color plane.

To fully understand the effect of color mixing, consider a one-pixel image with RGB components such that red and green are present but with no blue component:

```
img = zeros(1 ,1 ,3);
img(1 ,1 ,1) = 1; % red
img(1 ,1 ,2) = 1; % green
img(1 ,1 ,3) = 0; % blue
image(img);
axis('off');
```

This mixing of red and green only gives the color yellow. Note one other subtle but important issue in the above: We have used a double-precision image matrix. In that case, MATLAB expects the components to be in the range 0–1. However, if the RGB values were unsigned 8-bit integers (the usual case), then they would be scaled to the range 0–255.

3.7.3 Palette Color Representation

To reduce storage requirements, some storage formats and display systems use a palette-based system, which uses a color lookup table (LUT) to indirectly produce the colors from the samples as shown in Figure 3.23. Each pixel value is used to index a palette or table of colors in the video display hardware. Each entry contains one intensity level for each of the three primary colors. The palette must be chosen and optimized carefully for the given image, as the palette can represent only a finite number of colors. For 8-bit pixel values, this gives 2^8 possible colors as compared to 2^{24} for full-color systems.

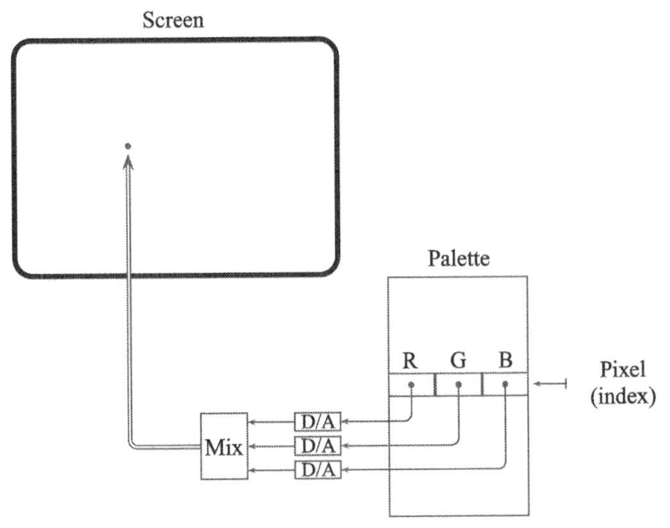

FIGURE 3.23 Using the color palette as a lookup table to produce a pixel.

Note that although using a palette-based system reduces the amount of memory required for the video display and reduces the transmission time on digital links by a factor of three, the palette itself is unique to the image being displayed. Therefore, it must be prepended to image files when stored or transmitted separately.

The MATLAB command **colormap**() is used to set the color palette (or color map) for a particular image figure. This function expects an array of length equal to the number of quantization levels. Each entry contains three elements, ranging from 0.0 to 1.0, to represent the intensity of each of the three primary colors.

As a simple example of using a palette, the following shows how to generate a random grayscale image using a matrix. An 8-bit (256-level) display is used, although the pixels are quantized to one of only four possible levels:

```
M = 64;
N = 64;
x = rand(M, N);
m = floor(255*(floor(x * 4)/3));
image(m([1:4], [1:4]));
colormap(gray(256))
axis('off');
```

The upper 4×4 pixels from this image matrix are extracted using matrix subscripts as follows:

FIGURE 3.24 A magnified view of a 4 × 4 pixel image with four gray levels. Of course, in a realistic display scenario, each pixel would be quite small—effectively merging with other nearby pixels to form a continuous image if viewed at a normal distance.

```
m([1:4], [1:4])
ans =
    255     255     0      170
    170     255     170    85
    0       0       85     255
    170     85      170    170
```

These values are then displayed using the current color map with the **`image`** `()` command and appear as shown in Figure 3.24. Obviously, the pixel size in this image is exaggerated. Note that this is a grayscale image—color images are usually processed by decomposing the colors into the separate constituent primary colors (red, green, and blue) as described earlier.

The following example shows how to display an image using a blue-only scale by zeroing out the red and green components:

```
x = 1:8:256;
m = x(ones(20 ,1), :);
image(m);
z256 = zeros(256,1);
cmap = [ z256 z256 [1:256]' ]/256;
colormap(cmap);
axis('off');
```

Monochrome or grayscale image representation is possible by setting each of the red, green, and blue components in the palette to equal values. For an 8-bit image, this gives a linear scale of intensity, ranging linearly from black (level = 0) to white (level = 255). The following MATLAB code illustrates this:

```
M = 32;
N = 128;
x = rand(M, N);
x = floor(x*256);
image(x);
colormap(gray(256));
axis('off');
set(gca, 'DataAspectRatio', [1 1 1]);
```

We can alter the earlier RGB image processing code to read images which contain a color palette. The following code reads in an image and checks the returned value cmap, which is the matrix of color map values (notice that this was ignored in the earlier example). For an 8-bit image, this matrix will be of the dimension 256×3 since there are 2^8 color index values, each with a red, green, and blue component.

```
FileName = 'MyImage.bmp';
[ImageMat cmap] = imread(FileName);
if( ~isempty(cmap) )
    disp('This image has a colormap');
end
[ImageHeight ImageWidth NumColorPlanes] = size(ImageMat);
image(ImageMat);
colormap(cmap);
set(gca, 'DataAspectRatio', [1 1 1]);
axis('off');
```

Note that in this case, we have used a bitmap image (bmp file extension). It should be made clear that the extension does not determine the presence of a palette—the extension determines how the data are internally stored (such as an uncompressed bitmap or a compressed JPEG). Both of these image types are able to encapsulate palette-based or RGB image data. A robust program will test for the presence of a color map and process the image accordingly.

3.8 ALIASING

Let us now again consider the problem of how fast to sample a given signal. If the sampling rate is not sufficiently fast, a problem called *aliasing* occurs. Put simply, this makes the signal "appear" to be slower than what it actually is. The sampling rate theorem states that we must sample at a rate which is greater than twice as fast as the highest frequency component present in order to be able to later reproduce the original waveform faithfully. This minimum sampling rate is usually called the *Nyquist rate*.[3] Put in another way, the sampling rate used dictates the highest frequency that can be faithfully represented, and we have the following fundamental principle:

> The *minimum sampling frequency* must be at least twice that of the highest frequency component present in the original signal.

Consider Figure 3.25. In simple terms, we might ask what the input signal is doing *between* samples. Intuitively, we might think that the signal must be smooth and "well behaved" between samples. For this reason, a low-pass filter is used before the sampling process as shown in Figure 3.26. This is done in order to remove any frequencies higher than half the sampling frequency $(f_s/2)$. This filter is termed an *anti-aliasing filter.* If this filter limits the highest frequency component in a given signal (called bandlimiting), then the Nyquist rate is applicable:

> The *Nyquist rate* is twice the bandwidth of the signal, and we must sample above this rate in order to ensure that we can reproduce the signal without error.

This may sound surprising and, in some ways, not really intuitive. If we think about this further, it might suggest that for a sinusoidal waveform, we only need two samples per complete cycle. But remember that it is a *lower bound*—in other words, we must sample *faster* than this.

If we turn the argument around, then a given sampling rate will dictate the highest frequency which we can reproduce after sampling. This is called the Nyquist or folding frequency.

> The *Nyquist frequency* or *folding frequency* is half the sampling rate and corresponds to the highest frequency which a sampled data system can reproduce without error.

So, to summarize these important points: The fundamental theorem of sampling states that we must sample at least twice as fast as the highest frequency component that we want to maintain after sampling. In practical terms, the sampling rate is chosen to be higher than this two-times rule above would suggest. For example, CD audio systems are designed to handle frequencies up to approximately 20 kHz. The sample rate is 44.1 kHz—a little more than double the highest expected frequency. The precise reasons for this will be examined in a later chapter, after the theory of filtering is developed.

[3]Sometimes, the terms Nyquist or Shannon sampling theorem are used, after Harry Nyquist and Claude Shannon, who developed the key underpinnings of the sampling and digital transmission theory used today.

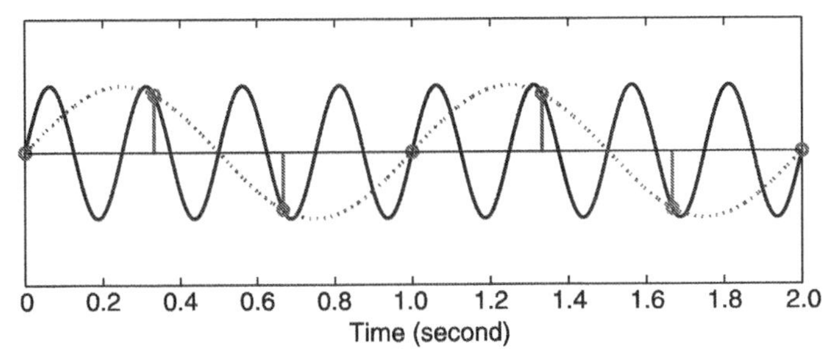

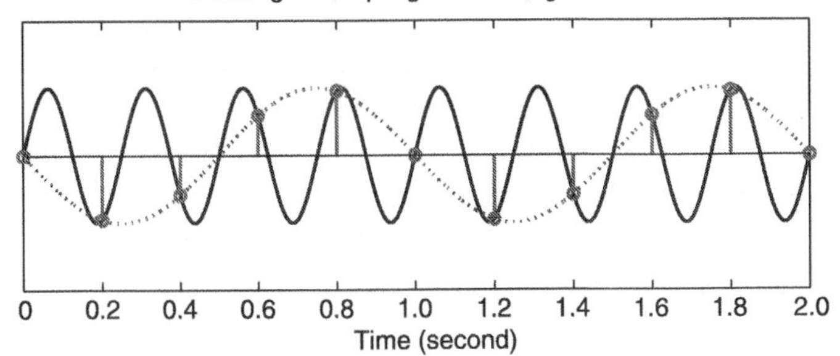

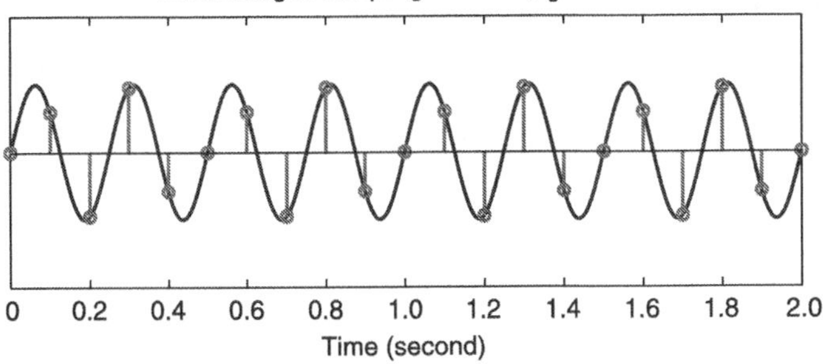

FIGURE 3.25 How aliasing arises. We have a 4-Hz sine wave to be sampled, so the Nyquist rate is 8 Hz, and we should sample above that rate. In the upper waveform, we sample below the input frequency. Clearly, we cannot expect to capture the information about the waveform if we sample at a frequency lower than the waveform itself. The waveform appears to have a frequency of $|f - f_s| = |4 - 3| = 1$ Hz. Next, we sample a little above the waveform frequency. This is still inadequate, and frequency is effectively "folded" down to $|f - f_s| = |4 - 5| = 1$ Hz. Finally, we sample at greater than twice the frequency ($f_s > 2f$ or $10 > 2 \times 4$ Hz). The sampled waveform does not produce a lower "folded-down" frequency in this case.

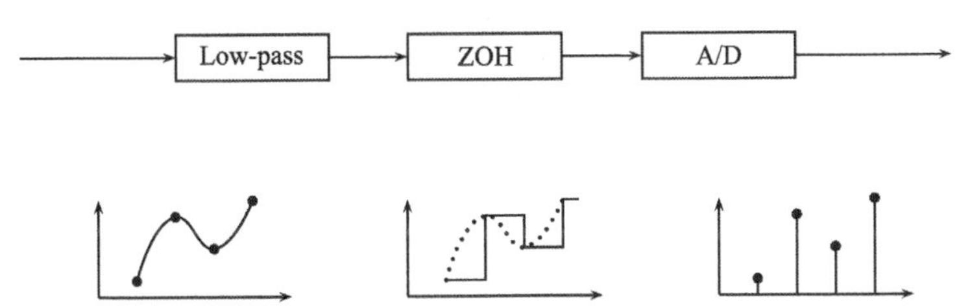

FIGURE 3.26 The positioning of an anti-aliasing filter on the input to a sampling system. It stops any signals above half the sampling rate from passing through—the signal is then said to be "band-limited." The resulting sample stream then contains enough information to *exactly* reconstruct the signal as it was after the low-pass filter.

3.9 RECONSTRUCTION

For a different reason, a *reconstruction filter* is used after the output of the D/A system, as depicted in Figure 3.27. This is also a low-pass filter—one that removes the high-frequency components. The need for this filter is a little more obvious: Consider the "stairstep" reconstructions that have been shown so far. We need to remove the "jagged" edges to reconstruct the original signal. This filter also has a cutoff frequency of $f_s/2$.

So far, we have the signal of interest being first band-limited with the anti-aliasing input filter, sampled at a sufficient rate, processed, output through a ZOH, and finally reconstructed with a low-pass filter. The obvious question is "how do we know this low-pass filtering on the output is the ideal arrangement?" In the process of answering this question, it becomes apparent that it is possible to calculate the values which the signal is *expected* to take on between the known samples. This process is termed "interpolation." We will investigate reconstruction and interpolation now and return to the mathematical underpinnings of reconstruction in Section 8.7.

3.9.1 Ideal Reconstruction

So far, we have used the ZOH as the output of the D/A system. The *ideal* function for interpolating between samples is[4]

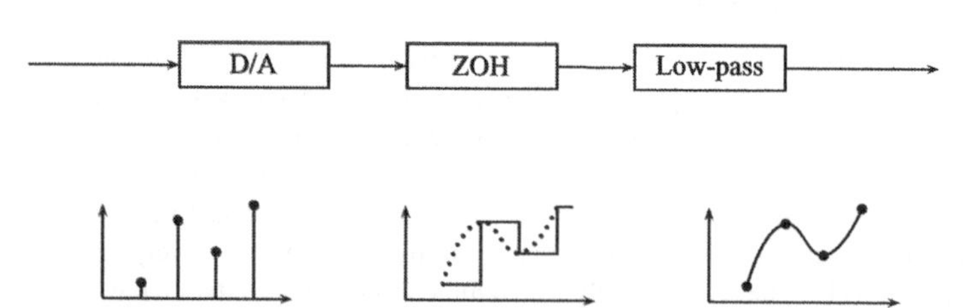

FIGURE 3.27 A reconstruction filter on the output of a sampled data system. Provided the signal was sampled at a rate greater than twice the highest frequency present in the original, the output of a reconstruction filter will be identical to the original signal. Note that the zero-order hold (ZOH) is not present as a separate component as such; it simply represents how a practical D/A converter works in reality.

[4]This will be derived in Section 8.7.

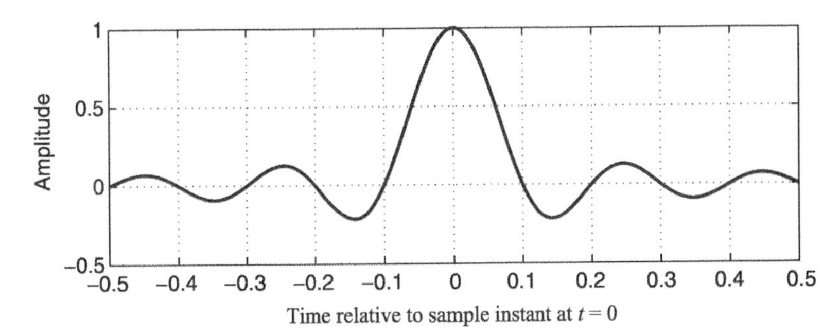

FIGURE 3.28 The "sinc" function $\sin(\pi t/T)/(\pi t/T)$ for the ideal reconstruction of a sample at $t = 0$, with a sampling period of $T = 0.1$.

$$\begin{aligned} h(t) &= \frac{\sin\frac{\pi t}{T}}{\frac{\pi t}{T}} \\ &= \operatorname{sinc}\frac{\pi t}{T}. \end{aligned} \tag{3.29}$$

This $(\sin x)/x$ or "sinc" function is shown in Figure 3.28, for the case of a sampling period $T = 0.1$.

Note that in the reconstruction function $h(t)$, the zero crossings are at nT, where n is an integer. This may easily be seen from the above equation for $h(t - nT)$—the sine function is zero when $\pi t/T$ is $0, \pi, 2\pi, \ldots$, which is whenever $t = nT$ for any integer n. What this means is that the reconstruction is done by a sample-weighted sinc function, where the sinc impulse from each sample does not interfere with the other sinc functions from other samples at each sampling instant. This can be seen in Figure 3.28, for the case where $T = 0.1$. In between the sampling instants, the sinc function provides the optimal interpolation.

To effect the ideal or perfect reconstruction, an infinite number of interpolation functions must be overlaid, one at each sample instant, and weighted by the sample value at that instant. The equation for each sinc function delayed to each sample instant $t = nT$ is

$$h(t - nT) = \operatorname{sinc}\frac{(t - nT)\pi}{T}. \tag{3.30}$$

To see this, consider Figure 3.29, which illustrates the steps involved in taking the original continuous signal and sampling it at discrete intervals. In this case, the samples are at intervals of $T = 0.1$ second. Using the conventional ZOH, this would be reconstructed as shown. However, using the sinc function interpolation, a weighted sinc function centered at each sampling instant, we can theoretically obtain perfect reconstruction. Figure 3.29 shows one sinc function, and this single function is translated in time and scaled in amplitude as shown in Figure 3.30. When all these

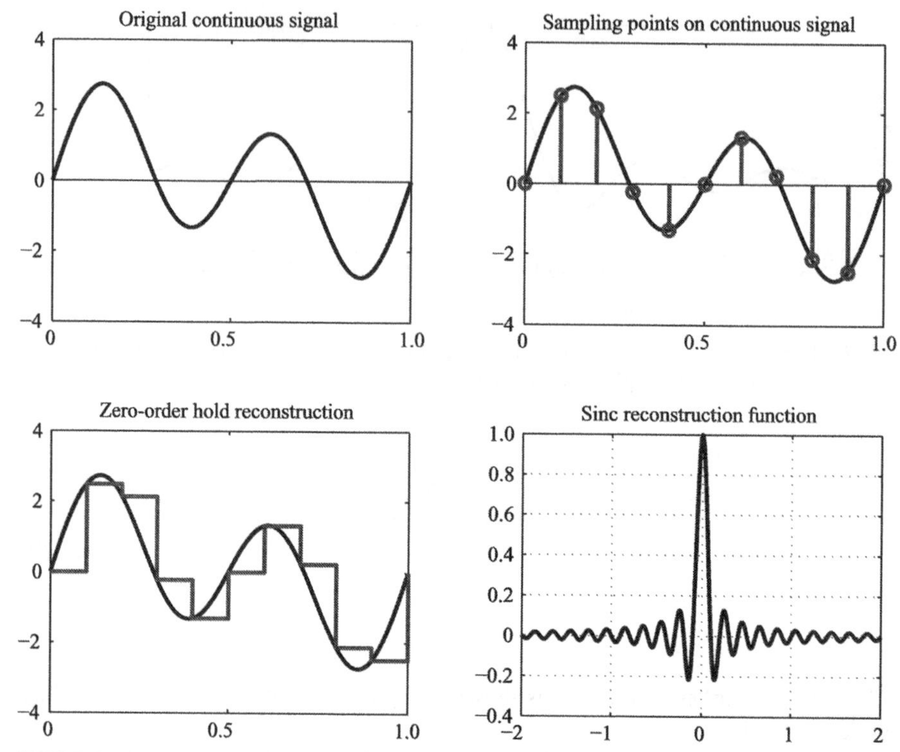

FIGURE 3.29 Illustrating the sampling of a signal. The original signal (upper left) is sampled at discrete-time instants. The reconstruction (lower left) can be done by holding each sample value until the next sample arrives. We can, however, reconstruct the waveform exactly if we use the sinc function. One sinc function centered on $t = 0$ is shown.

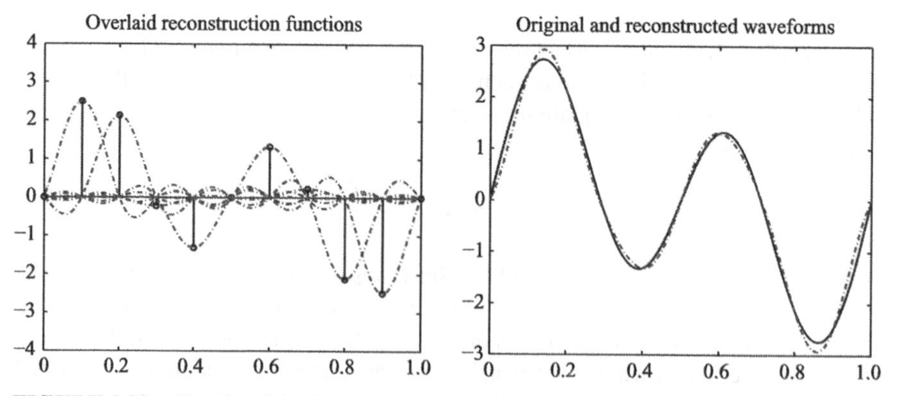

FIGURE 3.30 The sinc functions scaled and centered on each sample (left). The sum of these sinc functions equals the original (right), if they are allowed to extend for infinite time in either direction. Here, the sinc functions are truncated to show the discrepancy between the sum and the original waveform.

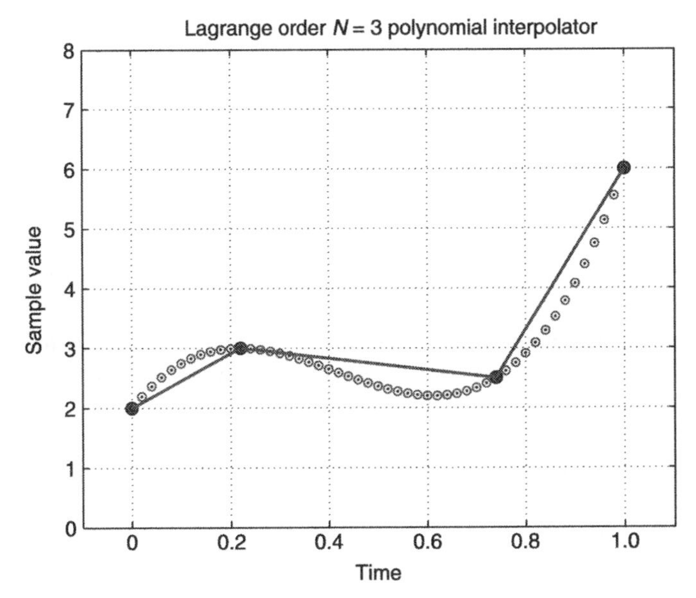

FIGURE 3.31 Lagrange interpolation for $N = 3$, with linear interpolation for comparison.

translated and scaled functions are added together, we can obtain an *exact* reconstruction. However, this is only valid for reconstruction functions extending to an infinite duration, in both positive and negative time (i.e., future and past). Obviously, this is not feasible in practice. Fortunately, we do not need to obtain perfect time reconstruction in order to capture sufficient information about the signal.

3.9.2 Polynomial Interpolation

Another interesting approach, rather than using a sinc function interpolation, is to use a polynomial to fit the wave at the sample points. The Lagrange interpolation function is ideal for this purpose.[5]

The problem is to estimate all the points $x(t)$ over continuous time t (or in practice, a large number of discrete points) given only samples $x(t_k)$ at discrete points t_k. An Nth-order polynomial interpolation of $x(t)$ at known points t_k requires the use of $N + 1$ samples. To see why this is so, consider a linear interpolation of the form

$$x(t) = at + b. \tag{3.31}$$

This equation has two unknowns, a and b, and hence requires two data points for a unique solution. Thus, two data points are required for this first-order interpolator. Interpolation of order zero is effectively a ZOH, as has been used previously. A third-order interpolation (requiring four points) is shown in Figure 3.31, with a first-order interpolation between each pair of points shown for comparison. Higher-order polynomial interpolations as $N \rightarrow \infty$ tend to a sinc function.

[5]Named after the French-Italian mathematician Joseph-Louis Lagrange.

The Lagrange interpolation is given mathematically by

$$x(t) = \sum_{k=0}^{N} L_k(t)x(t_k), \tag{3.32}$$

where $x(t_k)$ is the value of the sample at instant t_k and $L_k(t)$ is the interpolating polynomial term multiplied by each sample value, and is given by

$$L_k(t) = \frac{(t-t_0)\dots(t-t_{k-1})(t-t_{k+1})\dots(t-t_N)}{(t_k-t_0)\dots(t_k-t_{k-1})(t_k-t_{k+1})\dots(t_k-t_N)}. \tag{3.33}$$

Note that the terms on the numerator are of the form $(t - t_k)$, and the terms on the denominator are of the form $(t_k - t_i)$, with the single term $(t_k - t_k)$ removed from both the numerator and denominator. At the sample time $t = t_k$ (corresponding to the current "center" of interpolation), the interpolating function $L_k(t_k) = 1$ because the numerator will equal the denominator. At time instants corresponding to one of the other known samples, say, $t = t_k + 1$, one of the numerator terms will be zero, and thus the interpolating function $L_k(t_k + 1) = 0$. Therefore, the interpolation is guaranteed to pass through the known samples with weighting of one:

$$L_k(t_n) = \begin{cases} 1 & : \quad n = k \\ 0 & : \quad n \neq k \end{cases}. \tag{3.34}$$

3.10 BLOCK DIAGRAMS AND DIFFERENCE EQUATIONS

Once the samples have been acquired and quantized, they can be processed by a signal processing algorithm. This is done using a mathematical "model" for the processing. Such an input–output model is depicted in Figure 3.32. The input at sample instant n is denoted $x(n)$, and the output is $y(n)$. The transfer function from input to output characterizes the system: Given a known input, we can compute the corresponding output. This transfer function may be represented by its *impulse response*—that is, the response of the system when subjected to a single input of amplitude 1.0 at sample instant $n = 0$.

The output is represented as the set of amplitudes $y(n)$, as n takes on the integer values 0, 1, 2, 3, … . The impulse response is useful because the response of any linear system to *any* given input may be thought of as being the summation of

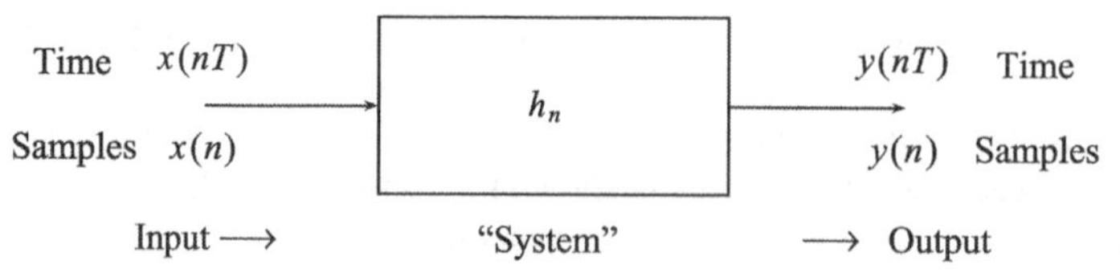

FIGURE 3.32 System block diagram: input $x(n)$, output $y(n)$, and impulse response coefficients h_n.

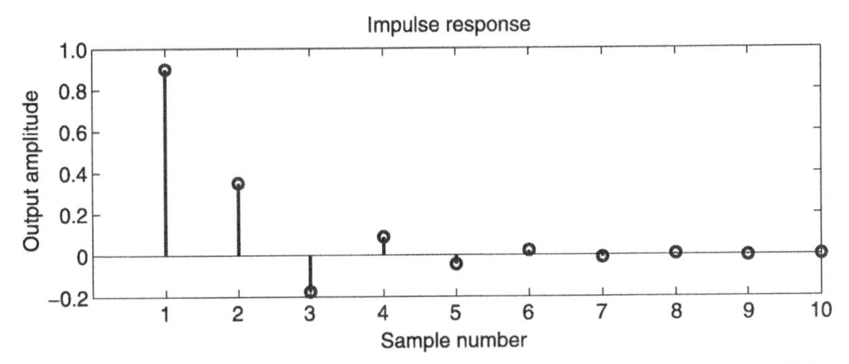

FIGURE 3.33 Impulse response samples h_n given the difference equation coefficients of Equation 3.35.

responses to as a series of sequential, scaled impulses. This is sometimes called the "principle of superposition." To see this, consider the simple *difference equation* given by

$$y(n) = 0.9x(n) + 0.8x(n-1) - 0.5y(n-1). \tag{3.35}$$

A realization of the impulse response of this equation in MATLAB is shown in the following. Note that MATLAB uses array indexes starting at 1, whereas the usual algorithmic description (as above) uses indexes starting at 0. Also, note that it is necessary to test for the proper indexing of the arrays. The resulting impulse response is shown in Figure 3.33.

```
NT = 10;
x = zeros(NT, 1);
y = zeros(NT, 1);
x(1) = 1;
for n = 1:NT
    y(n) = 0.9 * x(n);
    if(n > 1)
        y(n) = y(n) + 0.8 * x(n - 1);
        y(n) = y(n) - 0.5 * y(n - 1);
    end
end
stem(y);
```

Now consider input sequence

$$x(n) = \{0.6, -0.2, 0.8\}.$$

As shown in Figure 3.34, the response of the system (lower panel) is in fact the summation of the individual, scaled, and delayed impulse responses.

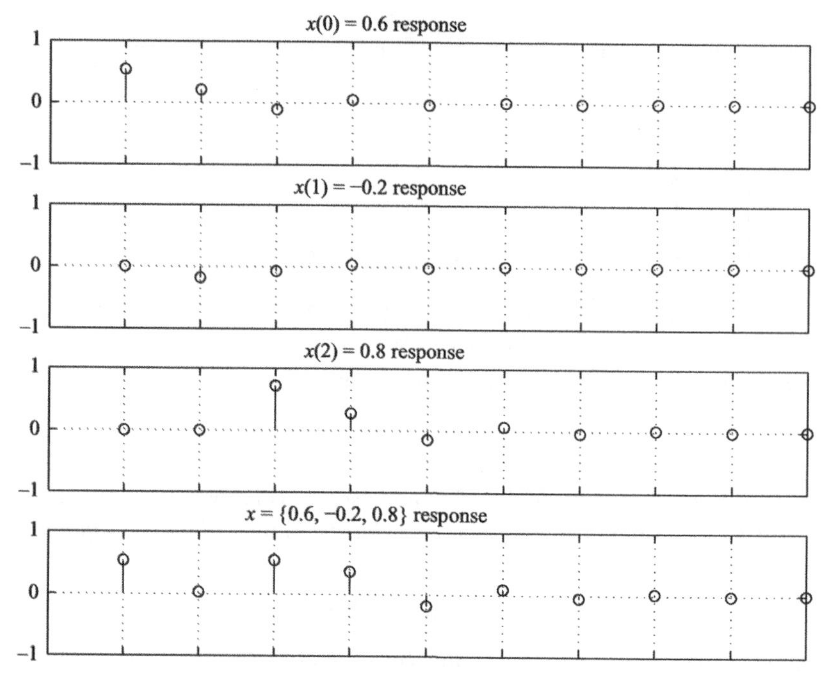

FIGURE 3.34 Visualizing the impulse response as the superposition of successive impulse responses.

Equation 3.35 is the difference equation for a specific system. To expand the difference equation to cater for all possible systems we may encounter, we need a more general equation for the case where there may be more coefficients of either the input or output. The following general form will do this:

$$\begin{aligned} y(n) = {} & (b_0 x(n) + b_1 x(n-1) + \ldots + b_N x(n-N)) \\ & - (a_1 y(n-1) + a_2 y(n-2) + \ldots + a_M y(n-M)). \end{aligned} \tag{3.36}$$

Effectively, the coefficient of the left-hand side $y(n)$ is $a_0 = 1$.

The implementation of this equation is shown in block diagram form in Figure 3.35. Note how the output is fed back to the input via the a_k coefficients (which may or may not be present). At least one of the b_k terms must be present in order to couple the input $x(n)$ into the system. The choice of a negative sign for the a coefficients is arbitrary (since the a's are constant coefficients). However, a negative sign is normally used to simplify later mathematical analysis.

In order to cater for difference equations of any arbitrary size, a more "general-purpose" coding provides for easier modification—both of the number of coefficients and their values. The MATLAB code for one possible method of realizing this is shown below. Note that although it appears more complex, it is more general and thus more easily modified for use with other systems.

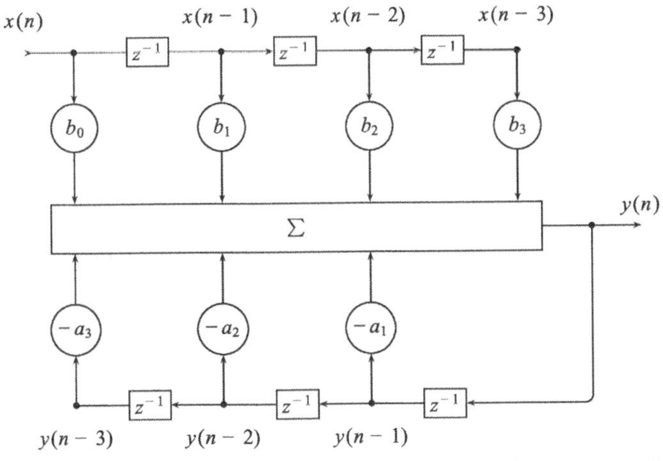

FIGURE 3.35 Block diagram representation of a difference equation. Four feed-forward b coefficients are shown, together with three feedback (a) coefficients. Of course, this could be increased to any arbitrary size.

Setting `x(1) = 1` in the code provides an impulse input to the system (i.e., sample $x(0) = 1$). The variable NT is the total number of input–output points to iterate over.

```
NT = 10;
b = [0.9 0.8];
N = length(b) - 1;
a = [1 0.5]; % a(1) = 1 always
M = length(a) - 1;
x = zeros(NT, 1);
y = zeros(NT, 1);
x(1) = 1;
for n = 1:NT
    y(n) = 0;
    for k = 0:N
        if( (n - k) > 0 )
          y(n) = y(n) + x(n - k) * b(k + 1);
        end
    end
    for k = 1:M
        if( (n - k) > 0 )
          y(n) = y(n) - y(n - k) * a(k + 1);
        end
    end
end
```

Noting the symmetry in the basic difference equation, the loops for multiplying the b and a coefficients are seen to be virtually identical. Effectively, $a_0 = 1$ because a_0 is implicitly the coefficient of $y(n)$, the output on the left-hand side of Equation 3.36. Also, the question may be asked as to why the a values are subtracted. The answer is simply that it is conventional, and since the a's are constant coefficients, it makes no difference whether they are defined as positive or negative in the difference equation.

Since the iteration of difference equations is common in discrete-time systems, MATLAB includes a built-in function to perform the above difference equation iterations. This function is called **filter()**, which takes, as expected, the equation coefficients a_k, b_k, and the input signal $x(n)$, and generates the output signal $y(n)$, as shown in the following example:

```
NT = 10;
b = [0.9 0.8];
a = [1 0.5];
x = zeros(NT, 1);
x(1) = 1;
y = filter(b, a, x);
stem(y);
```

3.11 LINEARITY, SUPERPOSITION, AND TIME INVARIANCE

In iterating difference equations, it is normal to assume zero initial conditions; that is, samples prior to $t = 0$ or sample values of index $n < 0$ are assumed to be zero. This is easily done in practice by initializing all variables and arrays to zero at the start-up of the iterations. This brings us to several underlying principles which underpin the operation of discrete-time systems and models: linearity, superposition, and time invariance.

Consider *linearity* first. Suppose we have a system defined as

$$y(n) = 0.9x(n) + 0.8x(n-1) - 0.4x(n-2) + 0.1x(n-3). \qquad (3.37)$$

We can iterate this difference equation to determine the output in response to an impulse as a specific case and extend that to any arbitrary input signal. Suppose we have an input sequence $x(n)$ as illustrated in Figure 3.36, which happens to consist of five constant samples followed by zeros. The output $y(n)$ of the system defined by this equation is shown. Now suppose we double the magnitude of the input. As shown, we expect this to translate to a linear multiplication of the output. This means the output is now $2y(n)$. In general, for a linear system, if we have an input sequence $x(n)$ giving an output sequence $y(n)$, then $kx(n) \rightarrow ky(n)$. Furthermore, if we have two separate inputs, say, $x_1(n)$ and $x_2(n)$, and the output of a system in response to

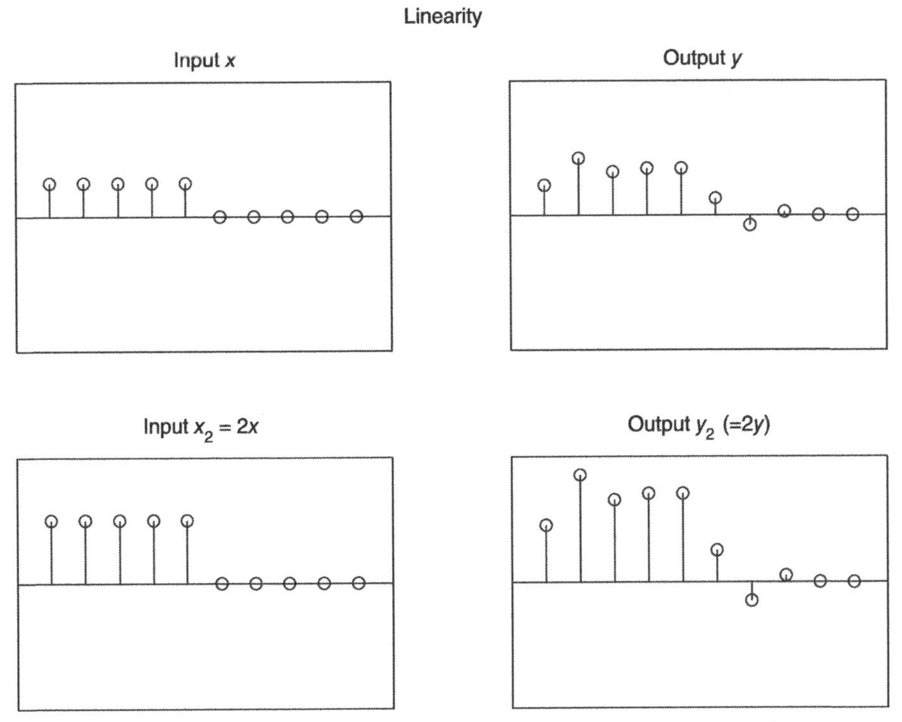

FIGURE 3.36 Illustrating the *linearity* of a system. Scaling the input scales the output accordingly, in a linear fashion. We can further extend this to multiple inputs to a system, which leads to the principle of superposition as illustrated in Figure 3.38.

each applied separately is $y_1(n)$ and $y_2(n)$, respectively, then the output of the combined inputs $x_1(n) + x_2(n)$ will equal $y_1(n) + y_2(n)$ if the system is *linear*.

We now examine what happens when we apply a given input at a different time. This is illustrated in Figure 3.37. We have an input $x(n)$ applied, which happens to be two positive samples in succession. The output $y(n)$ is shown. Now if we delay the $x(n)$ sequence and again apply to the linear difference equation, the output is clearly the same shape but delayed by a corresponding amount. In other words, if an input $x(n)$ gives an output $y(n)$, then delaying by N samples, we have $x(n - N) \rightarrow y(n - N)$. Such a system is said to be *time invariant*.

An essential consequence of linearity is that a linear system will also obey the *principle of superposition*. This is illustrated in Figure 3.38. We have an input $x(n)$, consisting of five nonzero samples, and the corresponding output $y(n)$. Now suppose we imagine the input $x(n)$ to be constructed from the sequences $x_1(n)$ and $x_2(n)$ as shown in the figure—in other words, $x(n) = x_1(n) + x_2(n)$. Each of these, applied to the system independently, gives the outputs $y_1(n)$ and $y_2(n)$, respectively. So, if we use the combined signal $x_1(n) + x_2(n)$ as input (which happens to equal $x(n)$), then the summation of the separate outputs is $y_1(n) + y_2(n)$, which is identical to the original $y(n)$. Thus, the separate inputs are able to be applied separately, and the

Time invariance

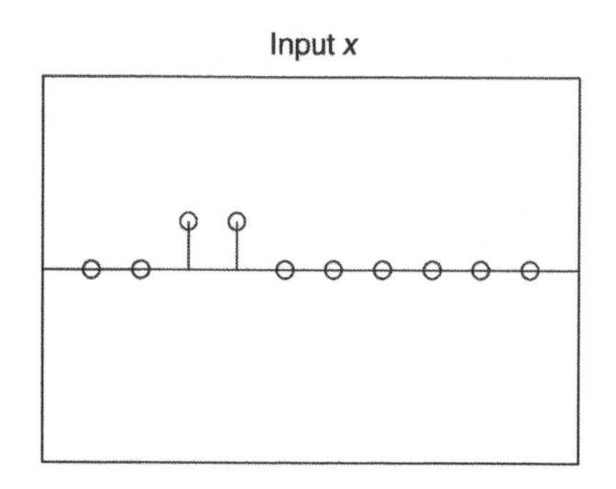

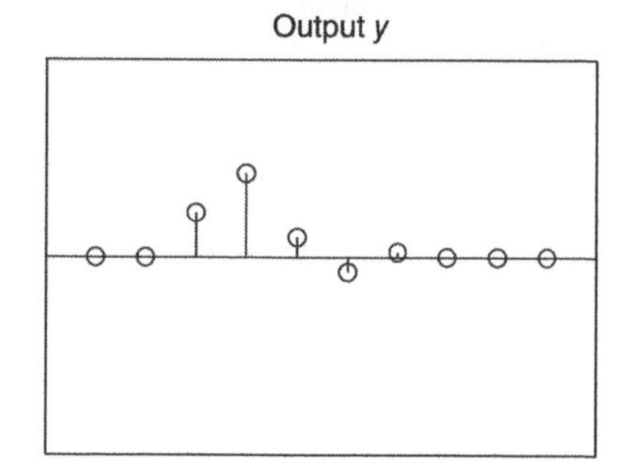

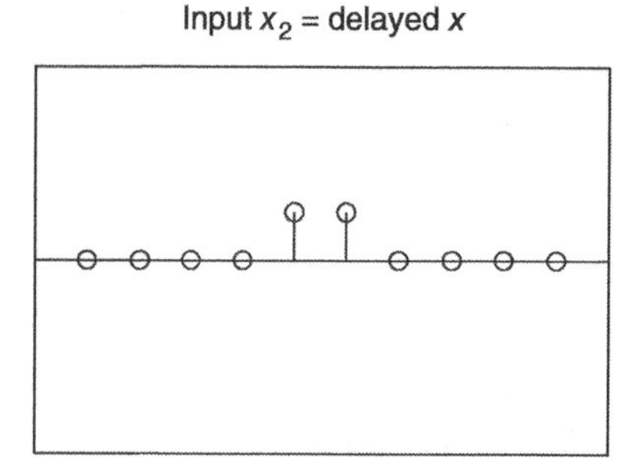

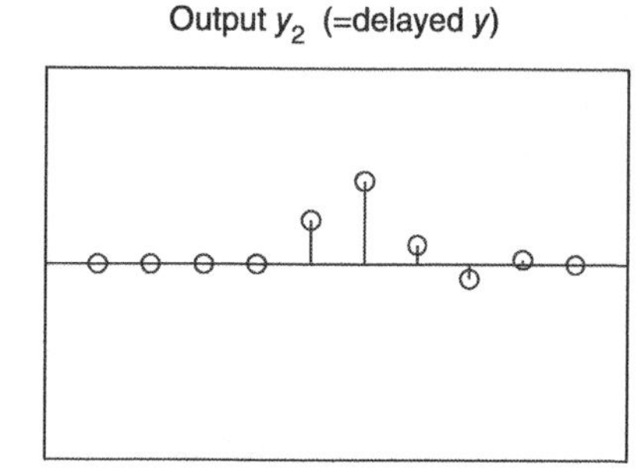

FIGURE 3.37 Illustrating the concept of *time invariance.* If the input is time shifted, the output of the system is correspondingly shifted in time.

output will be identical to the sum of the separate outputs as if each sequence was applied independently. Mathematically, this may be summarized by saying that if $x_1(n) \rightarrow y_1(n)$ and $x_2(n) - y_2(n)$, then $x_1(n) + x_2(n) \rightarrow y_1(n) + y_2(n)$.

So, let us look at a hypothetical example. Suppose we have a system whose characteristic is $y(n) = x^2(n)$. If we have an input to this system of $x(n) = x_1(n) + x_2(n)$, then the output would be $(x_1(n) + x_2(n))^2$. But if we applied each input separately and added them together, it would be $x_1^2(n) + x_2^2(n)$, and this is clearly *not* equal to $(x_1(n) + x_2(n))^2$. Hence, this is not a linear system.

To formalize this a little more, if the system performs an operation $f(\cdot)$, then for constants k_1 and k_2, a linear system obeys

$$f\{k_1x_1(n)+k_2x_2(n)\}=k_1f\{x_1(n)+k_2f\{x_2(n)\}. \tag{3.38}$$

Finally, there is a very important class of algorithms called *adaptive algorithms*, which do in fact have time-varying characteristics. Such algorithms are used to process speech, for example, which varies over time. If we examine a short enough time interval, the speech signal characteristics appear "approximately" the same, though over a longer interval they change. Adaptive algorithms are also used in the

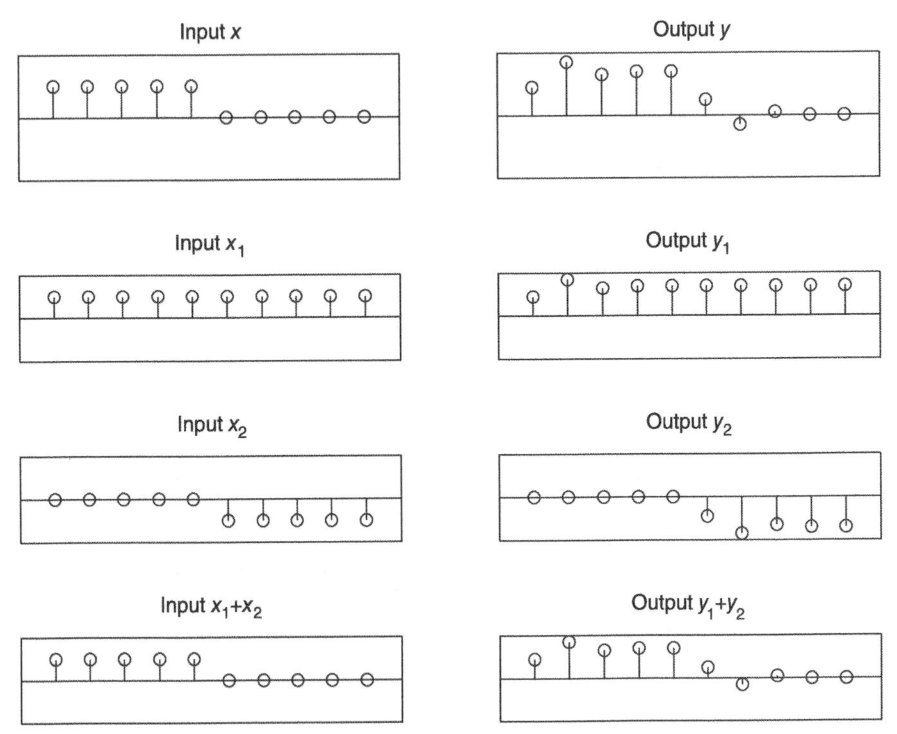

FIGURE 3.38 Illustrating the principle of *superposition.* If the input is decomposed into separate inputs which are applied separately, the output is effectively superimposed as the sum of all inputs considered separately. A system must be linear in order for this to occur.

form of adaptive filters, which iteratively adjust themselves according to some steering function over time and attempt to converge on a final solution according to the signals presented to them.

Most of our work will be concerned with linear, time-invariant (LTI) systems. Real signals and systems often only approximate the above requirements, but if the approximation is close enough, the LTI theory can be successfully applied.

3.12 PRACTICAL ISSUES AND COMPUTATIONAL EFFICIENCY

The previous examples have implicitly assumed that we iterate the difference equation over a fixed maximum number of samples. Realistically, the system output may in fact go on indefinitely, thus apparently requiring an infinite amount of memory. Inspection of the difference (Equation 3.36) shows that it is only necessary to store the previous N inputs, plus the present one, and M previous outputs.

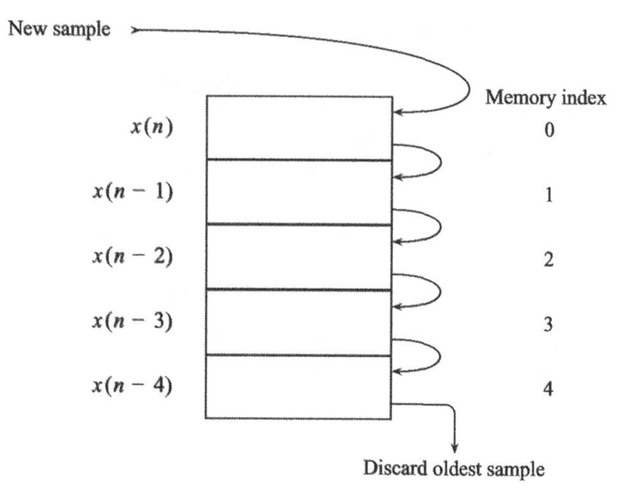

FIGURE 3.39 Delay-line buffering using a "stack-based" algorithm. The indexes are shown starting at zero, but it must be remembered that MATLAB indexes start at one.

The obvious practical solution is to use a "stack" architecture to store the previous inputs and outputs, as illustrated in Figure 3.39. At the end of each iteration, the oldest sample is discarded. Sample $x(n)$ is stored in vector element `x(1)`; sample $x(n-1)$ is stored in vector element `x(2)`, and so forth.[6] We can extend our MATLAB code for difference equations to implement this arrangement in a more general-purpose manner, as shown below. The variable `yplot` is only used for the purpose of saving the data over the iteration—in a practical system, it need not be saved. Note the segment labeled "age samples," which implements the stack in order to "push back" samples on each iteration.

One problem with the stack-based storage of past samples is that it requires a considerable amount of copying for high-order systems. Apart from the calculation of the difference equation itself, for N current and previous inputs and for M current and previous outputs, a total of $N+M$ sample copy operations are required. One solution is to employ special-purpose digital signal processors which have dedicated copy-and-multiply instructions.

```
NT = 10;
b = [0.9 0.8];
N = length(b) - 1;
a = [1 0.5];
M = length(a) - 1;
% input: impulse
x = zeros(N + 1, 1);
x(1) = 1;
```

[6]Remember that MATLAB starts array indexes at 1 and not 0. C starts indexes at 0.

```
% output
y = zeros(M + 1, 1);
yplot = zeros(NT, 1); % saves for plot
for n = 1:NT
    % difference equation
    y(1) = 0;
    for k = 0:N
        y(1) = y(1) + x(k + 1) * b(k + 1);
    end
    for k = 1:M
        y(1) = y(1) - y(k + 1) * a(k + 1);
    end
    yplot(n) = y(1); % save for plotting
    % age samples
    for k = N + 1:-1:2
        x(k) = x(k - 1);
    end
    for k = M + 1:-1:2
        y(k) = y(k - 1);
    end
end
```

Alternatively, the memory copying may be eliminated by using a circular buffer arrangement as shown in Figure 3.40. Instead of physical copying, a modulo-N index is used to "dereference" the actual samples required. A modulo variable is one which counts up to the modulo limit and then resets to zero. For example, counting modulo-5 gives the sequence 0, 1, 2, 3, 4, 0, 1, 2, 3, 4, 0, 1, 2, 3, 4, … .

The indexing of the required samples is calculated with a modulus of the number of samples stored. Referring to Figure 3.40, suppose there are five samples stored, $x(n)$ to $x(n-4)$. These are assigned physical (memory) indexes 0–4, numbering from the left-hand side. At time t_k, suppose the current sample $x(n)$ has physical index 3 (starting from zero). Sample $x(n-1)$ has index $3-1=2$; sample $x(n-2)$ is stored in location $3-2=1$, and so forth. At time sample t_{k+1}, sample $x(n)$ is stored in one location further on, in physical location 4. Sample $x(n-1)$ is stored in location $4-1=3$. At the subsequent time instant, time t_{k+2} in the diagram, the physical index is incremented to 5. However, since only samples up to $x(n-4)$ are required in this application, we do not need to use another new memory location.

It is possible to reuse the oldest sample location, which was sample $x(n-4)$ at time instant t_{k+1}, but which would otherwise have become sample $x(n-5)$ at time instant t_{k+2}. When the index for the new sample $x(n)$ is incremented, we simply test if it is equal to the number of samples (in this case, 5). If it is, then the index is simply reset to 0. Calculating the true physical index of previous samples is then slightly more difficult. As before, sample $x(n-1)$ is stored one location prior to the

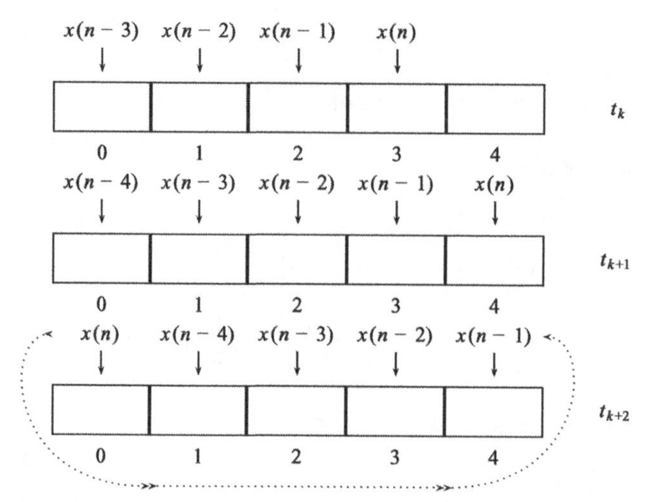

FIGURE 3.40 Circular buffering: "in-place" buffering using modulo buffer pointer arithmetic. Sample $x(n)$ represents the current sample at each time. At time t_{k+1}, we reuse the oldest sample space by "wrapping" the current sample pointer around.

current sample. This now corresponds to $0 - 1 = 1$. Sample $x(n-2)$ is stored at index $0 - 2 = -2$. Since memory location indices only start at 0, this represents a problem. The solution is simply to test whether the physical index thus calculated is negative, and if it is, simply add the number of samples to it. In this case, index -1 becomes $-1 + 5 = 4$. This tells us that sample $x(n-1)$ is now stored at location 4. As mentioned before, this type of "wrapping" calculation is termed "modulo arithmetic."

3.13 CHAPTER SUMMARY

The following are the key elements covered in this chapter:

- The nature of *binary representations*, including *integer*, fixed, and *floating-point* numbers, and how these are applicable to signal-processing problems
- Sampling and *quantization* of analog signals and their reconstruction
- A review of *sampling theory* including time sampling and difference equations
- Important concepts in linear systems, including *superposition*, *linearity*, and *time invariance*
- The role of *transfer functions* and *block diagram* representations, including an understanding of how to *implement* difference equations, and related efficiency issues

PROBLEMS

3.1. A CD can store approximately 650 MB of data and can store about 60 minutes of digital audio.

(a) Show that the effective data rate is approximately 1.4 Mbps.

(b) Calculate the approximate sample rate, byte rate, and bit rate for a sample rate of 44.1 kHz (answers: 88,200 samples per second; 176,400 bytes per second; 1,411,200 bps)

3.2. Check the correctness of the negative values in Table 3.1 using Equation 3.2.

3.3. What is the trade-off between fixed-point and floating-point number representations? How can it be that, using the same number of bits, a floating-point number can represent much larger numbers as well as very small numbers?

3.4. The code listing in Section 3.4.4 shows numbers with 20 digits of precision after the decimal point, but the last several digits appear to be runs of zeros. Explain why this is, in terms of the MATLAB value for `eps` and the fact that the values are stored with precision `single`.

3.5. It was shown that MATLAB evaluates log2 (double(intmax + 1)) to 31. Why not 32? What is the reason for +1?

3.6. The multiplication of binary numbers in Section 3.4.3 showed how to multiply two N-bit binary numbers. How many bits does the product require in order to store the largest possible unsigned result?

3.7. The number 11.5 is stored as a 32-bit floating-point value in IEE754 format. Show that the biased exponent is 130 decimal (binary 10000010) and that the fractional part is 1.4375 decimal (binary 01110000000000000000000). Write the entire 32-bit representation.

3.8. Use the floating-point CORDIC code given to compute $\cos(\pi/4)$ and $\sin(\pi/4)$. For 8-bit iterations, how close does the result come to the single-precision calculation? How close does it come using 16-bit calculations?

3.9. Reimplement the CORDIC code given to use only bitwise integer operations in MATLAB and verify its correct operation using the example given in the text.

3.10. Work through the MATLAB image processing examples in Section 3.7. Note the use of the various `set()` commands and, in particular, how to set the aspect ratio. Using the `whos` command, note the dimensions and data type of the image itself and the palette matrix where applicable.

3.11. Given samples at times $t_0 = 1, t_1 = 3$ of value $x(t_0) = 2x(t_1) = 5$, determine the Lagrange interpolator. Plot this using a small step size for t in order to verify that the line goes through the defined points $(t_0, x(t_0))$ and $(t_1\ x(t_1))$.

3.12. Given samples at times $t_0 = 1$, $t_1 = 3$, $t_2 = 4$ of value $x(t_0) = 2$, $x(t_1) = 5$, $x(t_2) = 4$, determine the Lagrange interpolator. Plot the known points above, together with the interpolated function. Use a small step size to plot the interpolated points and verify that the interpolated points go through the known data points.

3.13. Check that the three difference equation methods given in Section 3.10 produce identical results.

3.14. Find the difference equation and then the transfer function corresponding to the following implementation of an impulse response:

```
b = 1;
a = [1 -2*cos(pi/10) 1];
x = zeros(100, 1);
x(1) = 1;
y = filter(b, a, x);
stem(y);
```

3.15. Using the color mixing code in Section 3.7.2, display a one-pixel image of the following colors:

(a) Red only
(b) Green only
(c) Blue only
(d) Red + green
(e) Red + blue
(f) Green + blue
(g) Red + blue
(h) Red + green + blue
(i) Red + 0.5 × green
(j) 0.6 × red + 0.5 × green + 0.6 × blue

In each case, display the resulting image and verify the mixing of the colors.

3.16. Repeat the previous question on color mixing, but use an unsigned 8-bit image for the display. Does the rounding implied by the color weighting when scaling to integers make a perceptible difference? For example, $0.3 \times 255 = 76.5$, and this would be an integer value of either 76 (truncated) or 77 (rounded).

3.17. Using the color mixing code in Section 3.7.2 and the color mixing concepts as discussed, load an RGB image and convert it to a grayscale image. Take care to handle overflow issues correctly when adding the color components.

3.18. Using the palette-based image display code in Section 3.7.3, display the random image using the grayscale color map. Then, use another color map which is built into MATLAB, such as **hsv** or **bone**. What use could you think of for these "artificial" color maps?

3.19. The terms RMS, MSE, and SNR were used in Section 3.6.2 in connection with quantization. These terms are important in many aspects of signal processing besides quantization.

(a) Explain in words the meaning of RMS, MSE, and SNR. What physical characteristic do they represent?

(b) MSE is defined as

$$\text{MSE} = \frac{1}{N} \sum_N (y(n) - x(n))^2.$$

Explain each term of the equation.

(c) RMS is defined as

$$\text{RMS} = \sqrt{\frac{1}{N} \sum x^2(n)}$$

Explain each term of the equation.

(d) SNR in decibels is defined as

$$\text{SNR} = 10\log_{10}\left(\frac{\sum_n x^2(n)}{\sum_n e^2(n)}\right).$$

Explain each term of the equation.

3.20. The measurements 44, 46, 53, 34, and 22 are quantized to the nearest multiple of 4. Determine the quantized sequence, the MSE, and the SNR.

3.21. Consider Figure 3.25, which shows a 4-Hz signal. What would happen if we sampled at 8 Hz? Why do we need to sample *above* the Nyquist rate?

3.22. The following code generates 2 seconds of 200-Hz sine wave and plays it back. Change the frequency to 7,800 Hz and repeat the experiment. Can you explain the result? Why does the frequency appear to be the same? What other frequencies would give similar results and why?

```
fs = 8,000;
T = 1/fs;
tmax = 2;
t = 0:T:tmax; % sample points
f = 200;      % sound frequency
y = sin(2*pi*f*t);
sound(y, fs);
```

CHAPTER 4

RANDOM SIGNALS

4.1 CHAPTER OBJECTIVES

On completion of this chapter, the reader should be able to

1. differentiate between *random* and *deterministic* signals;
2. explain the fundamental principles of *statistics*, as they apply to signal processing;
3. utilize *statistical distributions* in understanding and analyzing signal processing problems; and
4. explain and implement *histogram equalization* and *median filtering*.

4.2 INTRODUCTION

Random noise signals may be represented mathematically using statistical methods. It is important to understand the nature of "random" systems, and that "random" does not necessarily mean "totally unpredictable." From a signal processing perspective, if we understand the nature of random fluctuations in a signal, we are in a better position to remove, or at least to minimize, their negative effects. To this end, some practical examples in image filtering and image enhancement are included in this chapter to demonstrate some practical applications of theory.

4.3 RANDOM AND DETERMINISTIC SIGNALS

A broad but useful categorization of signal types is into two classes:

1. *Random signals* are those which are not precisely predictable; that is, given the past history of a signal and the amplitude values it has taken on, it is not possible to precisely predict what particular value it will take on at certain instants in the future. The value of the signal may only be predicted subject to certain (hopefully known) probabilities.
2. *Deterministic signals* are those which *are* predictable. They may be described by equations, and from the recent past history of the signal, we can predict

Digital Signal Processing Using MATLAB for Students and Researchers, First Edition. John W. Leis.

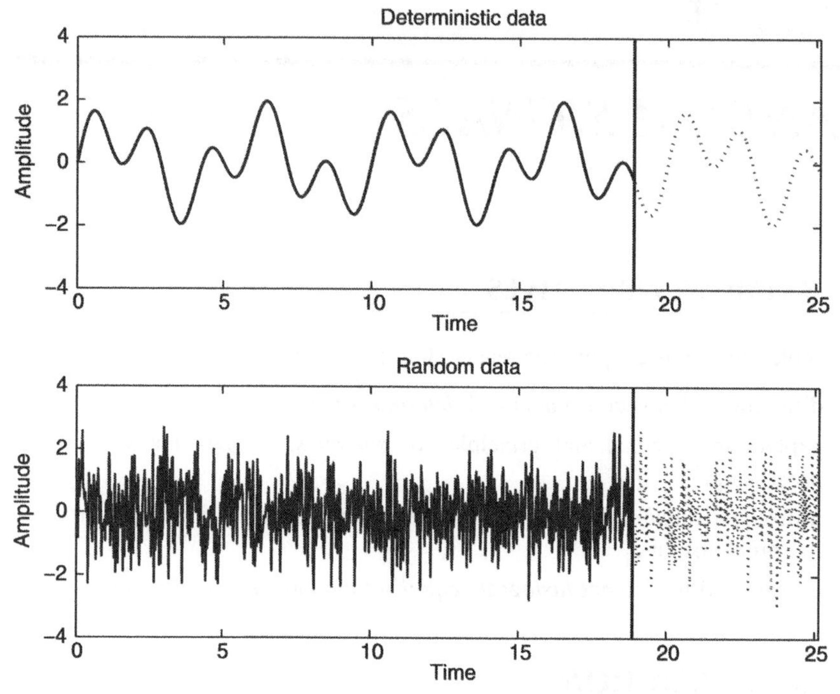

FIGURE 4.1 A comparison of random and deterministic signals. The question is: Are the dotted sections predictable, given only the previous observations (solid line)?

what amplitude values the signal will have at defined time instants. Subject to unavoidable quantization errors, arithmetic precision errors, and rounding errors, the prediction is exact and has no error.

Random signals can still be modeled—they are described by a probabilistic model, allowing us to state the *likelihood* of the signal taking on certain values in the future. We have encountered deterministic signals in the last chapter, where the difference equation was used to calculate new system output values.

Figure 4.1 illustrates the concepts of random and deterministic signals. It is immediately obvious that the signal in the upper panel is periodic (repeats itself over a certain interval or period, τ) and is therefore able to be predicted using a mathematical model. This mathematical description of the signal and the resulting equations are termed the *signal model*. Parameters may be determined for a particular model to describe a particular signal—in simple terms, consider a summation of sinusoidal waveforms:

$$x(t) = A_1 \sin(\Omega_1 t + \varphi_1) + A_2 \sin(\Omega_2 t + \varphi_2). \tag{4.1}$$

The parameters A_1 and A_2 represent amplitudes of each signal; Ω_1 and Ω_2 represent frequencies; and φ_1 and φ_2 are the phase shifts of each. This model would not adequately describe a signal which contains (say) three sinusoidal components, and the model would have to be extended to cater for that. The two-component model would be inaccurate.

For discrete systems, we can apply the general difference equation model:

$$\begin{aligned} y(n) = &(b_0 x(n) + b_1 x(n-1) + \cdots + b_N x(n-N)) \\ &- (a_1 y(n-1) + a_2 y(n-2) + \cdots + a_M y(n-M)). \end{aligned} \tag{4.2}$$

The constants N and M must be known, and the b and a parameters must be found (or approximated), in order to specify the signal model.

The signal shown in the lower plot of Figure 4.1 appears somewhat random and hence cannot be predicted. However, passive examination can be somewhat misleading. We need mathematical tools and algorithms to determine whether signals are in fact truly random. Random signals may be described by the likelihood of having a particular value. Such a statistical model also requires several parameters to be determined.

In reality, real-world signals are often best described by a combination of a base deterministic component together with a random or noise component. The simplest such formulation is an additive noise model:

$$x(t) = s(t) + \alpha v(t), \tag{4.3}$$

where $x(t)$ is the observed signal, $s(t)$ is the underlying periodic signal, and $v(t)$ is the additive random noise. The amount of additive noise is modeled by the parameter α.

In a similar way, a deterministic system is one whose underlying impulse response may be characterized by a mathematical model. In digital signal processing, this is usually a linear time-invariant (LTI) difference equation, as explained in Section 3.11.

Random signals will be considered in the following sections. In order to do this completely, it is necessary to first review some statistical theory, initially for one dimension but with an overview of multidimensional signals. Once the basic concepts have been developed, a practical application of the theory is then introduced. This involves enhancing a "noise-corrupted" image. This means that the signal data we have to process are no longer exactly the same as the original due to external factors beyond our control. These may be man-made or natural interference in the case of a transmission system, or perhaps errors introduced in storing the signal. A useful approach is to imagine the unwanted noise being described (or modeled) as a random signal added to the desired image data signal, as per Equation 4.3.

4.4 RANDOM NUMBER GENERATION

Random signal models are useful in real-world signal processing problems. Since we are considering digital signal processing, generating a random signal is equivalent to finding some method for generating random numbers. The first problem is the problem of randomness—what is truly random, after all? The operation of computer code is perfectly repeatable: Given a sequence of steps (an algorithm) in code and an initial value for all data used, running the code a second time produces *exactly* the same result as the previous time. Running it 100 times will produce 100 identical results. So, using computers to generate random numbers presents a problem. To get a feel for this, consider generating a sequence of random samples using MATLAB's **rand**() function:

```
x = rand(1000, 1);
subplot(2, 1, 1); stem(x(1:100));
subplot(2, 1, 2); stem(x(101:200));
```

Outwardly, the resulting samples will appear random—but are they? In fact, computer-based random number generators are termed *pseudorandom*. The sequence appears random but is in fact repeatable. Therefore, it is not truly random. If the random number generator is restarted using an initial starting point or "seed," the sequence is identical:

```
rand('seed', 0)
rand
ans =
    0.2190
rand
ans =
    0.0470
rand
ans =
    0.6789
```

The output on the first run is 0.2190, 0.0470, and 0.6789. On second and subsequent runs, the output is again 0.2190, 0.0470, and 0.6789. Thus, the samples are not, in the strictest sense, random. However, from one sample to the next, they are "random enough" for most signal processing purposes. Furthermore, it clearly depends on the initial starting point or seed for the random number generator. It is not uncommon to sample a high-speed system clock for this purpose.

4.5 STATISTICAL PARAMETERS

The simplest statistic to compute is the *mean*, which, mathematically, is in fact the *arithmetic mean*. For N signal samples, this is

$$\overline{x} = \frac{1}{N}\sum_{n=0}^{N-1} x(n). \tag{4.4}$$

This is what we commonly refer to as the *average*. Strictly speaking, this quantity is the *sample mean* because it is derived from a sample or subset of the entire population (not to be confused with the fact that a signal itself is sampled, in the signal processing sense). The symbol μ denotes the *population mean*, which is the mean of values if we have the entire set of possible samples available. Usually, we work with a subset of observed (sampled) values; hence, the sample mean is what we are referring to. If we take different subsets of samples, the sample mean calculated from each set itself forms a distribution which is representative of the population mean.

For a continuous (nonsampled) signal, the mean over a time interval, τ, is

$$\overline{x} = \frac{1}{\tau}\int_0^{\tau} x(t)\,dt. \tag{4.5}$$

Another parameter which will be shown in Section 4.11 to be very useful in designing image-restoring filters is the *median*. The median is the value for which half of the samples lie above it, and half lie below. For example, consider the sequence

6 4 8 3 4.

The (sample) mean is easily calculated to be 5. The median is found by first ordering the samples in ascending order:

3 4 4 6 8.

Crossing off 3 and 8, then 4 and 6, we are left with the median value of 4 (actually the second 4). This is the sample with index $(N-1)/2 = (5-1)/2 = 2$ in the ordered list.

Lastly, the *mode* or *modal value* is the most common value. It is found by counting the frequency of each sample. For example, the data used above have a frequency table which is easily derived:

Sample	0	1	2	3	4	5	6	7	8
Frequency	0	0	0	1	2	0	1	0	1

In this case, the mode is 4. Of course, the summation of the frequency counts must equal the total number of samples.

The *mean square* is defined as

$$\overline{x^2} = \frac{1}{N}\sum_{n=0}^{N-1} x^2(n). \tag{4.6}$$

A similar quantity, the *variance*, is defined as

$$\sigma^2 = \frac{1}{N}\sum_{n=0}^{N-1}\left(x(n)-\bar{x}\right)^2. \tag{4.7}$$

The variance, σ^2, and the standard deviation, σ, are related to the spread of the data away from the mean. The greater the variance, the greater the average "spread" of data points away from the mean. The limiting case—a variance of zero—implies that all the data points equal the mean.

4.6 PROBABILITY FUNCTIONS

Rather than one or two quantities (mean and variance) to characterize a signal, it is useful to have a more complete description in terms of the probability of observing the signal at or below a certain level. Many real-world signal processing applications (e.g., voice-based speaker identification) benefit from more advanced statistical characterizations.

A signal, $x(t)$, is shown in Figure 4.2. Suppose a particular level, X, is taken, where X is a value the random variable x can take on at some instant in time. Using the definitions in the figure, we could determine the proportion of time (in the whole observation record) that the signal spends under the nominated value X. Mathematically, this is

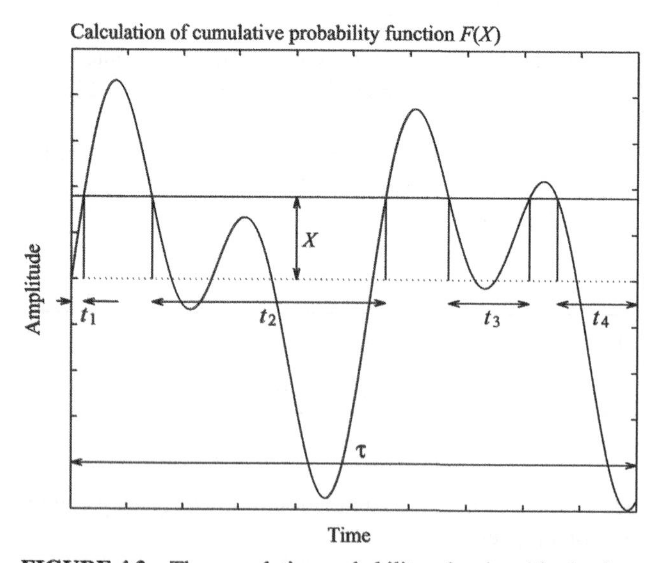

FIGURE 4.2 The cumulative probability of a signal is the time spent below a moving threshold, X. The value of this function must rise from zero (when X is most negative) to unity (when X is most positive). The shape of the curve is dictated by the nature of the time signal.

$$F(X) = \Pr\{x(t) \le X\}. \tag{4.8}$$

With reference to Figure 4.2, for a particular value X of signal $x(t)$ sampled over time interval τ,

$$F(X) = \frac{t_1 + t_2 + t_3 + t_4}{\tau}. \tag{4.9}$$

This is termed the *cumulative probability*. It is the probability of the signal being below a certain amplitude level. If we take equally likely random numbers in the interval [0, 1], what shape would the cumulative probability curve be? This can be demonstrated with the following MATLAB code:

```
N = 100000;
x = rand(N, 1);
dX = 0.01;
cumprob = [ ];
XX = [ ];
for X = -4:dX:4;
    n = length(find(x < X));
    cumprob = [cumprob n/N];
    XX = [XX X];
end
plot(XX, cumprob);
set(gca, 'ylim', [0 1.1]);
```

As might be expected, the value of $F(X)$, or the variable `cumprob` in the code, is found to be linearly increasing from 0 to 1. The random number generator is **`rand()`**, which generates *equally likely* random numbers in the range 0–1.

If we use a *normal* or *Gaussian* distribution, the values around the mean are more likely.[1] This is easily demonstrated by changing **`rand()`** above to **`randn()`**. The result is an "S-shaped" curve, gradually rising from 0 to become asymptotic to 1. Gaussian noise is commonly used as a statistical model for "real-world" noise sources.

A very much related concept is that of the *probability density* function (PDF). As shown in Figure 4.3, this is the likelihood that a signal at a given instant lies in a range from X to $X + \delta X$. From Figure 4.3, this is calculated over time interval τ as

$$f(X)\delta X = \frac{\delta t_1 + \delta t_2 + \delta t_3 + \delta t_4 + \delta t_5}{\tau}. \tag{4.10}$$

[1]Named after the German mathematician Johann Carl Friedrich Gauss.

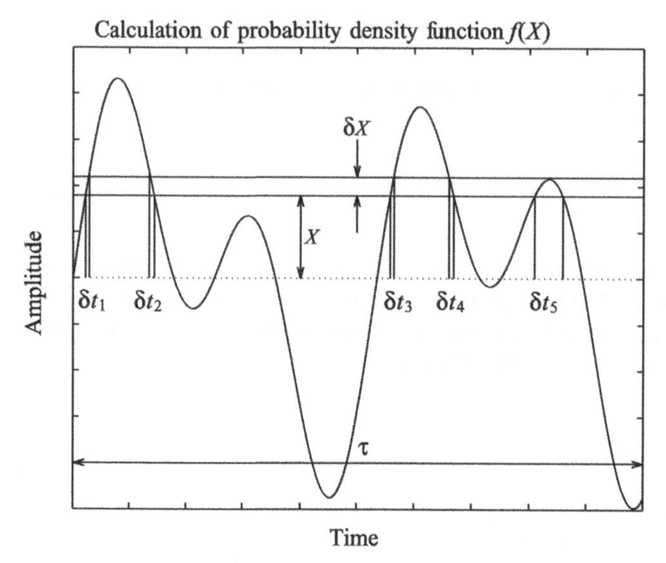

FIGURE 4.3 The probability density of a signal: time spent in an interval δX wide, at offset X. As the value of X sweeps from $-\infty$ to $+\infty$, the quantity of the signal present in the δX interval changes. Where the signal "spends" most of its time, the PDF will be largest. Conversely, values of X where the signal is not often spending time will result in a small PDF value.

Note that this is given the symbol $f(X)$ and that the definition is a *density*, which depends on the particular value of δX—hence, δX appears on the left-hand side of the definition.

Using the definition, it is relatively straightforward to write some MATLAB code to generate some random numbers in the interval [0, 1] and to determine their PDF, as follows:

```
N = 100000;
x = rand(N, 1);
dX = 0.01;
pdf= [ ];
XX = [ ];
for X = -4:dX:4;
    n = length(find((x > X) & (x < (X + dX))));
    prob = n/N;
    pdf = [pdf prob/dX];
    XX = [XX X];
end
plot(XX, pdf);
```

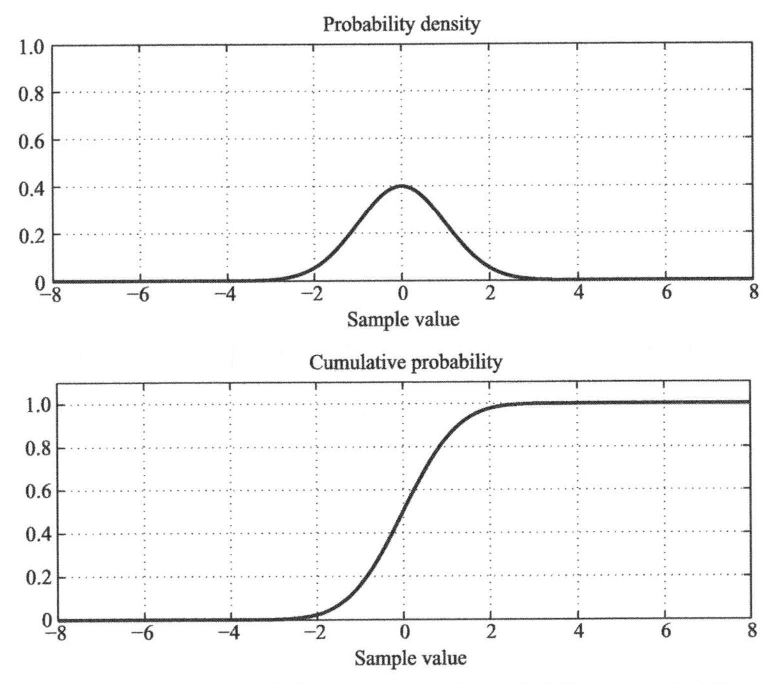

FIGURE 4.4 Relationship between cumulative probability and probability density, illustrated using a Gaussian distribution.

Intuitively, we would *expect* the PDF of a set of uniformly distributed random numbers to be a straight line—this is, after all, why they are called "uniformly distributed." But the above example shows that the distribution is not precisely uniform. This is because we are making two assumptions about a random process: that we have an infinite number of samples and that the interval of measurement is infinitely small. First, in the code example given, the number of samples is N, and ideally, we want the limit as $N \rightarrow \infty$. Second, the density interval δX is chosen to be an "arbitrarily small" number. Ideally, we want the limit as $\delta X \rightarrow 0$.

Changing the random number generator function **`rand()`** to a Gaussian source using **`randn()`** produces a "bell-shaped" curve for the PDF. The cumulative probability and probability density for a Gaussian source are illustrated in Figure 4.4. The "bump" in the Gaussian PDF shows the range of values which are more likely. In most practical applications, it is this observation—that some values are more likely than others—that is crucial to understanding the nature of the signal.

The PDF allows us to determine the probability of the signal being between X_1 and X_2—it is the *area* under the $f(X)$ curve between these limits:

$$\Pr\{X_1 \leq x(t) \leq X_2\} = \int_{X_1}^{X_2} f(X)dX. \tag{4.11}$$

Since the signal must exist between $\pm\infty$ with a probability of one, the total area must be equal to one:

$$\int_{-\infty}^{+\infty} f(X)\,dX = 1. \tag{4.12}$$

The relationship between $f(X)$ and $F(X)$ is also important. Since

$$\Pr\{x(t) \le X\} = F(X), \tag{4.13}$$

then it follows that

$$\Pr\{x(t) \le X + \delta X\} = F(X + \delta X). \tag{4.14}$$

Combining these two, we can say that

$$\Pr\{X \le x(t) \le X + \delta X\} = F(X + \delta X) - F(X). \tag{4.15}$$

The left-hand side of this expression is $f(X)\ \delta X$, and so $f(X)$ can be calculated from

$$f(X) = \frac{F(X + \delta X) - F(X)}{\delta X}. \tag{4.16}$$

This is the derivative of $F(X)$, and hence the rate of change of the cumulative probability equals the probability density, or

$$f(X) = \frac{dF(X)}{dX}; \tag{4.17}$$

that is, the slope of the $F(X)$ curve (lower curve) in Figure 4.4 equals the $f(X)$ curve (upper curve). Note the points where the slope is a maximum and also where it is equal to zero.

In reverse, the area under the PDF curve from $-\infty$ to some value X equals the cumulative density, or

$$F(X) = \int_{-\infty}^{X} f(x)\,dx; \tag{4.18}$$

that is, the area under the $f(X)$ curve (upper curve) in Figure 4.4 equals the $F(X)$ curve (lower curve).

The area starts off at zero (leftmost, or $X = -\infty$) and becomes equal to one, as shown by the $F(X)$ curve becoming asymptotic to one.

4.7 COMMON DISTRIBUTIONS

The Gaussian distribution has been encountered several times. It has a particular equation which describes the PDF in terms of the mean and variance:

$$f_{\text{Gaussian}}(x) = \frac{1}{\sigma\sqrt{2\pi}} \exp\left\{-\frac{(x - \bar{x})^2}{2\sigma^2}\right\}. \tag{4.19}$$

This equation can be used to determine the *mathematically expected* or most likely value of a random source, which is known to be Gaussian, provided its mean and

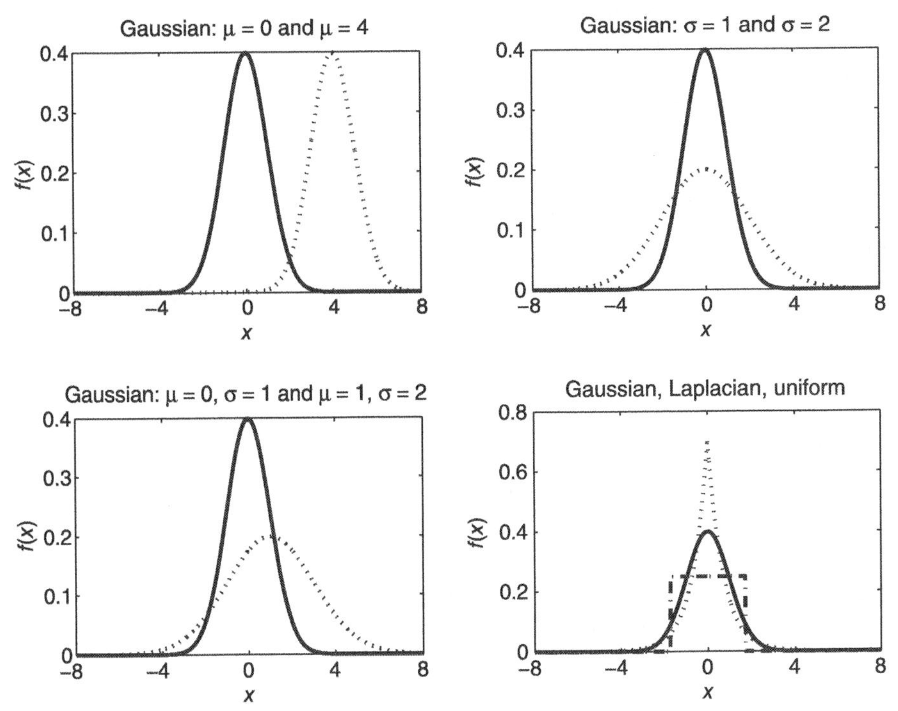

FIGURE 4.5 Probability density: effect of changing the distribution parameters. For a given distribution, we can change the mean and/or variance. Changing the mean moves the center peak, and changing the variance changes the spread of the distribution. The three types of distribution discussed in the text are also shown for comparison.

variance are known. Another common PDF signal model is the Laplacian distribution:[2]

$$f_{\text{Laplacian}}(x) = \frac{1}{\sigma\sqrt{2}} \exp\left\{-\frac{\sqrt{2}\,|x-\overline{x}|}{\sigma}\right\}. \tag{4.20}$$

Of course, the uniform PDF is simply

$$f_{\text{uniform}}(x) = \begin{cases} K = \dfrac{1}{x_{\max} - x_{\min}} & : \quad x_{\min} \le x \le x_{\max} \\ 0 & : \quad \text{otherwise} \end{cases}. \tag{4.21}$$

Figure 4.5 shows the effect of changing the mean of a Gaussian distribution. The overall shape is retained, but the center point is moved to coincide with the mean.

[2]Named after the French mathematician Pierre-Simon Laplace.

Figure 4.5 also shows the effect of changing the variance of a Gaussian distribution. The overall shape is retained but becomes somewhat more peaked as the variance tends to zero. A larger variance makes for a flatter curve—this is to be expected, as the area must equal one. The peak is still centered on the mean.

Note that the mean and variance of a signal may be calculated from the PDF alone. The mean of a signal, $\overline{x}$, is

$$\overline{x} = \int x f(x)\,dx. \tag{4.22}$$

This is really the summation of the signal value, weighted by the PDF. As the signal value gets further away from the mean, it is weighted less by the PDF. Similarly, the mean square in terms of the PDF is

$$\overline{x^2} = \int x^2 f(x)\,dx, \tag{4.23}$$

and the variance is

$$\sigma^2 = \int (x - \overline{x})^2 f(x)\,dx. \tag{4.24}$$

4.8 CONTINUOUS AND DISCRETE VARIABLES

The discrete-valued counterpart of the probability density is the *histogram*. The histogram is simply the count of how many samples have each possible value in the range. Generation of the histogram may be visualized as in Figure 4.6, where (for the sake of illustration) the histogram is shown on its side, with height proportional to the time the signal spends at a particular level.

Consider the image shown in Figure 4.7. The range of pixel values is from 0 to 255, but each pixel can only take on an integer value in this range. A pixel cannot have the value 67.2, for example. The histogram of this image is shown in Figure 4.8. For each possible pixel value of 0, 1, ... , 255, a count of the number of

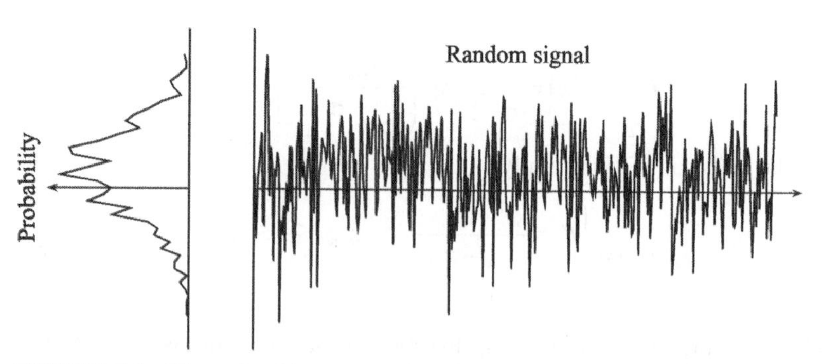

FIGURE 4.6 Illustrating histogram generation from raw samples. The signal (on the right) is analyzed and the fraction of time it spends at a certain level is plotted on the left-hand graph. This is then rotated on its side to become the histogram. The probability is this value scaled by the total number of values.

FIGURE 4.7 A test image. This is a grayscale (luminance only) image, having 256 levels. The size of the image is 1,024 × 768 (width × height).

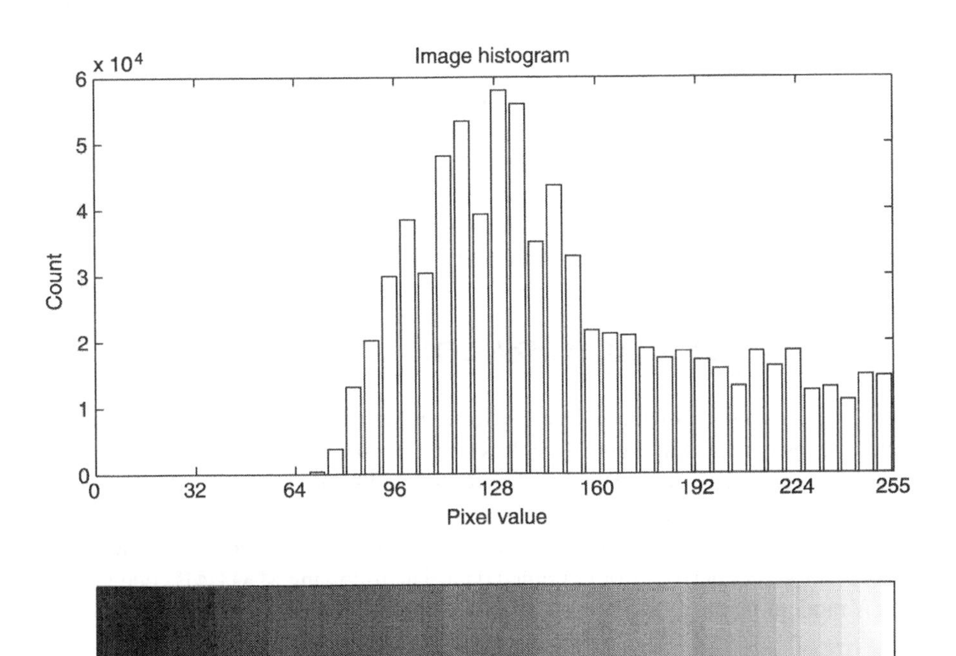

FIGURE 4.8 The pixel histogram of the test image shown in Figure 4.7. A grayscale bar is shown to indicate the relative brightness of each of the values. Thus, we see that there are few very dark pixels, a reasonable spread up to white, with a concentration in the middle gray region.

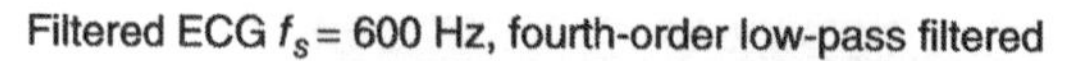

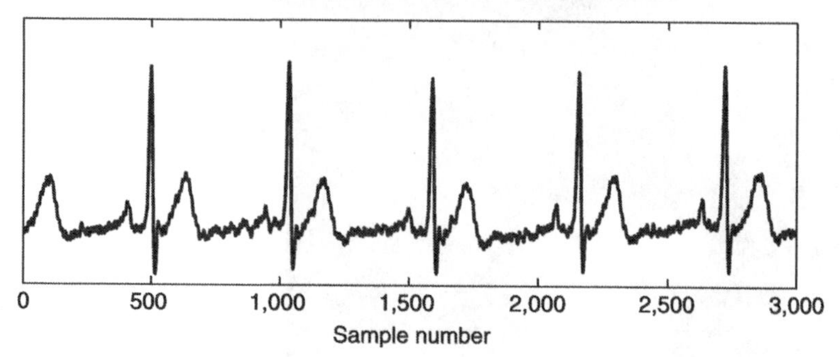

FIGURE 4.9 An electrocardiogram (heart waveform) sampled at 600 Hz.

pixels having that value is found. The bars in the histogram represent the count for that pixel. Thus, the summation of all histogram bars must equal the total number of source samples—in this case, the number of pixels in the image, or $1,024 \times 768 = 786,432$. This is the same reason that the area under the continuous variable PDF is equal to unity.

The modal (most common) value is easily seen in the histogram. However, the mean cannot be seen directly. If the histogram were uniform, then the mean would be precisely in the middle, as would the median. In the histogram of Figure 4.8, the presence of more, larger values will obviously "skew" the mean away from the halfway point.

As a second example, Figure 4.9 shows a representative resting heart electrocardiogram (ECG) waveform. A heart monitor could be developed, which samples a patient's heart and calculates the histogram of this parameter at certain intervals. An abnormal condition may be detected if the histogram deviates significantly from normal. In later chapters, it will be shown how the raw samples may be transformed into other parameters to give more flexibility in certain applications.

4.9 SIGNAL CHARACTERIZATION

Taking the histogram of the *difference* between adjacent pixels of the test image yields the result shown in Figure 4.10. It is tempting to use one of the models defined earlier to approximate this PDF. The Gaussian distribution is often used to approximate noise in a signal, and the Laplacian distribution may be used as a model for certain image parameters (as will be seen in Section 7.13). Figure 4.11 shows this concept for values derived from sampled data. The following MATLAB code generates N random numbers and plots their histogram:

```
N = 1000;
x = randn(N, 1);
[n, XX] = hist(x, 40);
bar(XX, n, 'w');
```

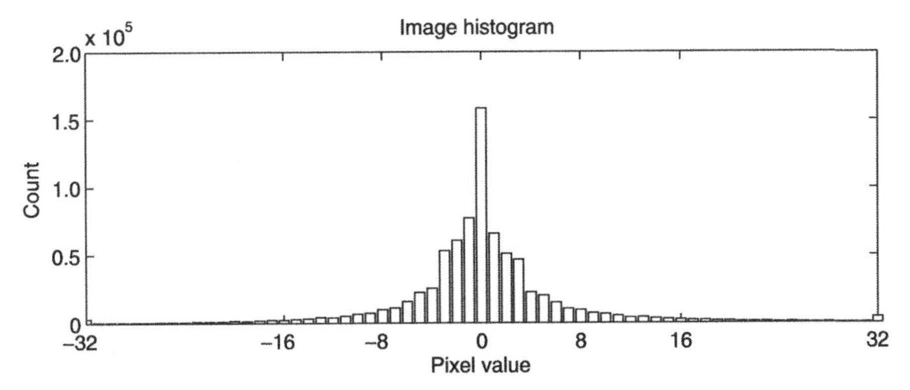

FIGURE 4.10 The image pixel difference histogram of the image shown in Figure 4.7.

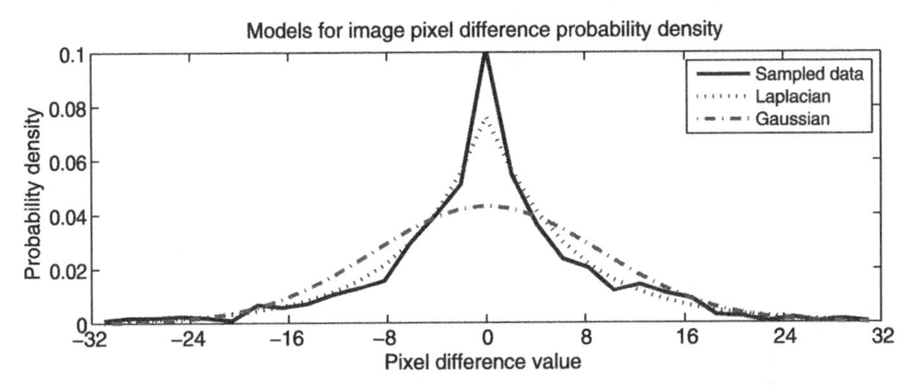

FIGURE 4.11 A PDF estimate—Gaussian and Laplacian distributions compared to the actual distribution of values derived from the sampled data.

Some signals are in fact represented as a vector of samples: several scalar samples taken at the same instant or somehow related. This could represent something as simple as two adjacent pixels in an image or perhaps a signal sampled simultaneously with several sensors. Each sample is then a vector of values, giving rise to a *joint probability*. This is illustrated in Figure 4.12, where the measured signal density and a two-dimensional Gaussian fit are shown. The parameters of the Gaussian density (means and variances for each dimension) may be known in advance, and the goodness of fit of some measured data may be compared to the Gaussian model, thus determining the likelihood that the new samples come from the same source.

4.10 HISTOGRAM OPERATORS

We will now consider how the histogram is useful in solving two problems of practical interest:

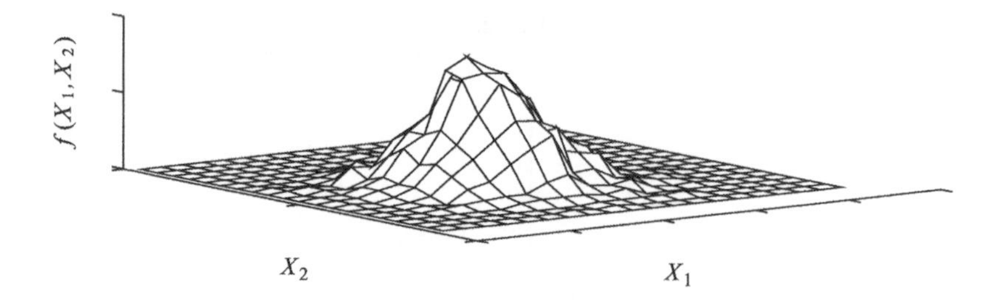

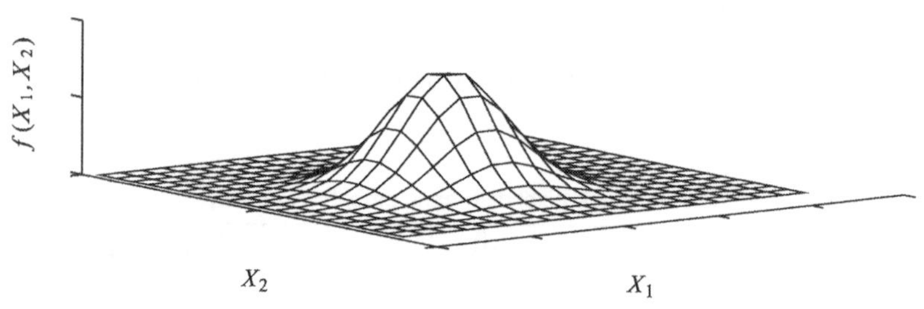

FIGURE 4.12 PDF estimate in two dimensions. The measured distribution (upper) is modeled by a distribution equation (lower).

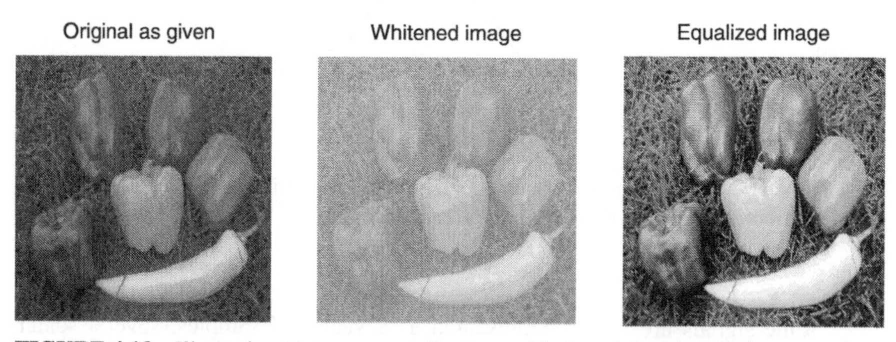

FIGURE 4.13 Illustrating histogram equalization, with the original image shown on the left. It is assumed that this image is not available. The image as presented is shown in the middle, and this is used as input to the histogram equalization algorithm. The output of the algorithm is shown on the right.

1. *Random sample generation* in order to generate random numbers with a known distribution
2. *Sampled data modification* to compensate for errors or natural conditions when the signal was sampled

4.10.1 Histogram Equalization

Consider Figure 4.13, which shows the original image, and the image which is available to use. The latter is clearly skewed toward the whiter end of the grayscale

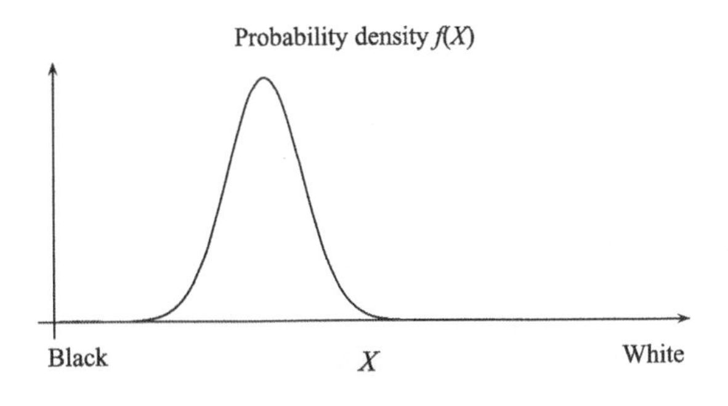

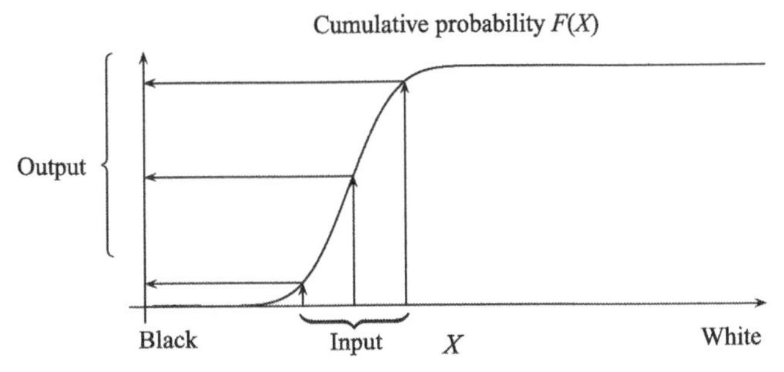

FIGURE 4.14 Viewing histogram equalization as a mapping of pixels using the cumulative distribution. The sampled values in the image (which occur over the range indicated) are mapped into the much wider range of values as shown. Of course, the mapping may well be different for differing density functions.

spectrum; this could be caused by the sampling process to begin with, excessive illumination, overexposure of the image sensor, or other causes. So, if this is all we have to work with, can we recover the original image or at least an approximation to it?

This scenario may be understood in terms of the PDF. In Figure 4.14, the PDF is shown skewed toward the black end of the luminance scale (toward zero).

The cumulative probability, shown below the PDF, may be used as a nonlinear mapping function: Pixels on the x axis are mapped to values on the y axis of the cumulative probability. Mathematically, this is a functional mapping of the form $y = f(x)$.

It is apparent that a small change in pixel value about the mean on the input will result in a large change in output luminance. The implementation of such an algorithm consists of two stages. The first pass through the data is needed to calculate the histogram itself. The second pass then uses the histogram as a lookup table, mapping each pixel of the input image in turn to a corresponding output pixel value, as determined by the histogram value at that pixel index.

The process as described ought to work for varying PDF mappings and nonlinearities. Figure 4.15 shows an image which has skews in both directions—toward black at one end and toward white at the other. The cumulative histogram of Figure 4.16 shows this. Only the middle image in Figure 4.15 is available to the algorithm,

FIGURE 4.15 Histogram equalization with skews in both directions. The image histogram is moved both toward the black end of the scale for darker pixels and toward the white end for lighter pixels, resulting in the middle image. Histogram equalization applied to this image results in the image shown on the far right.

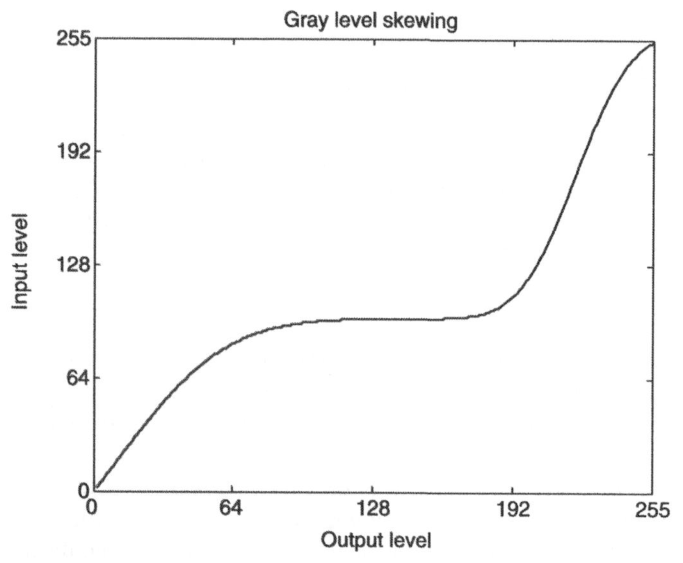

FIGURE 4.16 The distribution of pixels where lower- to mid-valued pixels are skewed toward black and higher-valued pixels skewed toward white.

and when the histogram equalization process is applied, the results are as shown in the right-hand image of the figure. Clearly, it is able to substantially improve the apparent visual quality of the image.

4.10.2 Histogram Specification

The other problem, that of random sample generation, is illustrated in Figure 4.17. Uniformly distributed random variables, which are comparatively easy to generate, are mapped from the y axis of the given cumulative probability function to the x

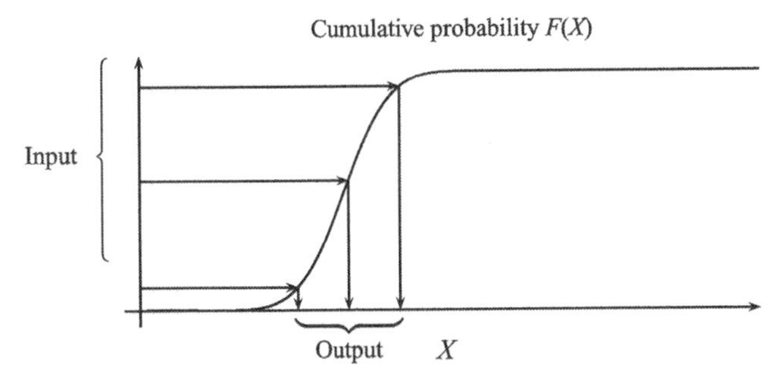

FIGURE 4.17 Histogram specification as a mathematical mapping. The input pixel values are mapped from that shown on the y axis to the much smaller range on the x axis. As with Figure 4.14, the mapping may well be different for differing density functions.

axis. Mathematically, this is an inverse function relationship of the form $x = f^{-1}(y)$. The result of this is a "concentrating" of the output values around the area where the cumulative probability increases at the greatest rate. If the given mapping $F^{-1}(X)$ represents a cumulative Gaussian curve, the random sample output will approximate a Gaussian distribution.

To understand why this works as it does, consider that in both generation and equalization, we wish to convert random variables u with a probability density $f_u(u)$ to random variables v with a probability density $f_v(v)$. For the mapping

$$u \to v, \tag{4.25}$$

we require the cumulative probability of the input variable u at some value U to equal the cumulative probability of the mapped variable v at some value V. By definition,

$$F_u(U) = \int_0^U f_u(u)\,du \tag{4.26}$$

and

$$F_v(V) = \int_0^V f_v(v)\,dv. \tag{4.27}$$

If we want the cumulative histograms to be equal, then

$$F_u(U) = F_v(V). \tag{4.28}$$

Hence, for any variable mapping $u \to v$,

$$v = F_v^{-1} F_u(u). \tag{4.29}$$

The generation of variables with a given distribution, say, Gaussian, from uniform variables is shown in Figure 4.18. We assume that uniformly distributed variables may be generated. u is distributed according to a uniform distribution; hence, the first stage of Equation 4.29, or $F_u(u)$, is a mapping $x \to y$. This is simply a constant scaling. The inverse function $F_v^{-1}(\bullet)$ is then applied. This is a mapping from $y \to x$.

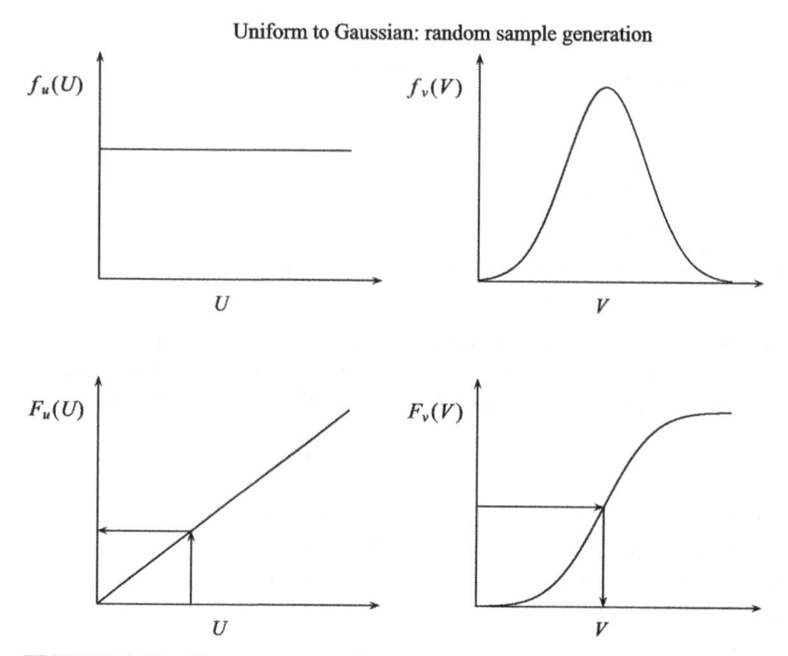

FIGURE 4.18 Histogram generation: uniform to Gaussian random sample conversion.

Application of Equation 4.29 also allows changing the distribution of the data. The distribution may in fact be measured, and not a "standard" distribution such a Gaussian. Figure 4.19 illustrates the process of changing from a Gaussian to a uniform distribution. U is distributed according to a nonuniform distribution; hence, the first stage of Equation 4.29, or $F_u(u)$, is a mapping $x \rightarrow y$. The inverse function $F_v^{-1}(\bullet)$ is then applied, which is a mapping $y \rightarrow x$. The difference between histogram equalization and the generation of samples from a known distribution is essentially just a reversal of the order of the projection functions.

4.11 MEDIAN FILTERS

As a simple example of statistical operations on samples, consider the problem of removing noise from an image. Figure 4.20 shows an image corrupted by so-called impulsive noise (also called "salt 'n' pepper" or "speckle" noise). This might be the result of noise on a communications channel, for example, or perhaps the image might be one frame from a historic film, which has suffered from degradation over time. Ideally, we would like to remove all noise and restore the original image.

A first thought might be to average each pixel out, with each pixel set to the average of its neighboring pixels. This causes the unwanted side effect of blurring the image, as illustrated in the figure. Another way may be to set a threshold and to replace pixels exceeding this threshold with the average of the neighboring pixels.

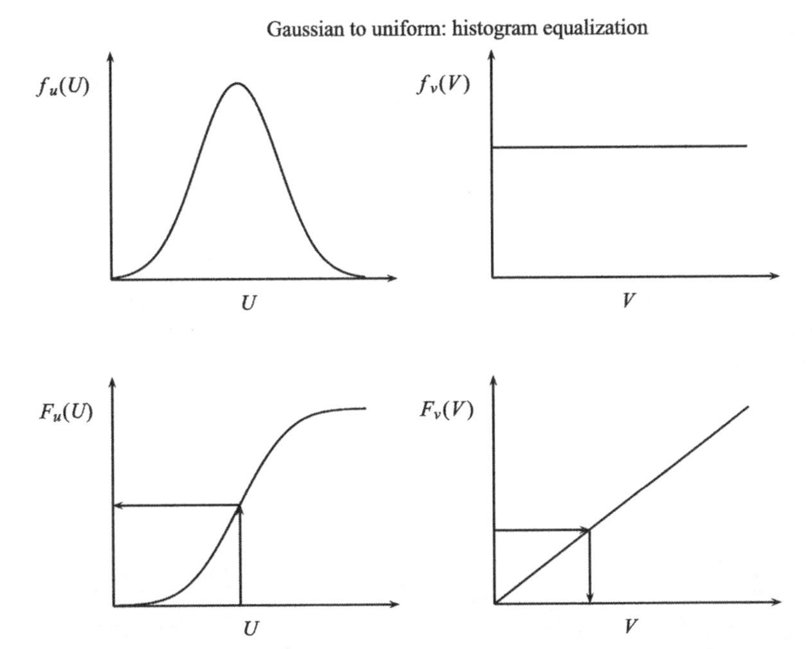

FIGURE 4.19 Histogram equalization: Gaussian (or other nonuniform) distribution to uniform sample conversion.

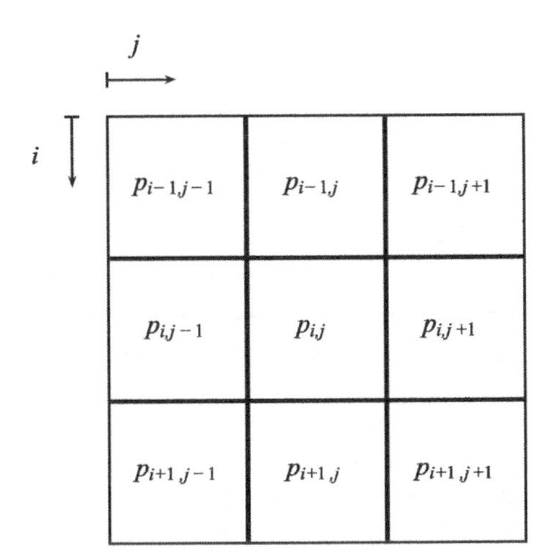

FIGURE 4.20 Illustrating a 3 × 3 image sub-block for the calculation of the median. Each square represents one pixel. The center pixel is the one currently being determined by the median filter algorithm.

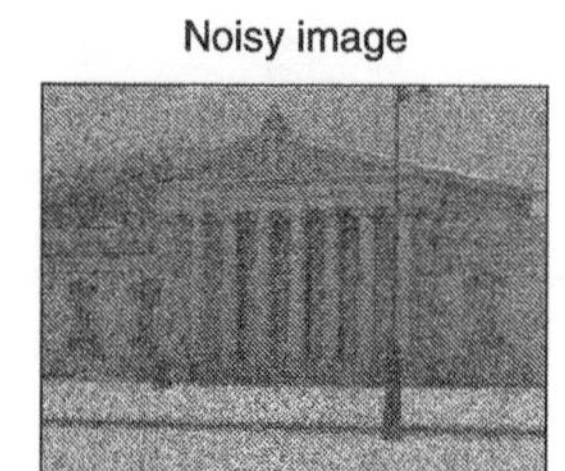

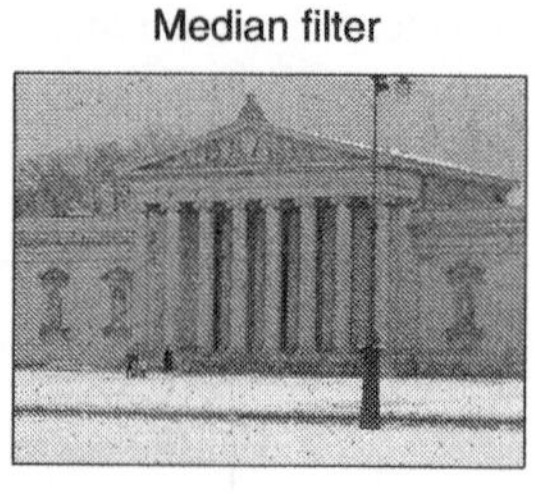

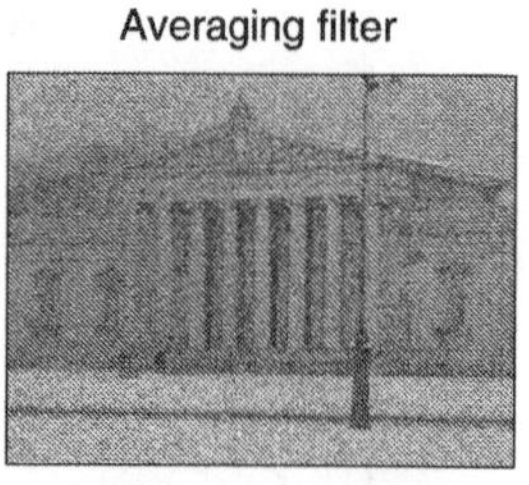

FIGURE 4.21 Restoration of an impulsive noise-corrupted image. The available image is corrupted by impulses (left). The median filter produces the center image, which is markedly improved. The averaging filter produces the right-hand image, which is noticeably smeared.

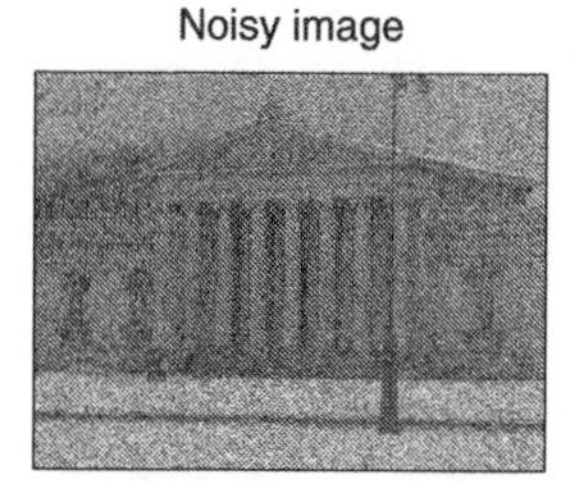

FIGURE 4.22 The effect of multiple iterations on the median filter. The noise-corrupted image is again on the far left, with the other images being the median filter restored images after one and two iterations, respectively.

The problem is then how to set the threshold: If set too low, too many genuine image pixels will be altered, thus altering the image; if set too high, the noise will not be removed.

One approach is to use a type of nonlinear filter: the *median filter*. In its simplest form, this simply sets each pixel to the median of its neighboring pixels. No thresholding is involved. To understand this, imagine the pixels segmented into matrices of size 3×3, as shown in Figure 4.21. Each pixel in the middle is replaced by the median value of those in the defined window.

As discussed previously, the median operator is simply the value of the middle pixel, when all pixels are ranked in order. For example, the values 7, 3, 9, 2, and 5 ranked in order are 2, 3, 5, 7, and 9. The median is the value occurring in the middle of the ordered list, or 5.

The results of applying a 3×3 median filter are shown in Figures 4.21 and 4.22. Most, though not all, of the visible noise has been removed. Importantly, the image itself has been largely preserved, with no immediately obvious degradation in other areas (such as loss of detail and edges). The figure also shows the result of a second pass over the image. Most of the residual impulses have been removed, with little degradation in the quality of the image itself. Median filters belong to a broader class of nonlinear filters termed "order statistic" filters.

4.12 CHAPTER SUMMARY

The following are the key elements covered in this chapter:

- a review of *statistics* as they apply to sampled data systems;
- the concepts of statistical *parameters*, *probability distributions*, and *histograms*;
- the similarities and differences between *continuous* and *discrete* distributions; and
- some applications of *median filters* and *histogram equalization*.

PROBLEMS

4.1. For the probability density and cumulative probability example code in Section 4.6, change the `rand()` to `randn()` and verify that the results are as expected.

4.2. For the sequence of signal samples 34, 54, 56, 23, and 78,

(a) calculate the mean, mean square, and variance; and

(b) subtract the mean from each number, and with the new sequence, calculate the mean, mean square, and variance. Explain your results.

4.3. For the sequence of samples $\{x\}$, write the equations defining mean, mean square, and variance. Then, let a new sequence $\{y\}$ be defined as $y(n) = x(n) - \overline{x}$; that is, generate the mean-removed sequence.

(a) Derive the equations for the mean, mean square, and variance of the sequence $\{y\}$.

(b) What do you conclude? Does the removal of the mean from each sample in a sequence have the expected effect on the mean? On the variance? What about the mean square?

4.4. For the sequence of samples $\{x\}$, write the equations defining mean, mean square, and variance. Then, let a new sequence $\{y\}$ be defined as $y(n) = \alpha x(n)$, that is, each sample multiplied by a real constant α (positive or negative).

(a) Derive the equations for the mean, mean square, and variance of the sequence $\{y\}$.

(b) What do you conclude? Are there any values of α for which this conclusion would not hold?

4.5. Repeat the above two problems using MATLAB, generating a sequence of random numbers for $\{x\}$. Try both `rand()` and `randn()` to generate the sequences.

4.6. The mean, mean square, and variance are all interrelated.

(a) For a sampled sequence $x(n)$, prove that $\sigma^2 = \overline{x^2} - \overline{x}^2$.

(b) How could the above result be used to simplify the computation of the variance of a set of sampled data?

4.7. The probability density and cumulative density were discussed in Section 4.6. The MATLAB code given in that section finds the probability by taking a sample value X and by determining the number of signal samples below X or within the range $X + \delta X$.

(a) Explain why this is not a particularly efficient approach, in terms of the number of passes that would have to be made through the data. For N input samples and M points in the probability curve, roughly how many passes would be required to determine all M output points?

(b) If we produce a running count of the M output data points, would we be able to use only one pass through the N data points?

4.8. Construct in MATLAB a histogram function `myhist()` for integer-valued data samples. Check that it works correctly by using randomly generated test data. Check your m-file against the output of MATLAB's **`hist()`** function.

4.9. Load an image in MATLAB (see Chapter 2 for details on how to do this). Convert the image to grayscale (luminance) values. Plot the histogram and cumulative histogram. Check that you have the correct results by ensuring that the final value in the cumulative histogram equals the number of pixels in the image.

4.10. Sketch the histogram and cumulative histogram as shown in Figure 4.14 for the cases were the image is too light, too dark, and finally where the image has peaks at both the light and dark ends of the spectrum. Explain how histogram equalization works in each case, with reference to the cumulative histogram.

4.11. The lower right-hand plot in Figure 4.5 includes the plot of a uniform distribution. For $\bar{x} = 0$, $\sigma = 1$, find the signal range and the constant scaling value in the PDF equation.

4.12. Verify that the area under a Laplacian distribution is equal to one.

4.13. The pixels in a 3×3 image sub-block are as follows:

$$\begin{matrix} 3 & 4 & 2 \\ 9 & ? & 1 \\ 3 & 7 & 3 \end{matrix}$$

(a) Estimate the unknown pixel in the center using both a *mean* and a *median* estimator.

(b) Repeat if the pixel of value 1 is changed to 25. Are both the mean and median affected? Why or why not?

4.14. A Rayleigh distribution has a cumulative distribution given by $F(x) = 1 - e^{-x^2/2\sigma^2}$ for $x \geq 0$.

(a) From $F(x)$, verify by differentiation that $f(x) = \frac{x}{\sigma^2} e^{-x^2/2\sigma^2}$ for $x \geq 0$.

(b) Starting with $f(x)$, verify by integration that $F(x)$ is valid.

(c) Sketch both $f(x)$ and $F(x)$.

CHAPTER 5

REPRESENTING SIGNALS AND SYSTEMS

5.1 CHAPTER OBJECTIVES

On completion of this chapter, the reader should be able to:

1. explain how sampled signals are *generated*, and be able to generate *sampled waveforms* using an equation.
2. explain the relationship between continuous and discrete *frequency*.
3. derive the z *transform* of a signal.
4. convert from a z transform to a *difference equation*.
5. determine the *stability* of a given system.
6. explain the relationship between a *pole–zero* plot and a system's *frequency response*.
7. explain what is meant by *convolution*, and how to calculate a system's response using convolution.

5.2 INTRODUCTION

We now turn our attention from analyzing a given sampled signal to synthesizing (generating) a sampled signal. Many applications depend on the ability to generate a signal: One important application area is to generate signals for communication systems. Once the theory for generating signals is introduced, it will be possible to analyze systems which alter a signal as it passes through a system. These two operations—analysis and synthesis—may be viewed as the inverse of each other, and much of the same theory applies equally well to either.

5.3 DISCRETE-TIME WAVEFORM GENERATION

Consider a sine wave as a point on a circle, sampled at discrete angular increments as shown in Figure 5.1. The path traced out on the horizontal axis is sinusoidal. As

Digital Signal Processing Using MATLAB for Students and Researchers, First Edition. John W. Leis.

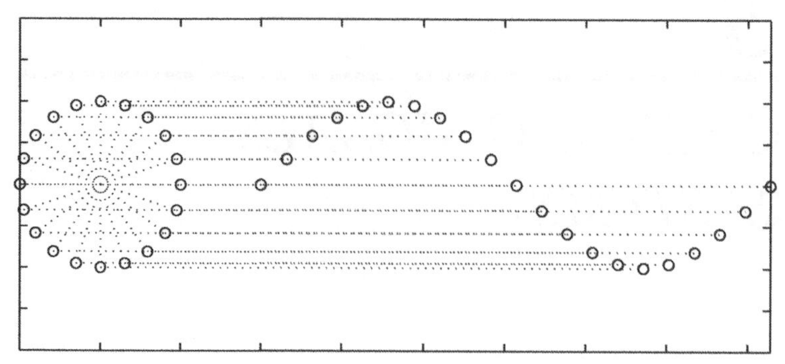

FIGURE 5.1 A sinusoidal waveform viewed as sampled points on a rotating circle. The amplitude corresponds to the radius of the circle, the phase corresponds to the starting point, the frequency is the rotational speed, and the sample rate of the waveform corresponds to the angle swept out on the circle between samples.

will be seen in Chapter 7, a fundamental principle of signal processing called "Fourier's theorem" states that periodic signals may be composed from a summation of sinusoids, of varying amplitude, phase and frequency.

Thus, the sine wave is of vital importance in signal processing. Firstly, we wish to generate a sine wave—or, more precisely, to generate the samples that constitute a sampled sine wave.

5.3.1 Direct Calculation Approach

The obvious approach to calculating the sample points for a sine wave is simply direct calculation. A sinusoidal waveform is mathematically described as:

$$x(t) = A\sin(\Omega t + \varphi), \tag{5.1}$$

where A is the peak amplitude of the waveform (i.e., a peak-to-peak value of $2A$), Ω is the frequency in radians per second, t is time, and φ is a phase shift (which determines the starting point in the cycle).

If we sample at a time $t = nT$, where n is the sample number, we generate the sequence of samples $x\,(n)$ at time instants nT:

$$x(n) = A\sin(n\Omega T + \varphi). \tag{5.2}$$

The direct calculation of this requires the evaluation of a trigonometric function, sin (). Defining:

$$\omega = \Omega T, \tag{5.3}$$

we have,

$$x(n) = A\sin(n\omega + \varphi). \tag{5.4}$$

From this, we can interpret ω as the angle swept out over each sample. So we now have the terms

T	sampling period
ω	"digital" frequency in radians per sample
Ω	"real" frequency in radians per second

"Real" time t is related to "sample time" (or sample number) n by:

$$t = nT$$

$$\text{seconds} = \text{sample number} \times \frac{\text{seconds}}{\text{sample}}.$$

"Real" frequency Ω radians per second is related to "sample-scaled frequency" ω radians per sample by

$$\omega = \Omega T$$

$$\frac{\text{radians}}{\text{sample}} = \frac{\text{radians}}{\text{second}} \times \frac{\text{seconds}}{\text{sample}}.$$

As always, radian frequency and Hertz (Hz) frequency are related by

$$\Omega = 2\pi f$$

$$\frac{\text{radians}}{\text{second}} = \frac{\text{radians}}{\text{cycle}} \times \frac{\text{cycles}}{\text{second}}.$$

From the equation $x(n) = A \sin n\omega$, we can see that the relative frequency ω is the phase angle swept out per sample. An example serves to demonstrate a waveform generated in this manner, with playback as an audio signal. We generate two signals: The first using the "analog" approach, where the waveform is $y_a = \sin 2\pi ft$, and time t stepped in small increments. The argument f is the frequency of the tone to generate (in this case, 400 Hz). Next, the "discrete" approach uses $y_d = \sin n\omega$, where n is the sample instant (an integer quantity), and ω is the "discrete" frequency in radians per sample, as explained above.[1]

[1]Note carefully the difference between Ω (capital Omega) and ω (lower-case omega).

```
fs = 8,000;
T = 1/fs;
tmax = 2;
t = 0:T:tmax;
f = 400;
Omega = 2*pi*f;
ya = sin(Omega*t);
sound(ya, fs);
pause(tmax);
n = 0:round(fs*tmax);
omega = Omega*T;
yd = sin(n*omega);
sound(yd, fs);
```

Note well the fundamental difference here: The "analog" approach uses small time increments δt, whereas the "discrete" approach uses sample index n. The calculation of the trigonometric function (sine) is computationally "expensive"—meaning that the calculation may take a long time to perform, compared with the sample period. If the sample period T is small enough, a processor may have difficulty keeping up. So, it is worthwhile to take a closer look at how the samples may be generated in practice.

5.3.2 Lookup Table Approach

An obvious simplification is to precompute the sinusoidal samples and store them in a *lookup table* (LUT). This works because the waveform is periodic, with only a finite number of samples for each period of the waveform. If the waveform were not periodic, this approach would not be feasible.[2] The operation of the lookup table is shown diagrammatically in Figure 5.2.

Once the values are precomputed and stored in the table, the index (or pointer) is simply stepped through the table. Each output sample sent to the D/A converter is the current entry in the table. When the pointer reaches the end of the table, it is simply reset to the start of the table again.

Amplitude scaling can be effected by a simple multiply operation on each sample as it is retrieved from the table. The frequency may be changed by stepping through not one, but several samples at a time. Finally, phase changes are effected by an abrupt change in the LUT index. To illustrate a phase change, suppose there were 256 samples in the table. The entire table represents 2π radians, or 360 degrees. A phase advance of 90° implies adding $(256 \times 90/360) = 64$ samples to the current position.

[2]Strictly speaking, we *could* generate nonperiodic signals in this way, but the signal would repeat itself at intervals corresponding to the table length.

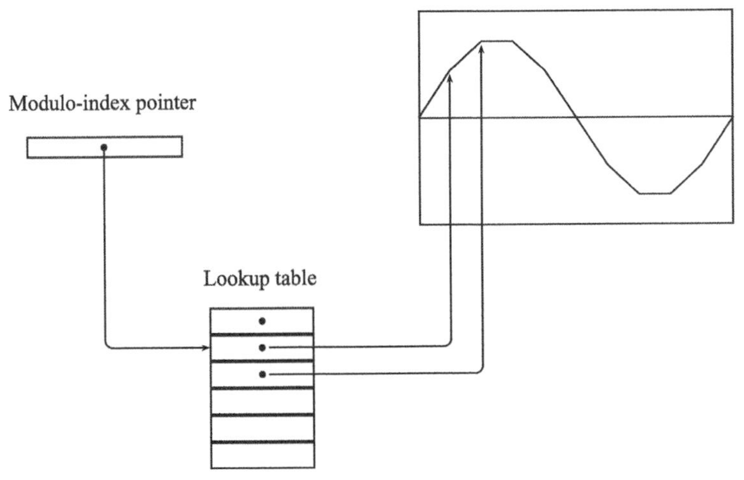

FIGURE 5.2 A lookup table for the generation of a sinusoidal waveform. The index pointer is incremented on every sample, to point to the next value in the table. After the pointer reaches the last sample in the table, it is reset back to the start of the table.

With a lookup table, the resolution or granularity of samples (and phase changes) is governed by the number of samples stored in the table. For example, a phase change of $\pi/3$ requires a noninteger increment of 42.67 samples in a table of 256 samples. The effective number of samples may be increased by mathematical interpolation between known samples (as shown in Section 3.9.2), although this obviously adds to the complexity.

This design is useful in applications such as signal generation for communications systems. An extension of this concept, called direct digital synthesis (DDS), is often used in communication systems. The approach used in DDS extends the method described, but rather than having a single step increment for each sample, a phase register adds a fixed offset to the table address according to the desired frequency so as to step through the table at the required rate (and thus generate the desired frequency). An address accumulator has more bits than is required for the LUT index, and only the higher-order bits of this accumulator are used to address the LUT. This enables finer control over the step size, and hence frequency generated.

5.3.3 Recurrence Relations

Another approach is to use the properties of trigonometric functions. The idea is to increment from the currently known position to the value of the sampled sinusoid at the next sample position. Applying the trigonometric expansions for sin $(x + y)$ and cos $(x + y)$, we have:

$$\cos(n\omega + \omega) = \cos n\omega \cos \omega - \sin n\omega \sin \omega \tag{5.5}$$

$$\cos(n\omega + \omega) = \cos n\omega \cos \omega - \sin n\omega \sin \omega. \tag{5.6}$$

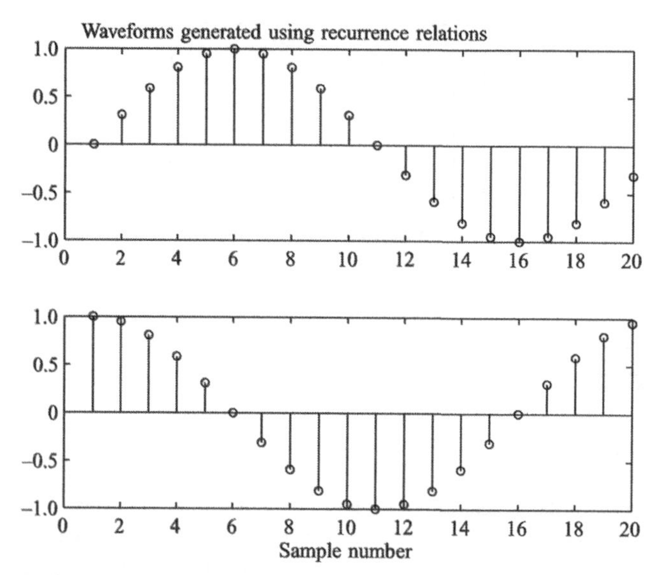

FIGURE 5.3 Quadrature sine and cosine waveforms, generated using recurrence equations.

We can set ω to be a small step size, and let $n\omega$ represent the current angle. Thus, the problem becomes one of progressing from the current sample value of $\sin n\omega$ to the next sample, $\sin(n\omega + \omega)$. The current angle $n\omega$ is varying, but the step size ω is fixed. Therefore, the values of $\sin\omega$ and $\cos\omega$ may be precomputed, as they are likewise fixed.

Using both of the equations above, $\sin n\omega$ and $\cos n\omega$ are known, and the values of $\sin(n\omega + \omega)$ and $\cos(n\omega + \omega)$ are easily computed using only multiplication and addition operations—no "expensive" (in computational terms) sine/cosine operations are required at each iteration. Note also that both sine and cosine waveforms (sometimes called "quadrature" waveforms) are inherently provided by the algorithm, as shown in Figure 5.3.

The following MATLAB code illustrates the implementation of a waveform generator using these recurrence relations. Note that, as explained above, there are no time-consuming trigonometric calculations (sin [] and cos []) within the main sample-generating loop.

```
N = 20;
w = 2*pi/N;
% running cos(w) and sin(w) values
cosw = cos(w);
sinw = sin(w);
% initial values
theta = 0;
```

```
cosnw = 1;
sinnw = 0;
% arrays to store resulting sine/cosine values
%(not needed in practice, only to show the
% computed waveform values in this example)
sinvals = [];
cosvals = [];
% note that no trig functions are used in this
% loop—only multiply and add/subtract
for n = 0:N-1
    % save values in array
    sinvals = [sinvals sinnw];
    cosvals = [cosvals cosnw];
    % recurrence relations for new values of
    % sin(nw + w) and cos(nw + w)
    newcos = cosnw*cosw - sinnw*sinw;
    newsin = sinnw*cosw + cosnw*sinw;
    % new values become current ones
    cosnw = newcos;
    sinnw = newsin;
end
figure(1);
stem(sinvals);
figure(2);
stem(cosvals);
```

5.3.4 Buffering, Timing, and Phase Errors

Once an appropriate method of generating the wave samples has been decided upon, it is necessary to think about how the samples are to be used in practice. If we are developing an audio or communications system, for example, the actual output of the sample values to the D/A hardware will likely be done by a separate digital signal processor chip. The reason for this is that the actual output of individual samples is usually done by interrupting the processor, which must then execute a routine to fetch the next sample from memory and send it to the hardware to generate the analog representation. If the sample rate is reasonably high, this constitutes a significant burden on the host CPU. For this reason, the underlying DSP processor subsystem does not usually deal with individual samples, but rather blocks of samples. The size of the block is necessarily a trade-off between lower processor requirements (using a larger block) and the amount of delay which can be tolerated (requiring a smaller block). It is the responsibility of the operating system and application using the signal to arrange for a block to be generated. There are two practical problems which arise in this scenario: The phase of the generated waveform, and the timing of gaps in the blocks. We look at each in turn.

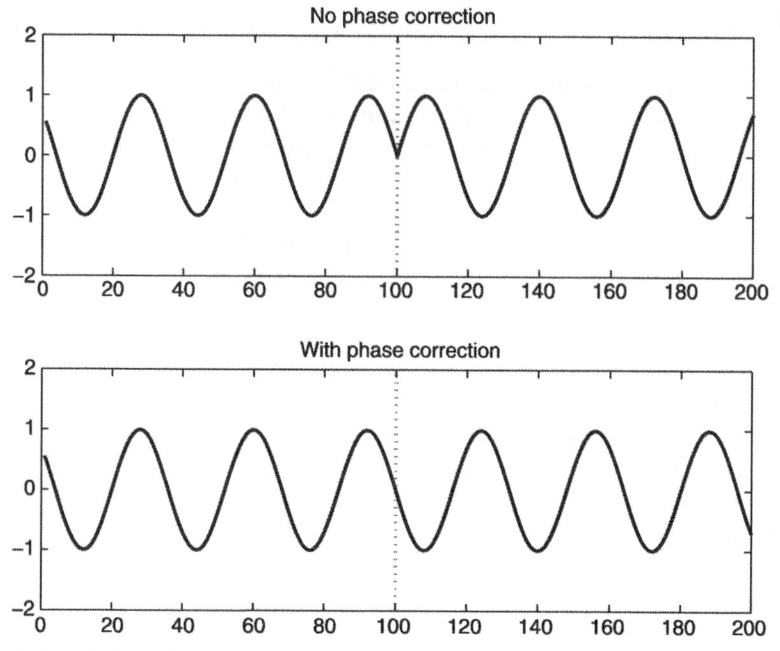

FIGURE 5.4 Illustrating the problem of matching the end of one block with the start of the next. Here, the block boundary occurs at index 100. In the upper plot, the subsequent block starts generating the sinusoid anew. The lower block starts the waveform using a phase which is adjusted so as to make the transition seamless. This phase is $\varphi = \theta + \omega$, where θ is the ending phase of the previous block, and ω is the phase increment per sample.

Suppose we have a block of output samples calculated and ready to transfer to the output device. We do this, and then start work on the next block. A problem arises if we start the next buffer using zero initial conditions. The blocks should seamlessly overlap, since the blocking is done only as a convenience to the hardware and software, and should not be visible in the output waveform. Figure 5.4 shows the situation which may arise. The middle of the upper plot is aligned with the block boundary. The waveform from block m may terminate at any amplitude, according to the frequency of the wave and the number of samples. As illustrated, it appears that the waveform abruptly ends at a particular phase. Block $(m + 1)$ then starts at zero phase.

What does this mean in practice? Consider an audio system which is required to generate a waveform. Some typical figures will help fix the idea. Suppose the sampling rate is 8 kHz, and we wish to generate a 250 Hz sinewave. Suppose the block size is 400 samples. This equates to 20 blocks per second. If we do not take into account the block boundaries, the situation depicted in Figure 5.5 arises. The discontinuities at the end of each block manifest themselves as an additional component at a rate equal to the block rate—in this case, 20 Hz. In audio systems, the noise thus introduced can be particularly objectionable at this frequency.

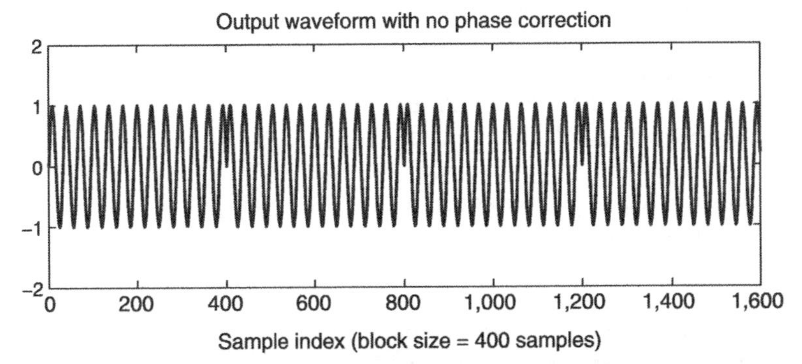

FIGURE 5.5 Showing multiple blocks and the discontinuity that can result at block boundaries. If this waveform were an audio waveform, the relative frequency of the blocks as compared with the waveform itself would mean that objectionable audible artifacts would be present.

The following shows how we initialize the system in order to generate such a waveform. The block buffer is `Bbuf`, and this is the buffer which is prepared by the main CPU and passed to the DSP subsystem. The buffer `Sbuf` contains all the sound samples. In practice, this would not be required, but since we are generating the entire waveform for subsequent playback, it is necessary for this demonstration. In this case, it stores 4 seconds of audio.

```
f = 250;        % frequency in Hz
fs = 8000;      % sampling frequency
B = 20;         % blocks/sec
L = 400;        % samples per block
D = 4;          % duration in seconds
% block buffer and output buffer
Bbuf = zeros(L, 1);
Sbuf = zeros(fs*D, 1); % = B*L*D
Omega = 2*pi*f;
T = 1/fs;
omega = Omega*T;
NBlocks = B*D;
```

The following shows how we generate the samples for each block of audio. Note that the phase increment per sample is ω radians per sample as per earlier theory, and in the code is denoted as `omega`. At the end of each individual block calculation, we copy the data into the final output buffer. In a real-time system, this step would not be necessary, and would be replaced with a system call to transfer the data block of L samples to the output DSP subsystem.

```
pSbuf = 1;
for b = 1:NBlocks
    % generate one block worth of samples, phase
    % independent
    for n = 1:L
        theta = (n - 1)*omega;
        x = sin(theta);
        Bbuf(n) = x;
    end
    % save the block in the output buffer
    Sbuf (pSbuf:pSbuf + L - 1) = Bbuf;
    % point to start of next output block
    pSbuf = pSbuf + L;
end
```

Now we address the problem of aligning the waveform, so that the transition between buffers is seamless. There are several ways to accomplish this, and the method below simply uses a phase start φ (phi) in the per-sample calculation. The phase angle at the end of the block is then stored for the start of the next block.

Since this phase would continue to increment, at some point it might overflow. But because any phase over 2π can be mapped back into the range 0–2π, we subtract the overflow as shown in the loop.

```
pSbuf = 1;
phi = 0;
for b = 1:NBlocks
    % generate one block worth of samples, correct phase
    % offset
    for n = 1:L
        theta = (n - 1)*omega + phi;
        x = sin(theta);
        Bbufpc(n) = x;
    end
    % phase for next block
    phi = theta + omega; % 1*omega at next sample
    % correct if over 2pi
    while( phi > 2*pi )
        phi = phi - 2*pi;
    end
    % save block in output buffer
    Sbufpc(pSbuf:pSbuf + L - 1) = Bbufpc;
    % point to start of next output block
    pSbuf = pSbuf + L;
end
```

The waveforms can be played back using **sound**(Sbuf, fs) and **sound**(Sbufpc, fs) for the phase-corrected approach. The resulting audio will exhibit an audible artifact at the block rate in the former case (in this case 20 Hz), which is not present in the latter case.

Another issue, which is caused by different factors but produces a similar type of waveform impairment, is the processing speed itself. In the preceding examples, we have calculated each buffer separately, and in practice would then pass the memory address of this buffer to the DSP subsystem. If we use the same block of memory, it would be necessary to wait until the output has been completed before preparing the new samples. This time would equate to LT, or 50 milliseconds, using the current example. So not only would this delay be mandatory, there would be a delay while the main processor calculated the L new samples, which may not be insignificant. This would again result in audible artifacts, since there would be a delay while the output buffer is prepared.

The solution to this problem is to use a technique termed *double-buffering*. In this approach, two buffers are used in an alternating fashion. While buffer A is being prepared, the samples in buffer B are output to the D/A converter. Once all samples in buffer B have been used, buffer A ought to be ready. All that is necessary is to pass the memory address of A to the output system, and start preparing buffer B (since it is no longer in use). Subsequently, the role of each buffer is again reversed: Buffer B is output while A is being updated. This ensures seamless transitions between buffers, and eliminates gaps between blocks.

5.4 THE *z* TRANSFORM

Another approach to waveform generation is to use a difference equation (as introduced in Section 3.10) to generate the output samples based on the present and past inputs, and possibly the past outputs.

This technique goes far beyond simple waveform generation; not only can we generate systems with an arbitrary response, but a similar approach may be used to analyze the behavior of systems in terms of time response and frequency response.

The principal tools required are the closely related *difference equation* and the *z transform*. Difference equations were encountered in Chapter 3. To recap, a general difference equation may be represented as:

$$\begin{aligned} y(n) = &\left(b_0 x(n) + b_1 x(n-1) + \cdots + b_N x(n-N)\right) \\ &- \left(a_1 y(n-1) + a_2 y(n-2) + \cdots + a_M y(n-M)\right). \end{aligned} \tag{5.7}$$

The coefficients b_k and a_k are constant coefficients, and determine the response of the system to an impulse or other input. N and M govern the number of past inputs and outputs respectively, which are used in the computation of the current output $y(n)$.

The closely related z transform is a very important mathematical operator when dealing with discrete systems. We start with a time function $f(t)$, which is sampled at $t = nT$. By definition, the z transform of $f(t)$ is denoted by $F(z)$, and calculated using the following formula:

$$F(z) = f(0)z^0 + f(1)z^{-1} + f(2)z^{-2} + \ldots$$
$$= \sum_{n=0}^{\infty} f(n)z^{-n}. \tag{5.8}$$

The time samples $f(n)$ are understood to mean $f(nT)$, with the sample period T being implicit in the fact that it is a sampled-data system. It may seem strange to have an infinite series and negative powers of z used. However, as will be shown in the following sections, the ideas embodied in this equation are fundamental to signal processing theory.

5.4.1 z^{-1} as a Delay Operator

Equation 5.8 has an interesting interpretation in the time domain. Using Figure 5.6 as a guide, let $f(n)$ delayed by one sample be $\tilde{f}(n)$. Thus, $\tilde{f}(n) = f(n-1)$.

By definition,

$$F(z) = f(0)z^0 + f(1)z^{-1} + f(2)z^{-2} + \cdots. \tag{5.9}$$

So,

$$\tilde{F}(z) = \tilde{f}(0)z^0 + \tilde{f}(1)z^{-1} + \tilde{f}(2)z^{-2} + \ldots \tag{5.10}$$
$$= \tilde{f}(0)z^0 + z^{-1}(\tilde{f}(1)z^0 + \tilde{f}(2)z^{-1} + \ldots)$$
$$= f(-1)z^0 + z^{-1}(f(0)z^0 + f(1)z^{-1} + \ldots) \tag{5.11}$$
$$= f(-1)z^0 + z^{-1}F(z).$$

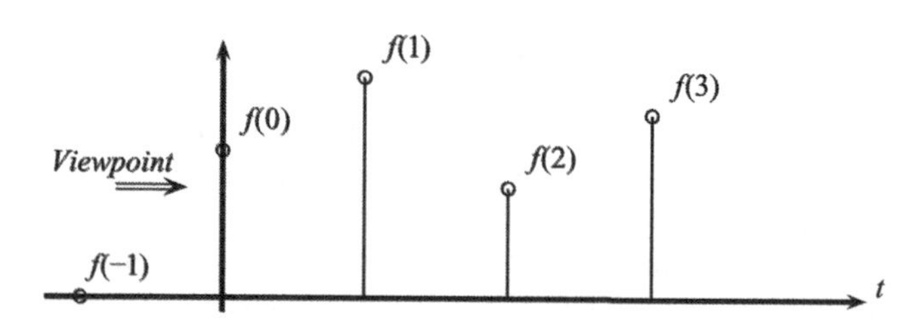

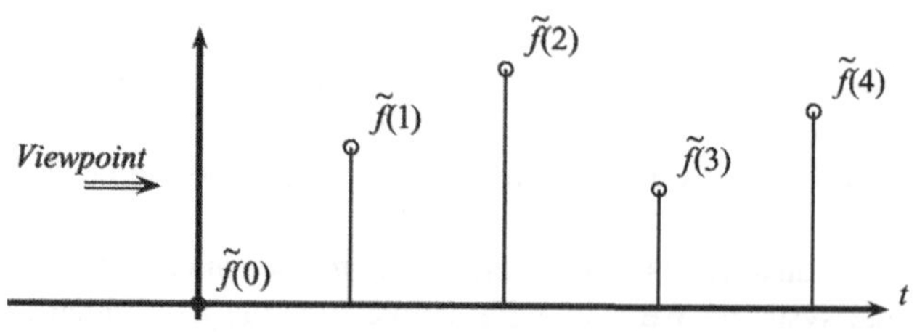

FIGURE 5.6 Viewing multiplication by z^{-1} as a delay of one sample. The upper diagram shows the original sequence $f(n)$; the lower shows $f(n)$ delayed by one sample.

If we assume zero initial conditions, $f(t) = 0$ for $t < 0$ and $f(-1) = 0$, and in that case, the above simplifies to:

$$\tilde{F}(z) = z^{-1}F(z). \tag{5.12}$$

In words, this conclusion may be expressed as:

Multiplying by z^{-1} corresponds to a **delay** of one sample.

In a similar way, it may be shown that multiplying by z^{-k} corresponds to a delay of k samples. Note that the term "delay" implies we are only referencing *past* samples—multiplying by z^{+k} corresponds to k samples in the *future*, a situation which is not normally encountered. Also, z^0 means a delay of 0 samples: Since $z^0 = 1$, the notation is consistent, as multiplying by 1 is expected to leave things unchanged, since $F(z) \cdot z^0 = F(z) \cdot 1 = F(z)$.

5.4.2 *z* Transforms and Transfer Functions

The z transform is useful because it allows us to write a *transfer function* for a given system. The transfer function is described in terms of the z equations, and it is iterated in a computer algorithm using the corresponding difference equation. The model may be depicted as in Figure 5.7. The transfer function is multiplicative:

$$Y(z) = X(z)H(z), \tag{5.13}$$

or, in words:

The output of a linear, time-invarant system is equal to the input $X(z)$ **multiplied by the transfer function** $H(z)$.

Or, equivalently,

$$\text{Linear, time-invarant system } H(z) = \frac{Y(z)}{X(z)}. \tag{5.14}$$

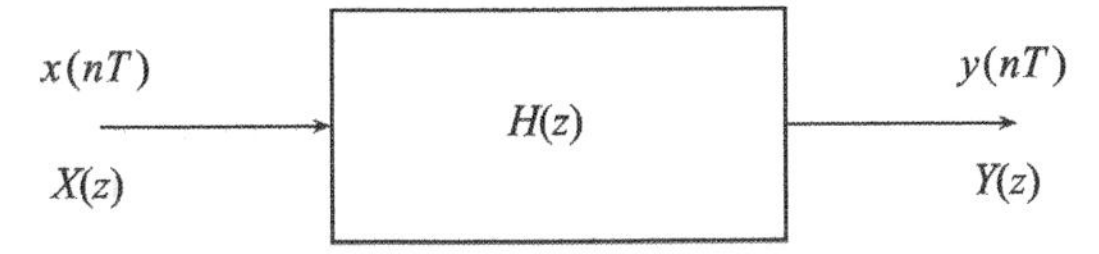

FIGURE 5.7 An input–output system model using z transforms. If the sampled input $x(n)$ is transformed into $X(z)$, then the output $Y(z) = X(z)\,H(z)$. From $Y(z)$, we can recover the samples $y(n)$.

or, in words:

$$\text{Transfer function} = \frac{output}{input}.$$

Note that the time-domain functions are *not* able to be expressed as a simple multiplication like the z transfer functions:

$$y(n) \neq h(n)x(n). \tag{5.15}$$

The output of a linear system may also be computed using the technique of *convolution*, which will be explained further in Section 5.9.

5.4.3 Difference Equation and z Transform Examples

To introduce the transfer-function analysis by way of example, consider a decaying exponential function. The time function is $f(t) = e^{-at}$, and therefore the sampled version is $f(nT) = e^{-anT}$. The function and the samples, shown in Figure 5.8, are generated using the following code:

```
a = 2;
N = 20;
T = 0.1;
% "continuous" time
tc = 0:0.01:N*T;
yc = exp(-a*tc);
% samples at nT
n = 0:N-1;
y = exp(-a*n*T);
plot(tc, yc, '-', n*T, y, 'o');
legend('Continuous system', 'Samples');
```

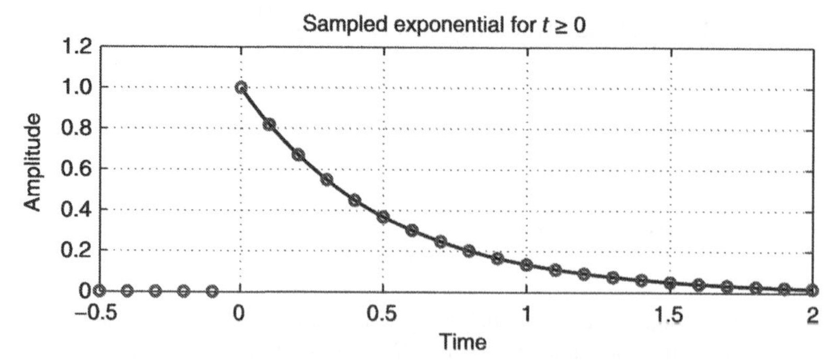

FIGURE 5.8 An exponential decay curve and its sampled version.

Applying the definition of a z transform to this function we get:

$$
\begin{aligned}
F(z) &= f(0)z^0 + f(1)z^{-1} + f(2)z^{-2} + \cdots \\
\therefore Z\{e^{-at}\} &= e^0 z^0 + e^{-aT} z^{-1} + e^{-2aT} z^{-2} + \cdots \\
&= 1 + e^{-aT} z^{-1}(1 + e^{-aT} z^{-1} + e^{-2aT} z^{-2} + \cdots) \\
&= 1 + e^{-aT} z^{-1} F(z) \\
F(z)(1 - e^{-aT} z^{-1}) &= 1 \\
F(z) &= \frac{1}{1 - e^{-aT} z^{-1}}.
\end{aligned}
$$

Note that the z transform is expressed as *negative* powers of z, because we can only access present or past samples. Multiplying by $\frac{z}{z}$, we can say that the z transform of e^{-at} is:

$$Z\{e^{-at}\} = \frac{z}{z - e^{-aT}}. \tag{5.16}$$

Now, we need to convert this z transform to a difference equation. Using

$$H(z) = \frac{Y(z)}{X(z)}, \tag{5.17}$$

we have:

$$\frac{Y(z)}{X(z)} = \frac{1}{1 - e^{-aT} z^{-1}} \tag{5.18}$$

Cross-multiplying

$$Y(z)(1 - e^{-aT} z^{-1}) = X(z). \tag{5.19}$$

We want the output $Y(z)$, so:

$$Y(z) = X(z) + e^{-aT} z^{-1} Y(z). \tag{5.20}$$

Converting to a difference equation is done by using the delay operator property of z:

$$y(n) = x(n) + e^{-aT} y(n-1). \tag{5.21}$$

Note how the delay-operator property of z has been used to transform from a transfer-function representation in z into a difference equation in terms of sample n. A z^{-1} in the z equation becomes a one-sample delay in the time (difference) equation:

$$
\begin{array}{ccccc}
Y(z) = & X(z) & + & e^{-aT} z^{-1} & Y(z) \\
\Updownarrow & \Updownarrow & & & \Updownarrow \\
y(n) = & x(n) & + & e^{-aT} & y(n-1)
\end{array}.
$$

TABLE 5.1 Calculating the Impulse Response for Sample Period $T = 0.1$ and Parameter $a = 2$

Sample number	Actual time	Impulse	
n	t	$x(n)$	$y(n) = x(n) + e - aT\, y(n - 1)$
0	0	1	1.00
1	0.10	0	0.82
2	0.20	0	0.67
3	0.30	0	0.55
4	0.40	0	0.45
5	0.50	0	0.37
6	0.60	0	0.30
7	0.70	0	0.25
8	0.80	0	0.20
9	0.90	0	0.17
10	1.00	0	0.14
⋮	⋮	⋮	⋮
18	1.80	0	0.027
19	1.90	0	0.022

Note that $y(n) = 0$ for $n < 0$, since we assume zero initial conditions.

The *impulse response* of this system is obtained by setting $x(n) = 1$ for $n = 0$, and $x(n) = 0$ elsewhere. Table 5.1 shows the calculation steps involved. It is important to realize that the $y(n)$ outputs are in fact identical to samples of e^{-at} sampled at $t = nT$.

This difference equation may be iterated to produce the sample values. This is coded in MATLAB as follows:

```
N = 20;
T = 0.1;
a = 2;
x = zeros(N, 1);
y = zeros(N, 1);
x(1) = 1;
for n = 1:N
    y (n) = x (n);
    if( n > 1 )
        y(n) = y(n) + exp(-a*T)*y(n - 1);
    end
end
stem(y);
```

A useful waveform in practice is that of a sine wave; it is often used as a test signal to measure the response of systems. For continuous time t, a sine wave is just:

$$f(t) = A \sin \Omega t. \tag{5.22}$$

The z transform of this will now be derived; the resulting difference equation will yield another method of generating sinusoidal waveforms.

The derivation from first principles can in fact be based on the previous result for $Z\,(e^{-at})$. Using the exponential of a complex number[3]:

$$e^{j\theta} = \cos\theta + j\sin\theta$$
$$\therefore e^{-j\theta} = \cos\theta - j\sin\theta.$$

The sine (and cosine) functions may be expressed by subtracting or adding as required:

$$e^{j\theta} - e^{-j\theta} = 2j\sin\theta$$
$$e^{j\theta} + e^{-j\theta} = 2\cos\theta.$$

Thus, sin() may be expressed as:

$$\sin\theta = \frac{1}{2j}(e^{j\theta} - e^{-j\theta}). \tag{5.23}$$

So,

$$\sin\Omega t = \frac{1}{2j}(e^{j\Omega t} - e^{-j\Omega t}). \tag{5.24}$$

Using the result of Equation 5.16 for the z transform of e^{-at}, this becomes:

$$\begin{aligned}
Z\{\sin\Omega t\} &= \frac{1}{2j}\left(\frac{z}{z - e^{+j\Omega T}} - \frac{z}{z - e^{-j\Omega T}}\right) \\
&= \frac{z}{2j}\left(\frac{(z - e^{-j\Omega T}) - (z - e^{+j\Omega T})}{(z - e^{+j\Omega T})(z - e^{-j\Omega T})}\right) \\
&= \frac{z}{2j}\left(\frac{e^{+j\Omega T} - e^{-j\Omega T}}{z^2 - z(e^{+j\Omega T} + e^{-j\Omega T}) + 1}\right). \\
&= \frac{z}{2j}\left(\frac{2j\sin\Omega T}{z^2 - 2z\cos\Omega T + 1}\right)
\end{aligned}$$

Now $\omega = \Omega T$ as seen before, hence:

$$Z\{\sin\Omega t\} = \frac{z\sin\omega}{z^2 - 2z\cos\omega + 1}. \tag{5.25}$$

[3]Also known as Euler's formula, after the mathematician Leonhard Euler.

The difference equation follows from the input/output model:

$$\frac{Y(z)}{X(z)} = \frac{z \sin \omega}{z^2 - 2z \cos \omega + 1}. \tag{5.26}$$

Since we require negative powers of z for causality,[4] we multiply the right-hand side by $\frac{z^{-2}}{z^{-2}}$ to obtain:

$$\frac{Y(z)}{X(z)} = \frac{z^{-1} \sin \omega}{1 - 2z^{-1} \cos \omega + z^{-2}}. \tag{5.27}$$

Cross-multiplying the terms on each side, we have:

$$\begin{aligned} Y(z)(1 - 2z^{-1} \cos \omega + z^{-2}) &= X(z)z^{-1} \sin \omega \\ \therefore Y(z) &= X(z)z^{-1} \sin \omega + Y(z)(2z^{-1} \cos \omega - z^{-2}). \end{aligned} \tag{5.28}$$

Hence,

$$y(n) = \sin \omega \, x(n-1) + 2 \cos \omega \, y(n-1) - y(n-2). \tag{5.29}$$

In terms of a difference equation, this is:

$$y(n) = (b_0 x(n) + b_1 x(n-1)) - (a_1 y(n-1) + a_2 y(n-2)), \tag{5.30}$$

with coefficients

$b_0 = 0$

$b_1 = \sin \omega$

$a_0 = 1$ (as always)

$a_1 = -2 \cos \omega$

$a_2 = 1.$

To produce a sine wave, we simply need to iterate the difference equation with a discrete impulse as input, as shown below. We can use the **`filter()`** function to do this, as introduced in Section 3.10.

5.5 POLYNOMIAL APPROACH

The previous section showed how to iterate a difference equation in order to determine the output sequence. It is particularly important to understand the relationship between difference equations and their transforms. The z transform of a linear system gives us the key to combining systems together to form more complex systems, since the z transforms in combined blocks are able to be multiplied or added together as necessary. We now give another insight into this approach.

Suppose we have a difference equation:

$$y(n) = 0.9x(n) + 0.8x(n-1) - 0.4x(n-2) + 0.1x(n-3). \tag{5.31}$$

[4]That is, the input causes the output, and we have no references to future samples.

If we iterate this for a step input, we have:

n	$x(n)$	$y(n)$
Sample Instant	Input	Output
0	1	0.9
1	1	0.9 + 0.8 = 1.7
2	1	0.9 + 0.8 − 0.4 = 1.3
3	1	0.9 + 0.8 − 0.4 + 0.1 = 1.4
4	1	0.9 + 0.8 − 0.4 + 0.1 = 1.4
5	1	0.9 + 0.8 − 0.4 + 0.1 = 1.4
6	1	. . .

Now let us take a different approach. The corresponding z transform is:

$$H(z) = 0.9 + 0.8z^{-1} - 0.4z^{-2} + 0.1z^{-3}. \tag{5.32}$$

If we subject this system $H(z)$ to an input $X(z)$, then the output $Y(z)$ must be the product $Y(z) = H(z)\,X(z)$. Let the input be a step function $x(n) = 1$. The z transform of a step input[5] is:

$$X(z) = \frac{1}{1 - z^{-1}}. \tag{5.33}$$

Hence, the output will be:

$$Y(z) = H(z)X(z) \tag{5.34}$$

$$= \frac{1}{1 - z^{-1}}(0.9 + 0.8z^{-1} - 0.4z^{-2} + 0.1z^{-3}). \tag{5.35}$$

So now, if we invert this function $Y(z)$ into $y(n)$, we should have the same result as if we iterated the difference equation for $H(z)$ using the input $x(n) = 1$. The above equation is effectively a polynomial division, so we can undertake a long division process to determine $Y(z)$. We first extend $H(z)$ with a (theoretically infinite) number of zeros:

$$H(z) = 0.9 + 0.8z^{-1} - 0.4z^{-2} + 0.1z^{-3} + 0z^{-4} + 0z^{-5} + \cdots. \tag{5.36}$$

We perform the polynomial long division as shown below.

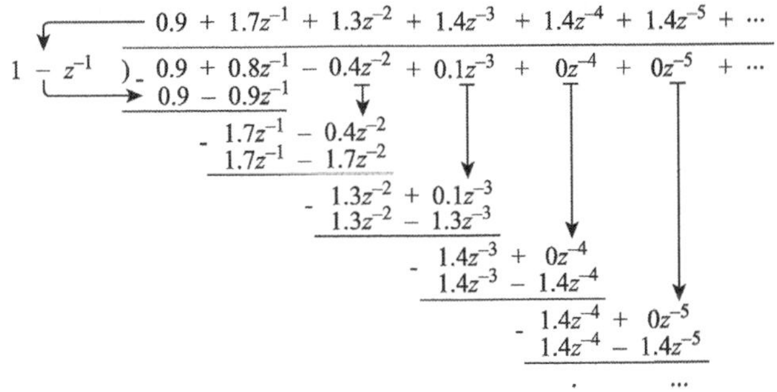

$$
\begin{array}{rl}
 & 0.9 + 1.7z^{-1} + 1.3z^{-2} + 1.4z^{-3} + 1.4z^{-4} + 1.4z^{-5} + \cdots \\
1 - z^{-1} \,) & 0.9 + 0.8z^{-1} - 0.4z^{-2} + 0.1z^{-3} + 0z^{-4} + 0z^{-5} + \cdots \\
- & 0.9 - 0.9z^{-1} \\
- & 1.7z^{-1} - 0.4z^{-2} \\
 & 1.7z^{-1} - 1.7z^{-2} \\
- & 1.3z^{-2} + 0.1z^{-3} \\
 & 1.3z^{-2} - 1.3z^{-3} \\
- & 1.4z^{-3} + 0z^{-4} \\
 & 1.4z^{-3} - 1.4z^{-4} \\
- & 1.4z^{-4} + 0z^{-5} \\
 & 1.4z^{-4} - 1.4z^{-5} \\
 & \cdots
\end{array}
$$

[5]See end of chapter problems.

At each successive stage, we choose and multiply the top coefficient (starting with 0.9) by the divisor $(1 - z^{-1})$ to give a product (in the first iteration, $0.9 - 0.9z^{-1}$), which is then subtracted from the numerator sequence. Then the next term is brought down (downward arrows) and the process repeated. The output is read off the top as the sequence 0.9, 1.7, 1.3, 1.4, 1.4, ...

Note that this is the *same result* as we get in iterating the difference equation. The coefficients are identical, and the z^{-k} terms show the relative position k. This corresponds to our earlier reasoning that z^{-k} is a delay of k samples.

5.6 POLES, ZEROS, AND STABILITY

The z domain equations provide a convenient way for converting between a sampled waveform and a difference equation. The transfer function in z is also invaluable for the analysis of systems. The following shows the relationship between a transfer function and the time domain, and shows a way to graphically visualize a system's expected response.

The *poles* and *zeros* of a system are quantities derived from its transfer function, and are fundamental to an understanding of how a system responds to a given input. They are defined as follows:

The Zeros of a system are those values of z which make the *numerator* $B(z)$ equal to zero.

The Poles of a system are those values of z which make the *denominator* $A(z)$ equal to zero.

Note that we are effectively solving for the roots of a polynomial, and hence the value (s) of z may be complex numbers. As a simple example, consider a system with one real pole at $z = p$. The transfer function is:

$$H(z) = \frac{z}{z - p}. \tag{5.37}$$

The corresponding difference equation is:

$$y(n) = x(n) + py(n-1). \tag{5.38}$$

The parameter p controls how much of the previous output is fed back. This recursive term is crucial to the stability of the system. Suppose initially $p = 0.9$. We may implement this using the difference equation coding approach outlined earlier, or using the MATLAB **filter()** command:

```
N = 20;
b = [1 0];
a = [1 - 0.9];
x = zeros(N, 1);
x(1) = 1;
y = filter(b, a, x);
stem(y);
```

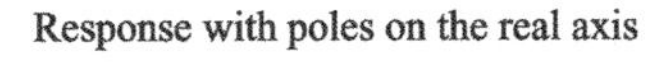

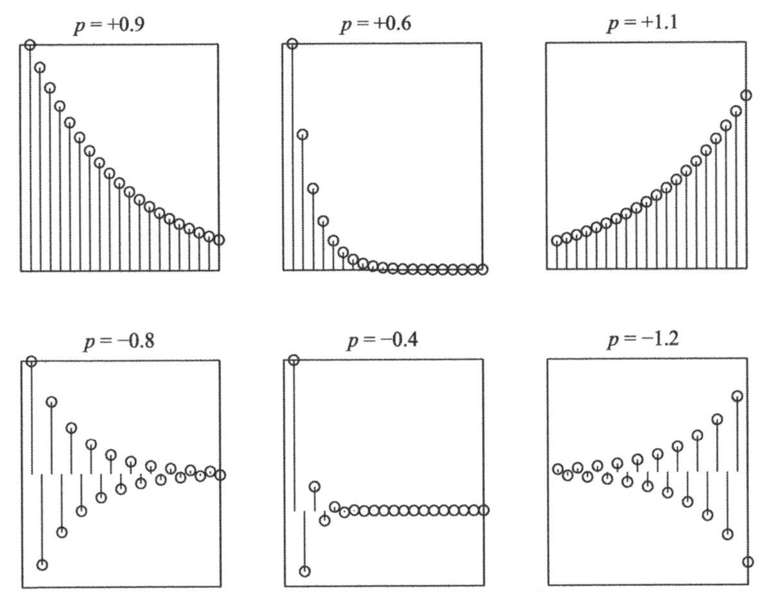

FIGURE 5.9 The impulse response of a system with poles on the real axis. Values of $|p| > 1$ lead to a diverging response. Negative values of p have an alternating +/− response. These effects are cumulative (as in the lower-right figure). That is, we have both alternation *and* divergence.

Figure 5.9 shows the family of impulse responses with poles in a variety of locations on the real axis. From this, some observations may be made:

1. Starting from the origin, as the pole location gets closer to ± 1, the response decays more slowly.
2. Poles at positive values of z show a smooth response, whereas poles at negative values of z oscillate on alternating samples.
3. Poles whose magnitude is less than 1 are stable, whereas poles whose magnitude is greater than one tend toward instability.

Now consider a system with complex poles. Because the coefficients of the time domain terms in the difference equation must be real, the coefficients of z when written out as a polynomial must in turn be real. This in turn means that the roots of the numerator and denominator polynomials in z (the zeros and poles respectively) must be complex conjugates. To give a simple second-order example, suppose a system has a transfer function:

$$G(z) = \frac{z^2}{(z-p)(z-p^*)}, \tag{5.39}$$

where * denotes complex conjugation. Let the pole p be at $p = 0.9e^{j\frac{\pi}{10}}$. That is, a radius of 09 at an angle of $\pi/10$. We may iterate the difference equation as shown previously using the coefficients derived from the polynomial expansion of the denominator of $G(z)$. Defining p as the value of the pole location, the **poly**() function may be used as shown below to expand the factored equation to obtain the denominator coefficients a. The numerator coefficients b are obtained directly, since the numerator is $1z^2 + 0z^1 + 0z^0$:

```
p = 0.9*exp(j*pi/10)
p =
     0.8560 + 0.2781 i
a = poly([p conj(p)])
a =
     1.0000 1.7119 0.8100
b = [1 0 0];
```

Note how the function **conj**() is used to find the complex conjugate. The pole locations may be checked using **roots**(), which is the complement to **poly**()

```
roots(a)
ans =
     0.8560 + 0.2781 i
     0.8560 - 0.2781 i
abs(roots(a))
ans =
     0.9000
     0.9000
angle(roots(a))
ans =
     0.3142
     -0.3142
```

Note that MATLAB uses i to display the complex number $j = \sqrt{-1}$, not j. By convention, the pole–zero plot of a system shows the poles labelled × and the zeros labelled o. We can extend the MATLAB code to show the poles and zeros, together with the unit circle, as follows:

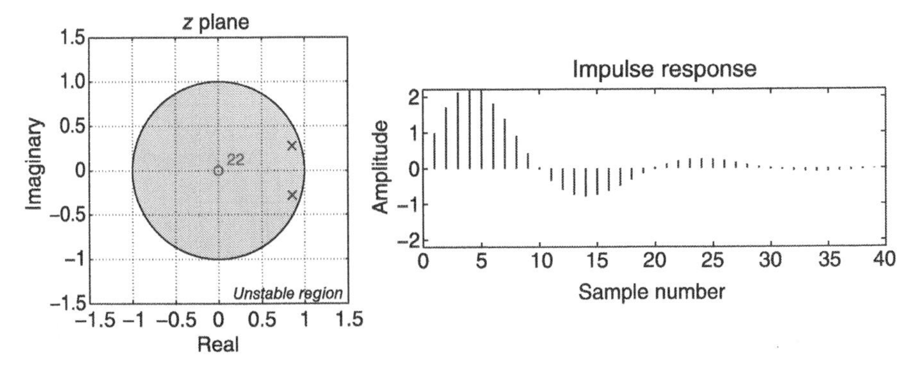

FIGURE 5.10 The pole–zero plot for a system with poles at $z = 0.9e^{\pm j\pi/10}$ and two zeros at the origin is shown on the left. The unit circle (boundary of stability) is also shown. The corresponding impulse response is on the right.

```
p = 0.9*exp(j*pi/10);
a = poly([p conj(p)]);
b = (1 0 0);
theta = 0:pi/100:2*pi;
c = 1*exp(j*theta);
plot(real(c), imag(c));
set(gca, 'PlotBoxAspectRatio', [1 1 1]);
zvals = roots(b);
pvals = roots(a);
hold on
plot(real(pvals), imag(pvals), 'X');
plot(real(zvals), imag(zvals), 'O');
```

Note that we need to be a little careful with simple zeros at the origin, and in fact the above has *two* zeros at $z = 0$, as shown in the context of Equation 5.39. Having more poles than simple zeros will only affect the relative delay of the impulse response, not the response shape itself.

The z plane pole/zero configuration resulting from this transfer function is shown in Figure 5.10. Iterating over 40 samples, the impulse response of this system is shown on the right-hand side. Note that it oscillates and decays as time progresses.

Figure 5.11 shows various responses with complex poles. From this, some observations may again be made:

1. Starting from the origin, as the pole gets closer to a radius of one, the response decays more slowly.
2. When the poles are *on* the unit circle, the response oscillates indefinitely.
3. When the poles are *outside* the unit circle, the system becomes unstable.

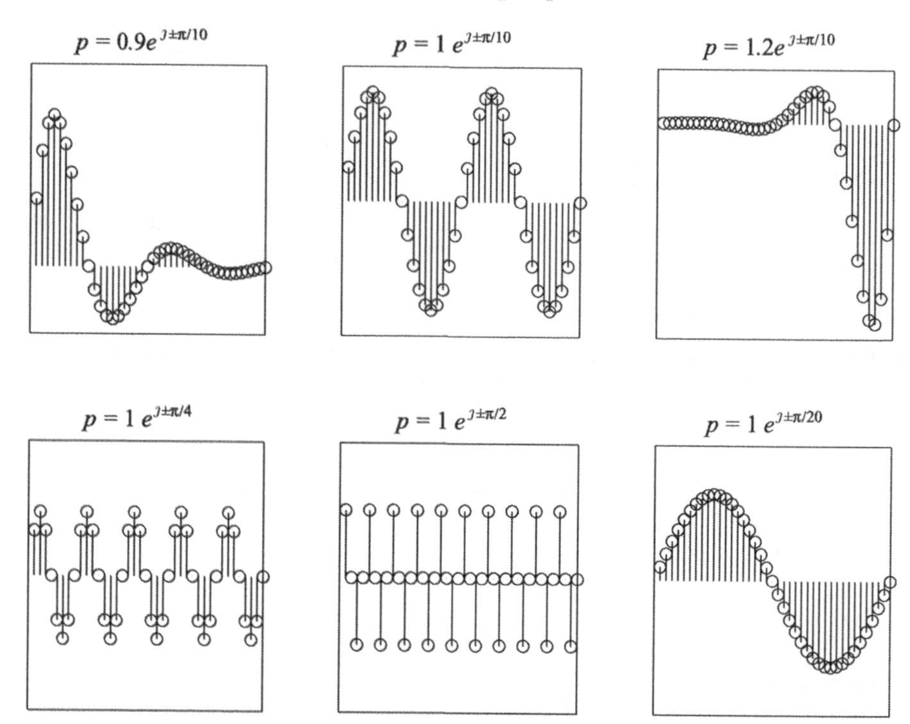

FIGURE 5.11 The impulse responses of a system with complex poles at various locations.

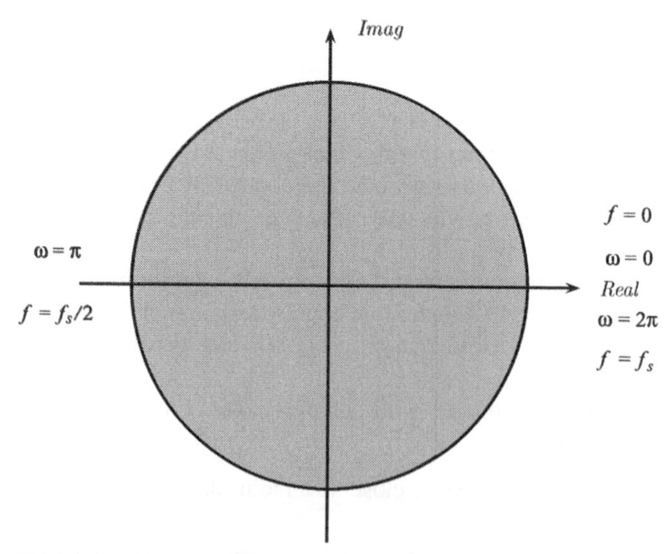

FIGURE 5.12 The z plane and its interpretation in terms of stability (the unit circle) and frequency (the angle of the poles relative to π).

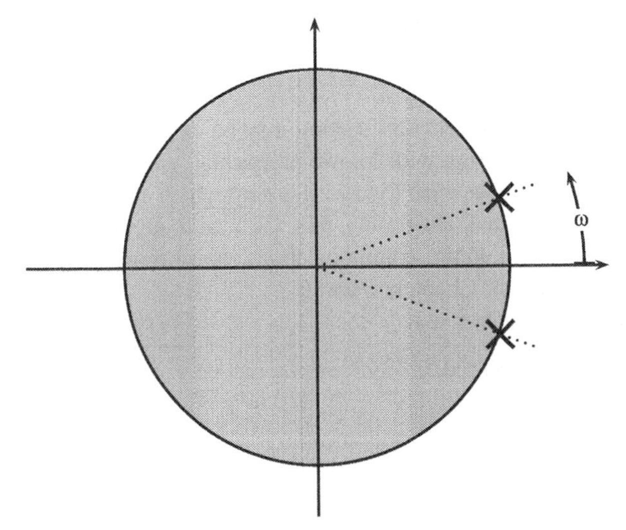

FIGURE 5.13 The poles of a marginally stable system. It is termed "marginally stable" because the poles are *on* the unit circle. When subjected to an impulse input, the corresponding difference equation will produce a sinusoidally oscillating output at frequency ω radians per sample.

4. When the magnitude of the poles is greater than one, the response expands.
5. When the magnitude of the poles is less than one, the response decays.
6. The angle of the pole controls the relative frequency of oscillation.

The previous arguments reveal that the relative angle of the poles controls the frequency of oscillation, and this may be visualized as shown in Figure 5.12. The positive real axis is the "zero frequency" axis, and the negative real axis is the "half sampling frequency" or $f_s/2$ axis. This is the highest frequency a sampled data system can produce, because it is constrained by the sample rate. The location of the poles for an oscillator—which is termed a *marginally stable* system—is shown in Figure 5.13. The relative frequency of oscillation ω is directly related to the angle of the poles, with an angle of π radians corresponding to a frequency of oscillation of $f_s/2$. The true frequency of oscillation is found by:

$$f_{\text{osc}} = \frac{\omega}{\rho}\frac{f_s}{2}. \tag{5.40}$$

This may be expressed in words as:

The true frequency of oscillation is equivalent to the angle of the upper-half z-plane pole, scaled such that an angle of π radians is equivalent to a real frequency of $f_s/2$ Hz.

5.7 TRANSFER FUNCTIONS AND FREQUENCY RESPONSE

The z domain notation introduced in the previous section may be extended further. Rather than simply generating waveforms with known frequency, amplitude and phase in response to an impulse input, we wish to have a system which can "shape" the frequency content of any given signal. Ultimately, this will lead to digital filters, which have numerous applications in audio, communications, noise removal, and elsewhere. Digital filters are covered in Chapters 8 and 9.

Suppose a system has a transfer function:

$$H(z) = 0.5z^{-4}. \tag{5.41}$$

The output is simply found by inference:

$$y(n) = 0.5x(n-4). \tag{5.42}$$

That is, a zero-frequency gain of 0.5 and a delay of four samples. Suppose the input is a sampled complex exponential of frequency ω radians,

$$x(n) = e^{jn\omega}. \tag{5.43}$$

Mathematically, the output is

$$\begin{aligned} y(n) &= 0.5e^{j(n-4)\omega} \\ &= \underbrace{e^{jn\omega}}_{\text{Input}} \cdot \underbrace{0.5e^{-4j\omega}}_{\text{Scaling}}. \end{aligned} \tag{5.44}$$

The multiplicative scaling consists of the magnitude term (0.5) and a phase delay of -4ω radians. Remembering that the input frequency was ω radians per sample, it is seen that the output frequency is the same as the input frequency.

In general, the output is equal to the input multiplied by a complex number representing the transfer function. Complex numbers have magnitude and phase—the magnitude is multiplied by the input amplitude, and the phase represents a time delay relative to the input frequency (recall that frequency is the rate of change of phase, $\Omega = d\varphi/dt$. The generalization of this result may be stated mathematically as:

$$H(e^{j\omega}) = H(z)\big|_{z=e^{j\omega}}. \tag{5.45}$$

or equivalently:

> In order to obtain the **frequency response** of a discrete (sampled-data) system, simply substitute
>
> $$z \to e^{j\omega}$$
>
> and evaluate for $\omega = 0 - \pi$.

This statement is quite fundamental to the understanding of the frequency response of digital systems, and we will now devote some time to the investigation of this result.

5.8 VECTOR INTERPRETATION OF FREQUENCY RESPONSE

If we factorize the transfer function numerator and denominator, it will have the general form:

$$H(z) = K \cdot \frac{(z-\beta_0)(z-\beta_1)(\cdots)}{(z-\alpha_0)(z-\alpha_1)(\cdots)}. \tag{5.46}$$

where K is a gain term, the β_m are the zeros and the α_n are the poles. The magnitude response is found by substituting $z = e^{j\omega}$ and finding the magnitude of the resulting complex number:

$$|H(e^{j\omega})| = K \cdot \left| \frac{(e^{j\omega}-\beta_0)(e^{j\omega}-\beta_1)(\cdots)}{(e^{j\omega}-\alpha_0)(e^{j\omega}-\alpha_1)(\cdots)} \right| \tag{5.47}$$

$$= K \cdot \frac{|e^{j\omega}-\beta_0||e^{j\omega}-\beta_1||\cdots|}{|e^{j\omega}-\alpha_0||e^{j\omega}-\alpha_1||\cdots|}. \tag{5.48}$$

To allow for a more intuitive geometric interpretation, consider each complex number as a vector in the z plane. The terms $|e^{j\omega}-\beta|$ and $|e^{j\omega}-\alpha|$ form a vector subtraction in the complex plane. Normal vector addition is carried out as shown in Figure 5.14, adding the vectors **a** and **b** "head to tail" to obtain the resultant **r**:

$$\mathbf{a} + \mathbf{b} = \mathbf{r}. \tag{5.49}$$

In the case of sampled-date systems in terms of z, the "vector" quantities are complex numbers, written in the form:

$$|e^{j\omega}-\alpha| = |1e^{j\omega} - |\alpha| e^{j\angle\alpha}|. \tag{5.50}$$

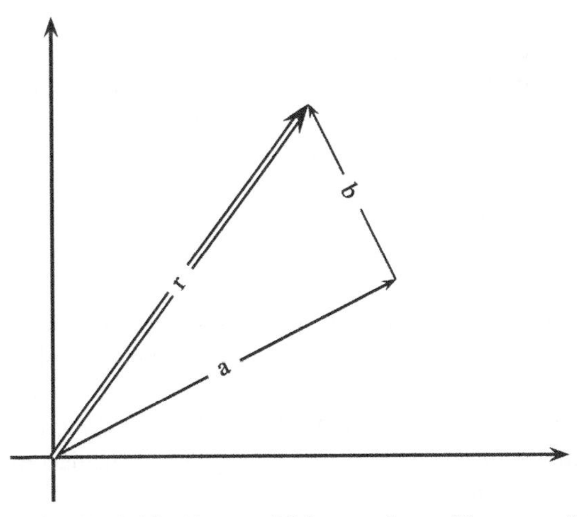

FIGURE 5.14 Vector addition **a** + **b** = **r**. The vector **b** is added to the head (arrow) of vector **a**, and the resultant vector **r** is drawn from the start of **a** to the end of **b**.

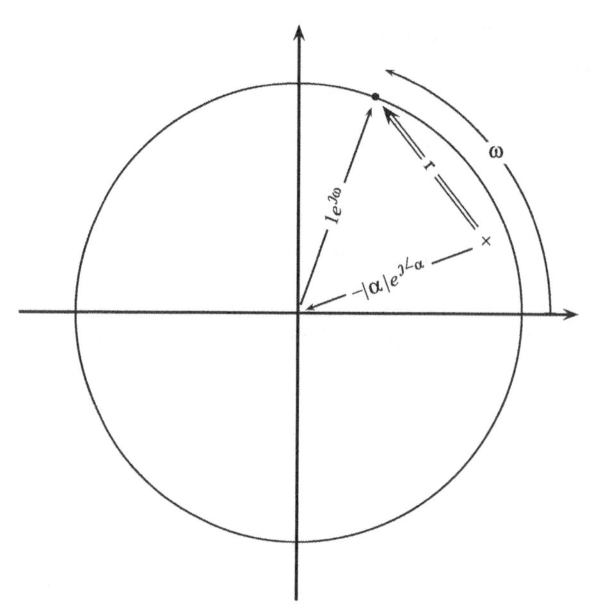

FIGURE 5.15 Frequency response as a vector evaluation. The specific frequency ω radians per sample defines a point on the unit circle. The vector from the origin to the pole is subtracted (reversed), leaving the resultant as the vector from the pole to the point on the unit circle.

The first term on the right-hand side defines a point at a radius of one, at an angle of ω. This angle changes as the frequency of interest changes. The second term is the complex point representing a pole (or zero) of the system. Obviously, this is constant for the system under consideration. Because it is subtracted, the equivalent vector is reversed. This is shown in Figure 5.15.

It may be seen that as the point on the unit circle approaches and then passes the angle of the upper pole, the resulting vector to that pole will pass through a minimum value. Thus, the response magnitude will be maximized, because (by definition) the pole is in the denominator. Effectively, the angle of evaluation ω is mapped into the horizontal axis of the frequency response curve (Figure 5.16), with the value of the response at that frequency shown as the vertical height of the curve. The resonant peak on the frequency response is seen to correspond to the pole angle in the pole–zero diagram on the complex plane.

This considers only one pole to be mapped into the frequency response. There are, of course, two poles to be considered, because as pointed out earlier, complex poles occur in conjugate pairs. However, since the conjugate pole is somewhat further away, its effect will be somewhat less. The net effect will be to draw the frequency down slightly, but for the purposes of approximation, it may be neglected in order to sketch an outline of the response.

As the angle ω goes from 0 to π, the top half of the unit circle in the z plane is swept out. As ω goes from π to 2π, the bottom half is covered. Remembering that

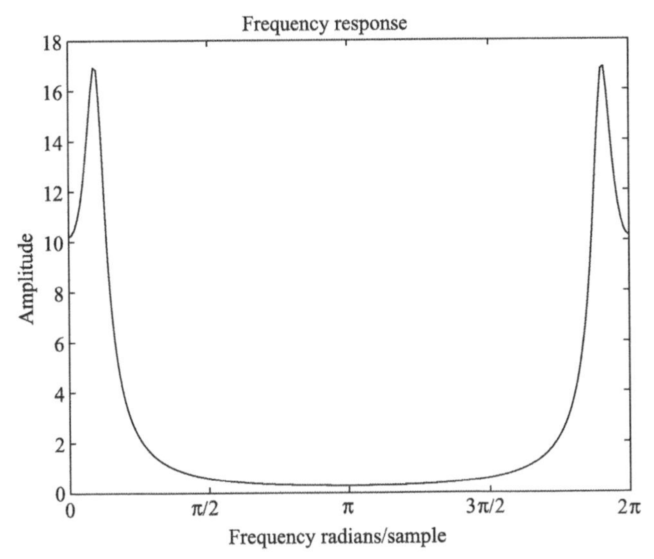

FIGURE 5.16 Frequency response of the system with poles at radius 0.9, angle $\pm\pi/10$.

poles (and zeros) occur in complex-conjugate pairs, the magnitude response is seen to be symmetrical: π to 2π is a mirror image of that from 0 to π.

The above gives a rough order-of-magnitude and shape estimate for the frequency response. We now show how the frequency response can be calculated using MATLAB. Of course, it simply depends upon the theory that the frequency response is given by evaluating the function $H(z)$ using $z = e^{j\omega}$. We will use the following example, which has two poles at radius r at angle θ,

$$H(z) = \frac{1z^2 + 0z^1 + 0z^0}{z^2 - (2r\cos\theta)z^1 + r^2z^0}. \tag{5.51}$$

As shown in the MATLAB code below, the key point to take care with is the power of z at each term in the numerator and denominator.

```
r = 0.9;                          % pole radius
theta = pi/10;                    % pole angle
omega = 0:pi/100:2*pi;            % frequency response range
% coefficients of z
b = [1 0 0];                      % b = numerator
a = [1 -2*r*cos(theta) r*r];      % a = denominator
% poles & zeros
syszeros = roots(b);
syspoles = roots(a);
fprintf(1, 'Pole angles: ');
fprintf(1, '% 2f ', angle(syspoles));
```

```
fprintf(1, '\n');
k = 0;
for omegak = omega
    k = k + 1;
    expvec = exp(j * [2:-1:0] * omegak);
    H(k) = sum(b. * expvec)./sum(a. * expvec);
end
plot(omega, abs(H));
set(gca, 'xlim', [0 2*pi]);
set(gca, 'xtick', [0:4]*pi/2);
```

5.9 CONVOLUTION

Convolution is a method of computing the response of a system to a given input. Convolution is based on mathematical principles, but has a very intuitive interpretation. In essence, the output of any linear, time-invariant system may be imagined as the superposition of a series of impulses. The net output is the summation of all these outputs, suitably scaled according to the magnitude of the input and aligned according to the time of the input (i.e., when the system "sees" each impulse).

Figure 5.17 shows the general input–output model of a system. The input to the system (the excitation waveform) generates an output. Obviously the output depends on the input and the nature of the system itself. This is where the transfer-function, and related difference-equation representations, come into play. Suppose we are given a difference equation:

$$y(n) = 0.8x(n) - 0.4x(n-1) + 0.2x(n-2). \tag{5.52}$$

Then the transfer function is:

$$\frac{Y(z)}{X(z)} = 0.8 - 0.4x^{-1} + 0.2z^{-2}. \tag{5.53}$$

The impulse response of this system is computed as we have seen previously:

n	$x(n)$	$y(n)$	
0	1	0.8×1	$= 0.8$
1	0	$0.8 \times 0 - 0.4 \times 1$	$= -0.4$
2	0	$0.8 \times 0 - 0.4 \times 0 + 0.2 \times 1$	$= 0.2$
3	0	$0.8 \times 0 - 0.4 \times 0 + 0.2 \times 0$	$= 0.0$
4	0	$0.8 \times 0 - 0.4 \times 0 + 0.2 \times 0$	$= 0.0$
⋮	⋮	*etc.*	

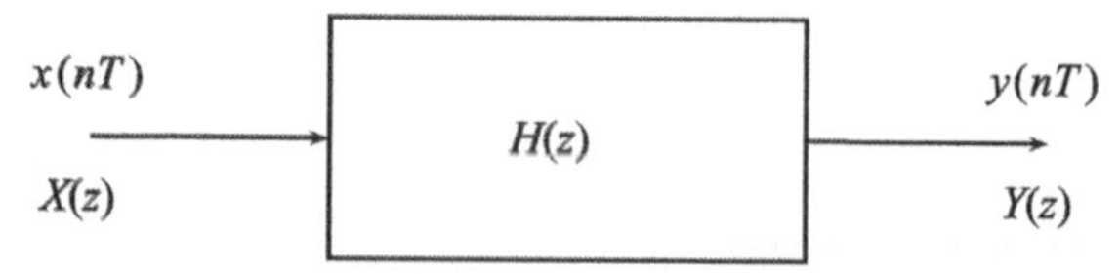

FIGURE 5.17 The input-output model of a sampled-data system.

h_k					0.8	−0.4	0.2					
$x(n)$	4	3	2	1	0	⇒						
$y(n)$					0							$\Sigma = 0$
h_k					0.8	−0.4	0.2	⇒				
$x(n)$		4	3	2	1	0	⇒					
$y(n)$					0.8	0						$\Sigma = 0.8$
h_k					0.8	−0.4	0.2					
$x(n)$			4	3	2	1	0	⇒				
$y(n)$					1.6	−0.4	0					$\Sigma = 1.2$
h_k					0.8	−0.4	0.2					
$x(n)$				4	3	2	1	0	⇒			
$y(n)$					2.4	−0.8	0.2					$\Sigma = 1.8$
h_k					0.8	−0.4	0.2					
$x(n)$					4	3	2	1	0	⇒		
$y(n)$					3.2	−1.2	0.4					$\Sigma = 2.4$
h_k					0.8	−0.4	0.2					
$x(n)$						4	3	2	1	0	⇒	
$y(n)$						−1.6	0.6					$\Sigma = -1.0$
h_k					0.8	−0.4	0.2					
$x(n)$							4	3	2	1	0	
$y(n)$							0.8					$\Sigma = 0.8$

FIGURE 5.18 Calculating the response of a system to a ramp input. The impulse response coefficients are h_n, the input is $x(n)$, producing an output $y(n)$.

It is clear that the output dies away after the input is removed. Specifically, if the system order is P (the highest power of z is z^{-P}), then the last nonzero output term will be P samples after the input is removed, regardless of whether the input is an impulse or not. More precisely, the impulse response lasts for $P + 1$ samples.

This particular system has only nonrecursive coefficients (coefficients of $x(n)$'s). After three output terms, the output is always exactly zero. Hence, this is termed a *finite impulse response* or FIR system.

Next, suppose the same system:

$$y(n) = 0.8x(n) - 0.4x(n-1) + 0.2x(n-2). \tag{5.54}$$

is subjected to a ramp input of 0, 1, 2, 3, 4. The output at each iteration is determined as shown by the sequence of operations in Figure 5.18. The output is seen to be calculated from the impulse response as follows:

1. Time-reverse the input sequence.
2. Slide the *time-reversed* input response coefficients past the impulse response.
3. Integrate (add up) the sliding sequence, and we have the output, or system response, at each sampling instant (shown in bold).

This operation is *convolution*—the input sequence is said to be *convolved* with the impulse response of the system. Note that what we have really done is the polynomial multiplication of input $X(z)$ and coefficients $H(z)$. The polynomial multiplication in the z domain is the same as convolution in the time domain. We can easily perform this in MATLAB using the **conv()** function:

```
h = [0.8 -0.4 0.2];
x = [0 1 2 3 4];
y = conv(h, x);
disp(y)
```

Note that this is actually polynomial multiplication. The convolution operation is in fact the same as multiplying two polynomials by shifting, multiplying, and aligning their coefficients. In this case, the polynomials are in fact the coefficients of the input $X(z)$ and the impulse response $H(z)$.

It also is a worthwhile exercise to use the above example to see what happens if we interchange $x(n)$ and h_k. That is, if we reverse the impulse response and hold the input sequence the same. The physical interpretation is of course that the input sequence is reversed, but mathematically, it does not matter whether the input or the impulse response is reversed.

So we now have a definition of convolution in a signal-processing sense:

> The output of a linear, time-invariant system is equal to the convolution of the input and the system's impulse response.

Mathematically, convolution is defined as:

$$y(n)=\sum_{k=0}^{P-1} h_k x(n-k). \tag{5.55}$$

where

h_k = impulse response coefficients

P = length of the impulse response

k = index to sum over

$y(n)$ = output of the system

Infinite impulse response (IIR) systems, on the other hand, have recursive terms in the difference equation. The output depends not only on the input, but on *past outputs* as well. A simple example is:

$$y(n)=x(n)-0.3y(n-1)+0.1y(n-2), \tag{5.56}$$

with transfer function

$$\frac{Y(z)}{X(z)}=\frac{1}{1+0.3z^{-1}-0.1z^{-2}}. \tag{5.57}$$

The *impulse response* of this system is:

n	$x(n)$	$y(n)$	
0	1	1.0	= 1.0
1	0	$0 - 0.3 \times 1.0$	= −0.3
2	0	$0 - 0.3 \times - 0.3 + 0.1 \times 1.0$	= 0.19
3	0	$0 - 0.3 \times 0.19 + 0.1 \times -0.3$	= −0.087
⋮	⋮	etc.	

This system has recursive coefficients (coefficients of $y[n]$'s). The output theoretically *never* reaches zero. Hence, it is termed an *infinite impulse response* (IIR) system.

Calculating some more terms in the impulse response, we get:

1 −0.3 +0.19 −0.087 +0.0451 −0.02223
+0.01117 −0.00557 +0.00279 +0.000697 ...

We cannot calculate the output of the IIR system directly using convolution, as we did with the FIR system. However, noting that the output terms die away, an FIR *approximation* could be written as:

$$\hat{y}(n) = x(n) - 0.3x(n-1) + 0.19x(n-2) - 0.087x(n-3). \tag{5.58}$$

where the circumflex or caret ("hat") above y means "approximation to." The corresponding transfer function is

$$\frac{\hat{Y}(z)}{X(z)} = 1 - 0.3z^{-1} + 0.19z^{-2} - 0.087z^{-3}. \tag{5.59}$$

The impulse response is then:

1 −0.3 +0.19 −0.087 0.0 0.0 0.0 0.0 0.0 ...

Some conclusions on convolution, FIR, and IIR systems are worth summarizing:

1. Theoretically, the FIR (nonrecursive) system needs an infinite number of terms to equal the IIR (recursive) system.
2. Depending on the accuracy required, it is possible to use a fixed-length FIR transfer function to approximate an IIR transfer function by truncating the impulse response.
3. Systems can have poles only, zeros only, or both poles and zeros. If it has zeros only, there will be one or more poles at the origin in order to make the system causal (i.e., have only zero or negative powers of z, and not depend on future samples).

These observations will be important later in the design and analysis of digital filters. We will return to convolution in Chapter 8, where it will be shown that the concept helps us understand the operation of digital filtering.

5.10 CHAPTER SUMMARY

The following are the key elements covered in this chapter:

- the importance of and relationship between *difference equations*, *transforms*, and the *impulse response*.
- the role of the z *transform* in signal processing.
- the concept of *poles and zeros*, and *system stability*.
- the concept of *convolution* and how it is used to calculate a system's output.

PROBLEMS

5.1. Show that the output from the difference equation $y(n) = 0.9x(n) + 0.8x(n-1) - 0.4x(n-2) + 0.1x(n-3)$ when subjected to an input 1, 1, 1, 1, 1, 0, 0, 0, . . . is as follows:

N	$x(n)$	$y(n)$
Sample Instant	Input	Output
0	1	0.9
1	1	$0.9 + 0.8 = 1.7$
2	1	$0.9 + 0.8 - 0.4 = 1.3$
3	1	$0.9 + 0.8 - 0.4 + 0.1 = 1.4$
4	1	$0.9 + 0.8 - 0.4 + 0.1 = 1.4$
5	0	$0.9 \times 0 + 0.8 \times 1 - 0.4 \times 1 + 0.1 \times 1 = 0.5$
6	0	$-0.4 + 0.1 = -0.3$
8	0	0.1
8	0	0
9	0	0
10	0	. . .

5.2. A system has poles at $r = 0.9$ and angle $\omega = \frac{\pi}{4}$.

(a) Determine the difference equation.

(b) Find the first eight terms resulting from an impulse input.

(c) Repeat if the radius is 0.95, 1.1, 2, and 4.

5.3. Convert the following difference equations into z transforms:

(a) $y(n) = x(n-2)$

(b) $y(n) = 0.1x(n) + 0.9x(n-1) - 0.3y(n-1)$

5.4. Starting with the definition of a z transform, show that a unit step has a transform $\frac{z}{z-1}$.

5.5. Find the z transform of te^{-at}, and the corresponding difference equation.

5.6. Section 5.4.1 showed how z^{-1} can be used to model a delay of one sample. What would z^{-k} mean? Prove that your inference is valid using a similar approach.

5.7. Using the example impulse response of Equation 5.54 and ramp input sequence of Figure 5.18 (Section 5.9), show that reversing the impulse response with the input fixed produces the same result as reversing the input sequence and keeping the impulse coefficients fixed. Check using the `conv()` function in MATLAB.

5.8. Enter the MATLAB code in Section 5.3.1 for generating a pure-tone waveform.

(a) Vary the frequency f, plot the waveform and listen to the result using the `sound()` command.

(b) Use the following approach to generate a frequency-modulated "chirp" signal whose frequency increases over time, and again listen to the result.

```
Fs = 8000;
dt = 1/Fs;
N = 8000;
t = 0:dt:(N-1)*dt;
f = 400;
fi = [1:length(t)]/length(t)*f;
y = sin(2*pi*t.*fi);
plot(y(1:2000));
sound(y, Fs);
```

5.9. Consider the phase compensation and audio buffering as described in Section 5.3.4. Implement the code given, both with and without phase compensation. Are the phase artifacts audible if compensation is not employed?

5.10. Again using the phase compensation and audio buffering as described in Section 5.3.4. Generate a sinusoidal waveform, but with a gap of D samples at the start of each buffer (i.e., introduce gaps in the continuous waveform). Start with $D = 1$, then progress to 10, 20, 100 samples. What length of gap is just noticeable?

5.11. Derive the z transforms and corresponding difference equations for the waveforms:

- $x(t) = \cos \Omega t$
- $x(t) = \sin(\Omega t + \varphi)$
- $x(t) = \cos(\Omega t + \varphi)$
- Implement the difference equations using MATLAB and check that each generates the expected wave form in each case.

5.12. Sketch the frequency response of a system with a pole at $z = 1$ and a pole at $z = -1$. Repeat for zeros at the same locations.

5.13. The image data compression standard *JPEG2000* uses a technique called *sub-band coding* to reduce the size of image data files, and hence the space taken to store the image (or, equivalently, the time taken to send the image over a communications channel). This method uses a pair of filters, called *quadrature mirror filters* (QMFs).

A very simple set of quadrature mirror filters $h_0(n)$ and $h_1(n)$ may be defined using the equations

$$F(z)=\frac{1}{\sqrt{2}}(1+z^{-1})$$
$$H_0(z)=F(z)$$
$$H_1(z)=F(-z)$$

(a) What are the corresponding difference equations for filters $H_0(z)$ and $H_1(z)$ that need to be implemented in this digital signal processing system?

(b) Use the pole/zero diagram to sketch the frequency responses of each filter, and hence explain how $H_0(z)$ and $H_1(z)$ form low-pass and high-pass filters, respectively.

(c) Show mathematically that the filter pair is energy preserving—that is

$$|H_0(\omega)|^2+|H_1(\omega)|^2=1$$

5.14. Low-bitrate speech coders often employ a noise-shaping filter, which involves changing the poles of a speech synthesis filter. The goal is to "shape" the response such that more quantizing noise is concentrated in the spectrum where there is greater energy, thus making it less perceptually objectionable.[6]

(a) Suppose we have a filter $G(z)=1/A(z)$ with poles at $z=(1/2)\pm \jmath(1/2)$. Determine the transfer function, the angle of the poles, and the radius of the poles.

(b) We employ a noise-shaping coefficient γ, and a noise-shaping transfer function $G(z/\gamma)$. Determine this transfer function, pole angle, pole radius, and gain.

(c) Show that the pole angle is unchanged in going from $G(z)$ to $G(z/\gamma)$. What happens to the pole radius?

(d) Using the above result, sketch the two frequency responses.

(e) Would this result hold in general for the same order of transfer function for other pole locations? Would it hold in general for higher-order transfer functions?

(f) In practice, the above approach is used in a feedback configuration, with the output filtered by $W(z)=A(z)/A(z/\gamma)$. Sketch the frequency response of $W(z)$, and show that the gain is reduced in the vicinity of the pole frequency.

5.15. Perform the convolution as tabulated in Figure 5.18 using multiplication of the corresponding z transforms. Does the result equal that obtained by the iterative method as in the table?

5.16. The output of a linear system is given by the product of the input and the transfer function, when all are expressed in the z domain. That is, $(z) = X(z)\,H(z)$. Recall that the z transform of h_n is defined as $H(z) = \Sigma_n h_n z^{-n}$, and that in the time domain, the output of a system is computed using the convolution as $y(n) = \Sigma_k h_k x(n-k)$.

[6] The precise details are beyond the scope of this chapter.

(a) Take the z transform of both sides of the convolution equation.

(b) By making the substitution $m = n - k$ and reversing the order of the summations, show that $Y(z) = X(z)\,H(z)$.

5.17. We are able to find the frequency response of a sampled-data system using the substitution $z = e^{j\omega}$ in the transfer function. Recall that the output of a discrete system is the convolution of the input and impulse response, given by $y(n) = \Sigma_k h_k x(n-k)$. Let the input to a system be a sampled complex sinusoid $x(n) = e^{jn\omega}$.

(a) Substitute this input into the convolution equation to derive an expression for $y(n)$.

(b) Using the definition of the z transform $H(z) = \Sigma_n h_n z^{-n}$, show that $y(n)$ may be written as the product of the sampled sinusoid and $H(z)$ when $z = e^{j\omega}$.

CHAPTER 6

TEMPORAL AND SPATIAL SIGNAL PROCESSING

6.1 CHAPTER OBJECTIVES

On completion of this chapter, the reader should be able to:

1. explain and use *correlation* as a signal processing tool.
2. be able to *derive* the *correlation equations* for some mathematically defined signals.
3. derive the operating equations for *system identification* in terms of correlation.
4. derive the equations for *signal extraction* when the signal is contaminated by noise, and explain the underlying assumptions behind the theory.
5. derive the equations for *linear prediction* and *optimal filtering*.
6. write down the basic equations used in *tomography* for reconstruction of an object's interior from cross-sectional measurements.

6.2 INTRODUCTION

Signals, by definition, are varying quantities. They may vary over time (temporal) or over an x–y plane (spatial), over three dimensions, or perhaps over time and space (e.g., a video sequence). Understanding how signals change over time (or space) helps us in several key application areas. Examples include extracting a desired signal when it is mixed with noise, identifying (or at least, estimating) the coefficients of a system, and reconstructing a signal from indirect measurements (tomography).

6.3 CORRELATION

The term "correlation" means, in general, the similarity between two sets of data. As we will see, in signal processing, it has a more precise meaning. Correlation in

Digital Signal Processing Using MATLAB for Students and Researchers, First Edition. John W. Leis.

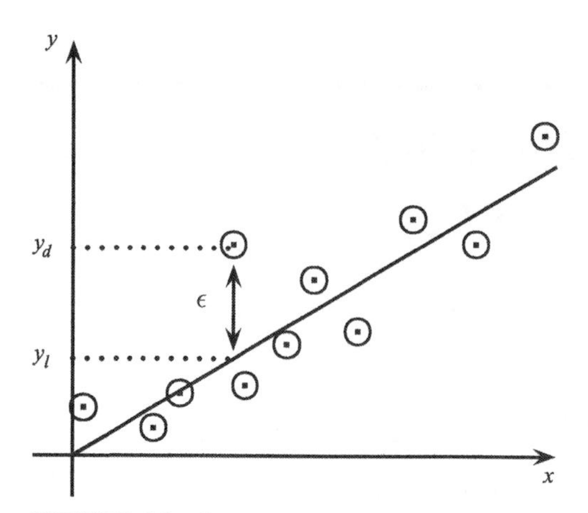

FIGURE 6.1 Correlation interpreted as a line of best fit. The measured points are show as ⊙, and the line shows an approximation to the points in the (x, y) plane. The error is shown between one particular point and the approximation; the goal is to adjust the slope of the line until the average error over all measurements is minimized ($\min \sum \varepsilon^2$).

signal processing has a variety of applications, including removal of random noise from a periodic signal, pattern recognition, and system parameter identification.

To introduce the computation of correlation, consider first a simpler problem involving a set of measured data points (x_n, y_n). We wish to quantify the similarity between them—or more precisely, to determine if they are correlated (if one depends in some way on the other). Note that this does not necessarily mean a causal relationship (that one causes another). These data points may represent samples from the same signal at different times, or samples from two different but related signals.

We may picture the two sets of samples as shown on the axes of Figure 6.1, on the assumption that there is a linear relationship between the two sets of measurements. Because the measurements are not precise, some variation or "noise" will cause the points to scatter from the ideal.

Mathematically, we wish to choose the value of b describing the line $y = bx$ such that the error ε between calculated (estimated) and measured points is minimized over all data points. The error is the difference between the measured data point y_d and the straight line estimation using an equation for y_l. So how do we determine b?

We can employ the tools of calculus to help derive a solution. Since the error may be positive or negative, we take the square of the error:

$$\begin{aligned} \varepsilon &= y_d - y_l \\ \therefore \varepsilon^2 &= (y_d - y_l)^2 \\ &= (y_d - bx)^2 \end{aligned} \tag{6.1}$$

The average squared (mean square) error over N points is:

$$E(\varepsilon^2) = \frac{1}{N} \sum_N (y_d - bx)^2. \tag{6.2}$$

For simplicity of notation, replace y_d with y. This is a minimization problem, and hence the principles of function minimization are useful. The derivative with respect to the parameter b is:

$$\frac{\partial E}{\partial b} = \frac{1}{N}\sum_{N} 2(y - bx)(-x). \tag{6.3}$$

Setting this gradient to zero,

$$\frac{1}{N}\sum_{N} 2\left(-xy + bx^2\right) = 0$$

$$\therefore \frac{1}{N}\sum_{N} xy = b\frac{1}{N}\sum_{N} x^2 \tag{6.4}$$

$$b = \frac{\sum_{N} xy}{\sum_{N} x^2}.$$

If the points x_n and y_n represent samples of signals $x(n)$ and $y(n)$, then b is simply a scalar quantity describing the similarity of signal $x(n)$ with signal $y(n)$. If $b = 0$, there is no similarity; if $b = 1$, there is a strong correlation between one and the other.

6.3.1 Calculating Correlation

The *autocorrelation* of a sampled signal is defined as the product

$$R_{xx}(k) = \frac{1}{N}\sum_{n=0}^{N-1} x(n)x(n-k), \tag{6.5}$$

where k is the *lag* or delay. The correlation may be normalized by the mean square of the signal $R_{xx}(0)$ giving:

$$\rho_{xx}(k) = \frac{R_{xx}(k)}{R_{xx}(0)}. \tag{6.6}$$

The subscript xx denotes the fact that the signal is multiplied by a delayed version of itself. For a number of values of k, the result is a set of correlation values—an *autocorrelation vector*. If we limit ourselves to one-way relative delays in the lag parameter (as in most practical problems), we have a vector

$$\mathbf{r} = \begin{pmatrix} R(0) \\ R(1) \\ \vdots \\ R(N-1) \end{pmatrix} \tag{6.7}$$

Note that the term "correlation" is used, and not "covariance." These terms are used differently in different fields. In signal processing, correlation is normally used as we have defined it here. However, it is often necessary to remove the mean of the sample block before the correlation calculation. Sometimes, the correlation is normalized by a constant factor, as will be described shortly. This is simply a constant scaling factor, and does not affect the shape of the resulting correlation plot.

To illustrate the calculation of autocorrelation, suppose we have an $N = 4$ sample sequence. For $k = 0$, we compute the product of corresponding points $x(0)x(0) + x(1)x(1) + x(2)x(2) + x(3)x(3)$, as shown in the vectors below:

·	$x(0)$	$x(1)$	$x(2)$	$x(3)$	·
·	$x(0)$	$x(1)$	$x(2)$	$x(3)$	·

For $k = +1$, we compute the product $x(1)x(0) + x(2)x(1) + x(3)x(2)$,

·	$x(0)$	$x(1)$	$x(2)$	$x(3)$	·
·	·	$x(0)$	$x(1)$	$x(2)$	$x(3)$

For $k = -1$, we compute a similar product of aligned terms:

·	$x(0)$	$x(1)$	$x(2)$	$x(3)$	·
$x(0)$	$x(1)$	$x(2)$	$x(3)$	·	·

and so forth, for $k = \pm 2, \pm 3, \ldots$. Extrapolating the above diagrams, it may be seen that the lag may be positive or negative, and may in theory range over

$$\text{lag } k \rightarrow -N+1, -N+2, \cdots, -1, 0, 1, \cdots, N-2, N-1$$

The continuous-time equivalent of autocorrelation is the integral

$$R_{xx}(\lambda) = \frac{1}{\tau} \int_{\tau} x(t)\, x(t-\lambda)\, dt, \tag{6.8}$$

where λ represents the signal lag. The integration is taken over some suitable interval; in the case of periodic signals, it would be over one (or preferably, several) cycles of the waveform.

Autocorrelation is the measure of the similarity of a signal with a delayed version of the *same signal*. *Cross-correlation* is a measure of the similarity of a signal with a delayed version of *another signal*. Discrete cross-correlation is defined as

$$R_{xy}(k) = \frac{1}{N} \sum_{n=0}^{N-1} x(n)\, y(n-k), \tag{6.9}$$

where, again, k is the positive or negative lag. The normalized correlation in this case is:

$$\rho_{xy}(k) = \frac{R_{xy}(k)}{\sqrt{R_{xx}(0) R_{yy}(0)}}. \tag{6.10}$$

The cross-correlation is computed in exactly the same way as autocorrelation, using a different second signal. For a simple $N = 4$-point example, at a lag of $k = 0$, we have $x(0)y(0) + x(1)y(1) + \cdots$ as shown:

·	$x(0)$	$x(1)$	$x(2)$	$x(3)$	·
·	$y(0)$	$y(1)$	$y(2)$	$y(3)$	·

At $k = \pm 1$, we have:

·	$x(0)$	$x(1)$	$x(2)$	$x(3)$	·
·	·	$y(0)$	$y(1)$	$y(2)$	$y(3)$
·	$x(0)$	$x(1)$	$x(2)$	$x(3)$	·
$y(0)$	$y(1)$	$y(2)$	$y(3)$	·	·

The process may be continued for $k = \pm 2, \pm 3, \ldots$. Multiplying out and comparing the terms above, it may be seen that autocorrelation is symmetrical, whereas in general cross-correlation is not.

The continuous-time equivalent of cross-correlation is the integral

$$R_{xy}(\lambda) = \frac{1}{\tau}\int_{\tau} x(t)\,y(t-\lambda)\,dt. \tag{6.11}$$

where λ represents the signal lag. The integration is taken over some suitable interval; in the case of periodic signals, over one cycle of the waveform (or, preferably, many times the period).

The cross-correlation (without normalization) between two equal-length vectors x and y may be computed as follows:

```
% illustrating one- and two-sided correlation
% set up data sequences
N = 4;
x = [1:4]';
y = [6:9]';
% two-sided correlation
ccr2 = zeros(2 * N - 1, 1);
for lag = -N + 1 : N - 1
    cc = 0;
    for idx = 1 : N
        lagidx = idx - lag;
        if((lagidx >= 1) && (lagidx <= N))
            cc = cc + x (idx) * y (lagidx);
        end
    end
```

```
    ccr2(lag + N) = cc;
end
disp(ccr2)
% one-sided correlation
ccr1 = zeros(N, 1);
for lag = 0 : N - 1
    cc = 0;
    for idx = lag + 1:N
        lagidx = idx - lag;
        cc = cc + x(idx) * y(lagidx);
    end
    ccr1(lag + 1) = cc;
end
disp(ccr1)
```

Note that we need nested loops: the outer loop to iterate over the range of correlation lags we require, and an inner loop to calculate the summation over all terms in each vector. Also, no attempt has been to optimize these loops: the loops are coded exactly as per the equations. It would be more efficient in practice to start the indexing of the inner **for** loop at the very first nonzero product term, rather than utilize the **if** test within the for loop as shown for the two-sided correlation.

The difference in computation time (and/or processor loading) may be negligible for small examples such as this, but for very large signal vectors, such attention to efficiency pays off in terms of reduced execution time.

This slide-and-sum operation should remind us of convolution, which was introduced in Section 5.9. The key difference is that in convolution, we reverse one of the sequences at the start. Since MATLAB has the **conv**() function available, it must (implicitly) perform this reversal. Now, in computing the correlation, we do not require this reversal. Thus, if we reverse one of the sequences prior to convolution, we should negate the reversal within **conv**(), and thus obtain our correlation. This is illustrated in the following, where we use the "flip upside-down" function **flipud**() to time-reverse the vector:

```
% using convolution to calculate correlation
x = [1:4]';
y = [6:9]';
cc = conv(x, flipud(y));
disp(cc)
```

Note that we need to take particular care to use **flipud**() or **fliplr**() (flip left-right) as appropriate here, depending upon whether the input vectors are column vectors or row vectors.

6.3.2 Extending Correlation to Signals

The calculation examples in the preceding section are for very short vectors. When considering signal processing applications, the vectors are generally quite long (perhaps thousands of samples), and so it helps to visualize matters not as discrete points, but as waveforms. Now that we have the calculation tools at our disposal, we can extend our understanding to longer vectors computed automatically.

Figures 6.2 through 6.4 show how the correlation product is computed for sinusoidal vectors as signals. One important practical aspect is the fact that the data window is finite. This gives rise to a decreasing correlation function, as illustrated.

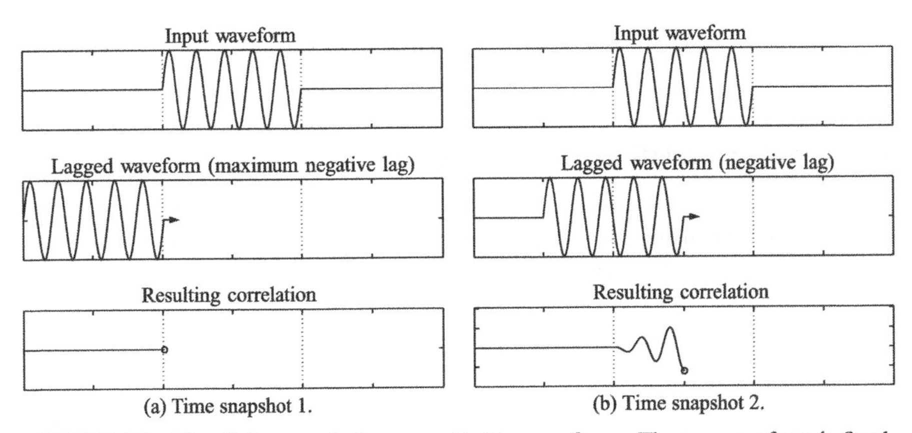

(a) Time snapshot 1. (b) Time snapshot 2.

FIGURE 6.2 Visualizing correlation as applied to waveforms. The top waveform is fixed, the middle is the moving waveform, and the bottom panel shows the resulting correlation of the waveforms. In panel a, we see the initial situation, where there is no overlap. In panel b, as the second waveform is moved further on, the correlation is increasing.

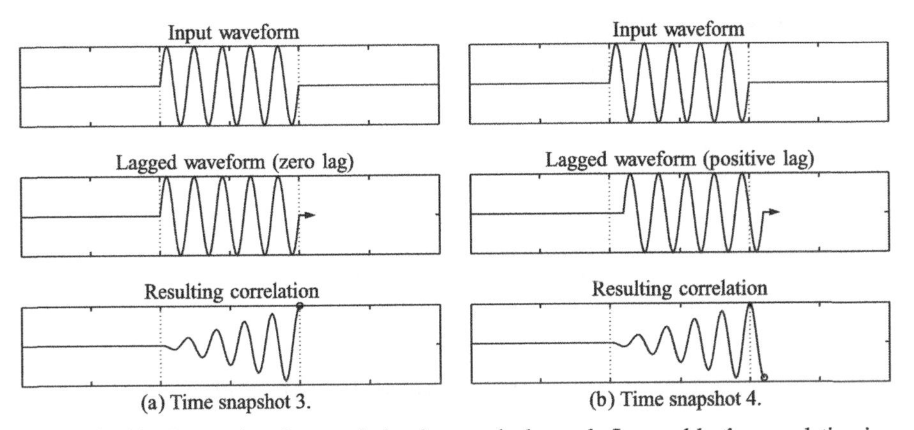

(a) Time snapshot 3. (b) Time snapshot 4.

FIGURE 6.3 In panel a, the correlation has reached a peak. In panel b, the correlation is just starting to decline.

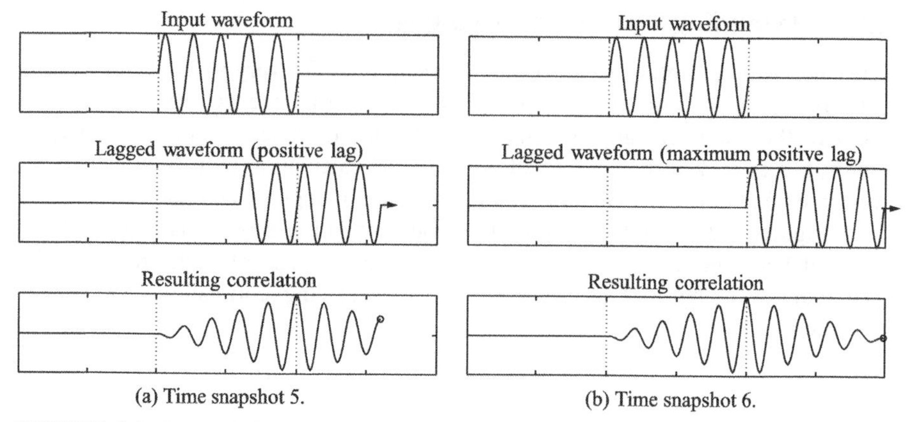

(a) Time snapshot 5. (b) Time snapshot 6.

FIGURE 6.4 In panel a, the correlation is declining further. In panel b, the correlation is almost back to zero.

6.3.3 Autocorrelation for Noise Removal

Suppose we have a sine wave, and wish to compute the autocorrelation. In MATLAB, this could be done using the code shown previously for correlation, but with a sine wave substituted for the input sample vector.

```
t = 0:pi/100:2*4*pi;
x = sin(t);
```

Note that the result has the form of a cosine waveform, tapered by a triangular envelope (as shown in Fig. 6.4). This is due to the fact that, in practice, we only have a finite data record to work with.

Mathematically, we can examine this result as follows. Using the definition of continuous autocorrelation

$$R_{xx}(\lambda)=\frac{1}{\tau}\int_{\tau}x(t)x(t-\lambda)\,dt. \tag{6.12}$$

If the sine wave is

$$x(t)=A\sin(\Omega t+\varphi), \tag{6.13}$$

then,

$$x(t-\lambda)=A\sin(\Omega(t-\lambda)+\varphi). \tag{6.14}$$

So,

$$R_{xx}(\lambda)=\frac{1}{\tau}\int_{0}^{\tau}A^2\sin(\Omega t+\varphi)\sin(\Omega(t-\lambda)+\varphi)\,dt. \tag{6.15}$$

Noting that τ is one period of the waveform, and that $\tau=2\pi/\Omega$, this simplifies to:

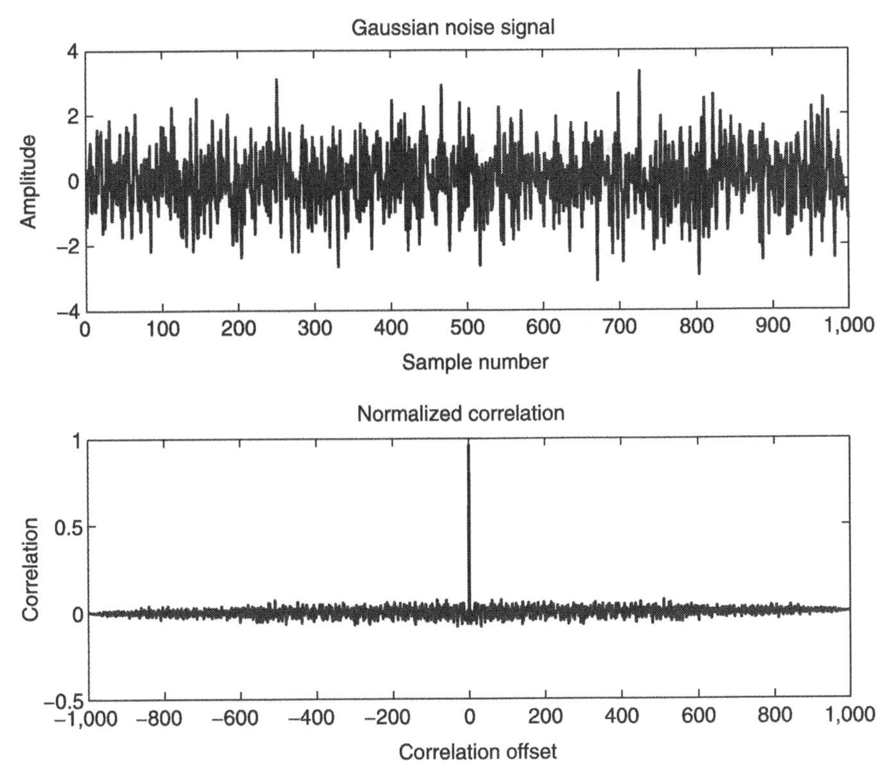

FIGURE 6.5 Autocorrelation of random Gaussian noise. This shows the autocorrelation function itself, which in theory is zero for all nonzero offsets. We also know that for additive signals, the correlations are effectively added together.

$$R_{xx}(\lambda) = \frac{A^2}{2}\cos\Omega\lambda. \tag{6.16}$$

This result can be generalized to other periodic signals. The autocorrelation will always retain the same period as the original waveform. What about additive random signals, which are often a good model for real-world signals?

If we have a Gaussian noise signal, its autocorrelation may be found in MATLAB using a random data source generated using `x = randn(1,000, 1);` The result, shown in Figure 6.5, shows that there is only one strong component, at a lag of zero. The autocorrelation at lag zero is simply the mean-square of the signal, since we are multiplying each sample by itself.

However, at lags other than zero, we are multiplying two random values. On average, the summation of these (the autocorrelation at that point) will equal zero. Thus, the autocorrelation at lag zero of zero-mean white Gaussian noise is equal to the variance:

$$R_{\text{Gaussian}}(k) = \begin{cases} \sigma^2 & : \ k = 0 \\ 0 & : \ \text{elsewhere} \end{cases}. \tag{6.17}$$

In practice, only an approximation to the ideal case exists. This is simply because we have a finite—not infinite—number of samples to average over.

6.3.4 Correlation for Signal Extraction

We now have two useful pieces of information: first, that the autocorrelation of a periodic signal preserves the periodicity; second, that autocorrelation of a random signal is zero at all lags except for a lag of zero.

Suppose we have a zero-mean noise-corrupted signal $y(n)$, which is comprised of a (desired) periodic signal $x(n)$ with additive, uncorrelated noise $v(n)$. The observed signal may be written

$$y(n) = x(n) + v(n). \tag{6.18}$$

Mathematically, the autocorrelation of the observed signal is

$$\begin{aligned} R_{yy}(k) &= \frac{1}{N}\sum_N (x(n)+v(n))(x(n-k)+v(n-k)) \\ &= \frac{1}{N}\sum_N (x(n)x(n-k)+x(n)v(n-k) \\ &\quad + v(n)x(n-k)+v(n)v(n-k)) \end{aligned} \tag{6.19}$$

$$\begin{aligned} &= \frac{1}{N}\sum_N x(n)x(n-k) + \frac{1}{N}\sum_N x(n)v(n-k) \\ &\quad + \frac{1}{N}\sum_N v(n)x(n-k) + \frac{1}{N}\sum_N v(n)v(n-k) \\ &= R_{xx}(k) + \cancel{R_{xv}(k)}^0 + \cancel{R_{vx}(k)}^0 + \cancel{R_{vv}(k)}^0 \end{aligned} \tag{6.20}$$

Thus, we have an equation for the signal cross-correlation in terms of the various individual signal-noise cross-correlation terms. As we have shown, it is simply an additive sum. If the noise $v(n)$ is uncorrelated with itself then

$$R_{vv}(k) = 0 \quad \text{for all } k \neq 0$$

If the noise is uncorrelated with the periodic signal, then for zero-mean signals the cross-correlation ought to equal zero, regardless of the lag chosen:

$$\begin{aligned} R_{xv}(k) &= 0 \quad \text{for all } k \\ R_{vx}(k) &= 0 \quad \text{for all } k \end{aligned}$$

Hence, all terms but the first in Equation 6.20 cancel out, leaving only the autocorrelation of the signal itself.

A sample input signal is shown in Figure 6.6. Only the noisy waveform is assumed to be available, and so determining the period in the time domain is likely to be quite error-prone. Figure 6.7 shows the autocorrelation of the clean signal (upper panel) and the autocorrelation of the noisy signal (lower). In theory, the noise only affects the correlation at a lag of zero. Hence, the periodicity of the original noise-free signal may be determined with greater accuracy.

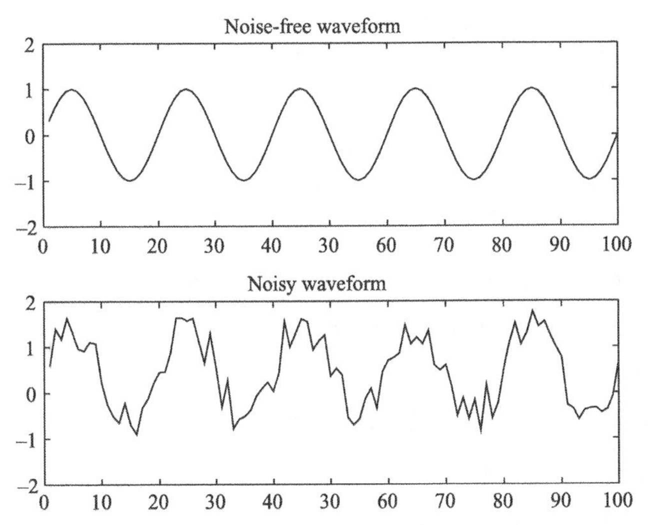

FIGURE 6.6 Time-domain waveforms: clean (upper) and noise-corrupted (lower). Ideally, we wish to determine the frequency of the clean signal from samples of the noisy signal only. The presence of noise makes this difficult to do.

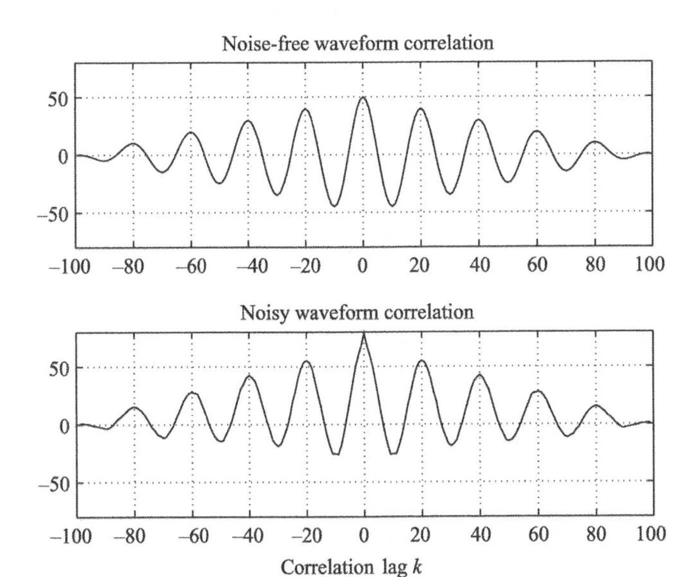

FIGURE 6.7 Correlation waveforms: clean (upper) and noise corrupted (lower). Apart from the addition at zero lag, the waveforms are nearly identical. The zero crossings of the correlation waveform are now well-defined, making the frequency determination much more reliable.

6.3.5 Correlation in System Identification

In many real-world applications, we wish to determine the transfer function of a system. Examples include telecommunications (where, e.g., the transfer function of a channel may be determined upon initialization of a system) and audio processing of room acoustics.

Signal processing techniques may be applied to determine the impulse response of a system. This leads to a characterization of the frequency response, and frequency compensation may be performed in the form of equalizing or pulse-shaping filters.

The impulse response of a system, and hence its FIR coefficients, may obviously be found by subjecting the system to an impulse input. However, in most practical applications, it is not feasible to subject a system to an impulse input. Hence, an alternative is required.

To begin, consider the output of a system subjected to an input $x(n)$ as:

$$y(n) = b_0 x(n) + b_1 x(n-1) + b_2 x(n-2) + \cdots. \tag{6.21}$$

Now multiply each side by $x(n)$ to obtain

$$y(n)x(n) = b_0 x(n)x(n) + b_1 x(n-1)x(n) + b_2 x(n-2)x(n) + \cdots.$$

Summing over a large number of samples N, we have:

$$\begin{aligned}
\sum_N y(n)x(n) &= \sum_N \big(b_0 x(n)x(n) + b_1 x(n-1)x(n) \\
&\quad + b_2 x(n-2)x(n) + b_3 x(n-3)x(n) + \cdots\big) \\
&= \sum_N b_0 x(n)x(n) + \sum_N b_1 x(n-1)x(n) \\
&\quad + \sum_N b_2 x(n-2)x(n) + \sum_N b_3 x(n-3)x(n) + \cdots \\
&= b_0 \sum_N x(n)x(n) + b_1 \sum_N x(n-1)x(n) \\
&\quad + b_2 \sum_N x(n-2)x(n) + b_3 \sum_N x(n-3)x(n) + \cdots.
\end{aligned}$$

The summation terms may be recognized as scaled correlations. If $x(n)$ is Gaussian noise, then all terms on the right-hand side except $\Sigma_N x(n)x(n)$ will be zero, leaving

$$\sum_N y(n)x(n) = b_0 \sum_N x(n)x(n). \tag{6.22}$$

And so b_0 is found as

$$b_0 = \frac{\sum_N y(n)x(n)}{\sum_N x(n)x(n)} \tag{6.23}$$

Similarly, multiplying both sides by $x(n-1)$ and summing over a large number of samples N, the system coefficient b_1 is found as

$$b_1 = \frac{\sum_N y(n)x(n-1)}{\sum_N x(n-1)x(n-1)} \tag{6.24}$$

Continuing on, the terms b_2, b_3, b_4, … may be found in a similar manner.

Thus, the system FIR model may be approximated subject to the assumptions detailed regarding the autocorrelation of the input and cross-correlation of the input and output. It is important to understand the assumptions made at the outset for developing the model, since these are also the model's limitations.

6.4 LINEAR PREDICTION

We now turn to the problem of estimating the future values of a sequence. The solution to this problem requires the correlation values—not the single correlation values as discussed earlier, but rather a matrix of correlation values. This problem occurs often in signal processing, and may be found in signal coding and compression, filtering, and other application areas.

To derive the problem and its general solution, suppose we have samples of a sequence $x(n)$, and want to predict the value at the next sample instant $\hat{x}(n)$. The prediction error is:

$$e(n) = x(n) - \hat{x}(n). \tag{6.25}$$

One obvious way to form the estimate $\hat{x}(n)$ is to base the estimate on past samples. If we estimate one sample ahead from the most recent sample, we could form the linear estimate $w_0 x(n-1)$, where w_0 is a constant to be determined. In this case, the estimate and error are:

$$\begin{aligned} \hat{x}(n) &= w_0 x(n-1) \\ e(n) &= x(n) - \hat{x}(n) \\ &= x(n) - w_0 x(n-1). \end{aligned} \tag{6.26}$$

The w_0 may be thought of as a multiplying "weight," and for more than one w, the prediction of the output is formed by the linear weighted sum of past samples. So we need to determine the value of the constant parameter w_0. The instantaneous squared error is

$$e^2(n) = (x(n) - w_0 x(n-1))^2. \tag{6.27}$$

The error is squared because it could be positive or negative, and the squaring operation also allows us to form the derivative easily when finding the optimal solution mathematically. This error value is only for one sample, but what we really need is the minimum error averaged over a block of N samples. This is sometimes called the "cost function." So the average squared error is

$$\begin{aligned} \overline{e^2} &= \frac{1}{N} \sum_N e^2(n) \\ &= \frac{1}{N} \sum_N (x(n) - w_0 x(n-1))^2 \\ &= \frac{1}{N} \sum_N (x^2(n) - 2x(n) w_0 x(n-1) + w_0^2 x^2(n-1)). \end{aligned} \tag{6.28}$$

To minimize the cost function with respect to the parameter w_0, we take the derivative

$$\frac{d\overline{e^2}}{dw_0} = \frac{1}{N}\sum_N \left(0 - 2x(n)x(n-1) + 2w_0x^2(n-1)\right). \tag{6.29}$$

To find the minimum average error, we set the derivative equal to zero

$$\frac{d\overline{e^2}}{dw_0} = 0.$$

Expanding Equation 6.29,

$$\frac{1}{N}\sum_N x(n)x(n-1) = w_0^* \frac{1}{N}\sum_N x^2(n-1), \tag{6.30}$$

solving this for the optimal predictor coefficient w_0^* yields:

$$w_0^* = \frac{\frac{1}{N}\sum_N x(n)x(n-1)}{\frac{1}{N}\sum_N x^2(n-1)}. \tag{6.31}$$

We assume that the summation has to be taken over a sufficiently large number of samples to form the prediction with reasonable accuracy. We can use our previous definition of correlation to simplify the above:

$$R(0) \approx \frac{1}{N}\sum_N x^2(n-1) \tag{6.32}$$

$$R(1) \approx \frac{1}{N}\sum_N x(n)x(n-1). \tag{6.33}$$

Hence, the optimal first-order predictor parameter w_0^* is:

$$w_0^* = \frac{R(1)}{R(0)}. \tag{6.34}$$

For a better estimate, we could try to extend the prediction by using a second-order term. Following similar reasoning, for a second-order predictor, we have the estimate based on a linear combination of the last two samples

$$\begin{aligned}
\hat{x}(n) &= w_0x(n-1) + w_1x(n-2) \\
\therefore e(n) &= x(n) - \hat{x}(n) \\
&= x(n) - (w_0x(n-1) + w_1x(n-2)) \\
\therefore e^2(n) &= [x(n) - (w_0x(n-1) + w_1x(n-2))]^2.
\end{aligned} \tag{6.35}$$

Again, over a sufficiently large number of samples, the average squared error is

$$\overline{e^2} = \frac{1}{N}\sum_N e^2(n) \tag{6.36}$$

$$= \frac{1}{N}\sum_{N}\left(x(n)-\left(w_0 x(n-1)+w_1 x(n-2)\right)\right)^2. \tag{6.37}$$

Once again, to minimize the average error with respect to the predictor parameters w_0 and w_1, we take derivatives. This time, we need partial derivatives with respect to each parameter in turn. We can employ the chain rule of calculus

$$\left(\frac{\partial e}{\partial w}=\frac{\partial e}{\partial u}\frac{\partial u}{\partial w}\right)$$

to find the derivative

$$\frac{\partial \overline{e^2}}{\partial w_0}=\frac{1}{N}\sum_{N}\{2[x(n)-(w_0 x(n-1)+w_1 x(n-2))][-x(n-1)]\}. \tag{6.38}$$

Setting this derivative to zero $\left(\partial \overline{e^2}/\partial w_0 = 0\right)$, as we would with any minimization problem, it is possible to determine an expression for the optimal predictor w_0^*:

$$\frac{1}{N}\sum_{N}\{2[x(n)-(w_0^* x(n-1)+w_1^* x(n-2))]\times[-x(n-1)]\}=0. \tag{6.39}$$

Expanding the summation through individual terms gives

$$\begin{aligned}\frac{1}{N}\sum_{N}x(n)x(n-1) &= w_0^*\frac{1}{N}\sum_{N}x(n-1)x(n-1)\\ &\quad + w_1^*\frac{1}{N}\sum_{N}x(n-1)x(n-2).\end{aligned} \tag{6.40}$$

Using autocorrelation functions as before,

$$R(1)=w_0^* R(0)+w_1^* R(1). \tag{6.41}$$

Similarly, by taking the partial derivative with respect to w_1 yields

$$R(2)=w_0^* R(1)+w_1^* R(0). \tag{6.42}$$

Thus, we have two equations in two unknowns. In matrix form, these equations are:

$$\begin{pmatrix}R(1)\\R(2)\end{pmatrix}=\begin{pmatrix}R(0) & R(1)\\R(1) & R(0)\end{pmatrix}\begin{pmatrix}w_0^*\\w_1^*\end{pmatrix}, \tag{6.43}$$

which is:

$$\mathbf{r}=\mathbf{R}\mathbf{w}^*, \tag{6.44}$$

where $\mathbf{r}$ is a vector of correlation coefficients starting at $R(1)$, $\mathbf{R}$ is a matrix of correlation coefficients, and $\mathbf{w}^*$ the vector of optimal predictor coefficients, all with terms as defined in Equation 6.43.

Taking the z transform of the predictor Equation 6.35, we have:

$$\begin{aligned}E(z) &= X(z)-\hat{X}(z)\\ &= X(z)-\left(w_0 X(z)z^{-1}+w_1 X(z)z^{-2}\right)\\ &= X(z)\left(1-\left(w_0 z^{-1}+w_1 z^{-2}\right)\right)\end{aligned} \tag{6.45}$$

$$\therefore \frac{X(z)}{E(z)} = \frac{1}{1-\left(w_0 z^{-1} + w_1 z^{-2}\right)}. \tag{6.46}$$

This gives us the required transfer function for the optimal predictor. To test this, we generate a data sequence of N samples using a second-order system with complex poles, as shown in Listing 6.1. This gives us a benchmark, since the optimal filter ought to approach the coefficients of the system which generated the signal.

The poles are defined and the coefficients calculated and stored in a. From this, the system excitation (here a simple random sequence) produces the observed system output x. Then we calculate the correlations and form the correlation matrix, and finally solve for the coefficients.

Note how the output estimate is generated using the linear predictor coefficients. We use a filter as defined by Equation 6.46, with coefficients determined using Equation 6.43.

Listing 6.1 Testing the Linear Prediction Theory

```
N = 1000;
r = 0.9                          % pole angle
omega = pi/10;                   % pole radius
p = r * exp(j * omega);
a = poly([p conj(p)]);
roots(a)
e = randn(N, 1);                 % system input
x = filter(1, a, e);             % response to input
% calculate autocorrelations
R0 = sum(x . * x)/N;
R1 = sum(x(1:N - 1) . * x(2:N))/N;
R2 = sum(x(1:N - 2) . * x(3:N))/N;
% autocorrelation matrix & vector
R = [R0 R1; R1 R0];
r = [R1; R2];
% optimal predictor solution
w = inv(R) * r;
% optimal predictor parameters as a filter
ahat = [1 ; -w];
% predicted sample values
xhat = filter(1, ahat, e);
a =
          1.0000         -1.7119          0.8100
ahat' =
          1.0000         -1.7211          0.8287
```

6.4.1 Geometrical Interpretation

We now digress to examine a geometrical interpretation of the linear prediction problem. It gives the same result, but using an entirely different approach. The insight gained will be useful when studying optimal filtering and other signal processing problems.

This method does not use a calculus-based minimization approach, but rather relies on the *orthogonality* of vectors. Consider again a simple second-order predictor given by

$$\hat{x}(n) = w_0 x(n-1) + w_1 x(n-2). \tag{6.47}$$

If the prediction were perfect, we could write a matrix equation for each predicted sample as follows.

$$\begin{pmatrix} x(2) \\ x(3) \\ \vdots \\ x(N-1) \end{pmatrix} = \begin{pmatrix} x(1) & x(0) \\ x(2) & x(1) \\ \vdots & \vdots \\ x(N-2) & x(N-3) \end{pmatrix} \begin{pmatrix} w_0 \\ w_1 \end{pmatrix}. \tag{6.48}$$

This may seem a little strange, but remember that the product of an $N \times 2$ matrix and a 2×2 matrix is permissible (since the inner dimensions are both 2), and the result is the product of the outer dimensions, or $N \times 2$ (strictly, as we have defined the terms here, $(N - 2) \times 2$, because we assume the first two samples are not able to be predicted).

Here we have made the assumption that the first predictable sample is $x(2)$, since we need the two prior samples $x(1)$ and $x(0)$ to form a valid prediction. This may not always be the case; as we will see later, the output quantity (here the prediction) may be formed differently, using successive overlapping blocks. The above equation may be written as a matrix-vector product

$$\mathbf{y} = \mathbf{M}\mathbf{w}, \tag{6.49}$$

where the prediction $\mathbf{y}$ is defined to be a vector

$$\mathbf{y} = \begin{pmatrix} x(2) \\ x(3) \\ \vdots \\ x(N-1) \end{pmatrix}. \tag{6.50}$$

The matrix $\mathbf{M}$ has columns formed by the input sequence, aligned to the predictor coefficients

$$\mathbf{M} = \begin{pmatrix} x(1) & x(0) \\ x(2) & x(1) \\ \vdots & \vdots \\ x(N-2) & x(N-3) \end{pmatrix}. \tag{6.51}$$

The predictor coefficients are in the form of a column vector

$$\mathbf{w} = \begin{pmatrix} w_0 \\ w_1 \end{pmatrix} \tag{6.52}$$

Unless we are able to perfectly predict the next sample, there will be no values of w_0 and w_1 that satisfy equation 6.49 exactly. This is akin to the earlier example of the straight-line fit from measured or experimental data with error—unless the error is zero, the points will not lie precisely on the line of best fit.

Given this matrix interpretation, the quantity we seek to minimize is the distance between two vectors **y** and **Mw**. This is the Euclidean difference defined as

$$\|\mathbf{y} - \mathbf{M}\mathbf{w}\|. \tag{6.53}$$

This is illustrated in Figure 6.8. The vector **y** lies above the plane, and the vector formed by the matrix-vector product **Mw** lies on the plane. The quantity we need to minimize is really the distance $\|\mathbf{y} - \mathbf{M}\mathbf{w}\|$, and this can only occur when it forms a vector perpendicular to the plane. So if we define the optimal solution as the one satisfying this criterion, the optimal value **w*** is

$$\mathbf{w}^* = \begin{pmatrix} w_0^* \\ w_1^* \end{pmatrix}. \tag{6.54}$$

Referring to Figure 6.8, **y** is fixed in space, and as **w** varies, the product **Mw** ranges over the plane shown. It is said to form a subspace (the column space) of **M**.

If **y** – **Mw** is to have a minimum length, it must be perpendicular to the plane. Let **w*** be such that **y** – **Mw*** is perpendicular to the plane. Now, if two vectors are orthogonal, we know that their inner or vector dot product must be zero. So the inner (vector dot) product of **y** – **Mw** with **Mw** must be zero.

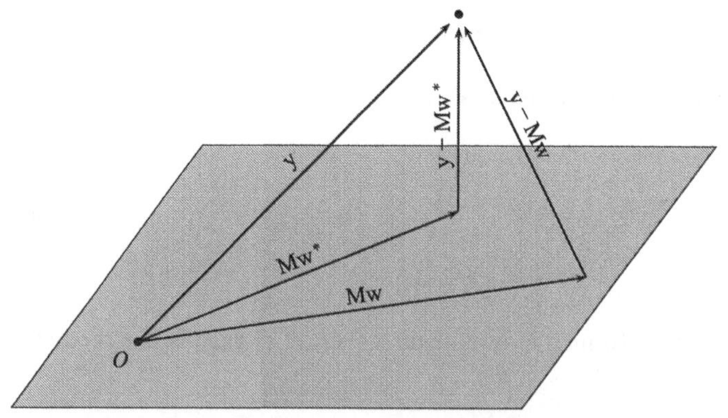

FIGURE 6.8 Illustrating the orthogonality of vectors. The vector y is the desired output vector. We need to reach it as closely as possible using a vector Mw, where M is formed from the input samples (measured), and w is the vector of filter weights (variable). The closest match is when the vector difference y – Mw is perpendicular to the plane defined by Mw. In that case, the vector Mw* is the closest in a "least-squares" sense, and w* is the optimal weight vector.

That is

$$(\mathbf{Mw}) \cdot (\mathbf{y} - \mathbf{Mw}^*) = 0 \tag{6.55}$$

$$\therefore (\mathbf{Mw})^T (\mathbf{y} - \mathbf{Mw}^*) = 0. \tag{6.56}$$

In the second line, we have used the fact that $\mathbf{a} \cdot \mathbf{b} = \mathbf{a}^T\mathbf{b}$. Since for any matrices **A** and **B**:

$$(\mathbf{AB})^T = \mathbf{B}^T \mathbf{A}^T, \tag{6.57}$$

where T represents the transposition operator (write rows as columns, write columns as rows), the orthogonal solution reduces to:

$$\mathbf{w}^T \mathbf{M}^T (\mathbf{y} - \mathbf{Mw}^*) = 0 \tag{6.58}$$

$$\mathbf{w}^T \left(\mathbf{M}^T \mathbf{y} - \mathbf{M}^T \mathbf{Mw}^* \right) = 0. \tag{6.59}$$

For this to hold, as long as $\mathbf{w}^T$ is nonzero, the term in brackets $(\mathbf{M}^T\mathbf{y} - \mathbf{M}^T\mathbf{Mw}^*)$ must be zero, and so:

$$\mathbf{M}^T \mathbf{Mw}^* = \mathbf{M}^T \mathbf{y}. \tag{6.60}$$

Thus, the unique solution is:

$$\mathbf{w}^* = \left(\mathbf{M}^T \mathbf{M} \right)^{-1} \mathbf{M}^T \mathbf{y}. \tag{6.61}$$

The quantity $\mathbf{M}^+ = (\mathbf{M}^T\mathbf{M})^{-1}\mathbf{M}^T$ is also called the pseudoinverse. The following shows how to extend the previous linear-prediction example to implement this matrix-based solution.

```
y = x(3:N) ;
M = [x( 2:N - 1)    x(1:N - 2)] ;
w = inv(M' * M) * M' * y
w =
   1.7222
  -0.8297
```

This is essentially a different route to the same end point. The numerical result as described will be slightly different to the earlier numerical result using derivatives, due to the way the starting points for the sequence have been defined.

6.5 NOISE ESTIMATION AND OPTIMAL FILTERING

We now extend the above and apply the basic concepts to the problem of removing noise from a signal, where we have access to an estimate the noise. Such an approach is called a Wiener filter.[1] Consider the problem where we want to minimize the noise in a signal, and we are able to sample the noise component separately. One received channel will comprise the signal plus noise, the other will consist predominantly of

[1]Named after Norbert Wiener.

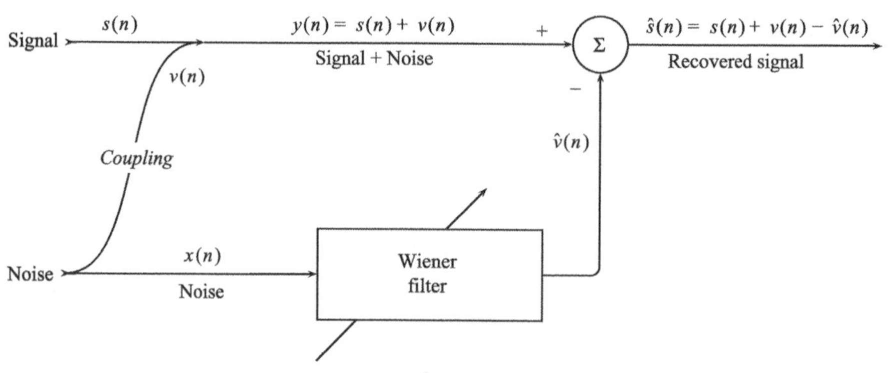

FIGURE 6.9 Wiener filtering for noise cancellation. The filter weights are updated for a block using both the noisy signal and the (approximately) noise-only signal. It is assumed that we do not have access to the clean signal, only the noise-contaminated signal. Furthermore, the noise signal is only an estimate, and we use its statistical properties to determine the optimal filter.

noise only. Of course, the noise component is not precisely, sample-by-sample, identical to that received by the signal sensor. If it were, the problem would reduce to a simple sample-by-sample subtraction.

The situation is illustrated in Figure 6.9. The noise source is correlated to some degree with the measured signal, and we represent this coupling via a linear transfer function. If we can estimate the linear transfer function, we can estimate the noise as received by the signal sensor, and thus subtract it out.

In the diagram, the output signal $\hat{s}(n)$ at instant n is

$$\hat{s}(n) = y(n) - \mathbf{w}^T \mathbf{x}(n), \tag{6.62}$$

where the weight vector $\mathbf{w}$ is

$$\mathbf{w} = \begin{pmatrix} w_0 \\ w_1 \\ \vdots \\ w_{L-1} \end{pmatrix}, \tag{6.63}$$

and the sample vector $\mathbf{x}_n$ is defined as the block of L samples starting at instant n

$$\mathbf{x}_n = \begin{pmatrix} x(n) \\ x(n-1) \\ \vdots \\ x(n-L+1) \end{pmatrix}. \tag{6.64}$$

As before, we square the output to get an indicative error power

$$\hat{s}^2(n) = (y(n) - \mathbf{w}^T \mathbf{x}(n))^2 \tag{6.65}$$

$$= y^2(n) - 2y(n)\mathbf{w}^T \mathbf{x}(n) + (\mathbf{w}^T \mathbf{x}(n))^2. \tag{6.66}$$

Now:

$$\begin{aligned}\hat{s}(n) &= y(n) - \hat{v}(n) \\ &= (s(n) + v(n)) - \hat{v}(n) \\ &= s(n) + (v(n) - \hat{v}(n)).\end{aligned} \tag{6.67}$$

Hence:

$$\begin{aligned}\hat{s}^2(n) &= (s(n) + v(n) - \hat{v}(n))^2 \\ &= s^2(n) + 2s(n)(v(n) - \hat{v}(n)) + (v(n) - \hat{v}(n))^2.\end{aligned} \tag{6.68}$$

Taking the average or mathematical expectation:

$$\begin{aligned}E\{\hat{s}^2(n)\} &= E\{s^2(n)\} + 2E\{s(n)v(n)\} - 2E\{s(n)\hat{v}(n)\} + E\{v(n) - \hat{v}(n))^2\} \\ &\approx E\{s^2(n)\} + E\{(v(n) - \hat{v}(n))^2\}.\end{aligned} \tag{6.69}$$

For perfect cancellation, the noise estimate $\hat{v}(n)$ equals the noise $v(n)$, and the last term on the right is zero. Since $E\{s^2(n)\}$ is fixed, minimizing the difference between noise and noise estimate corresponds to minimizing the output power $E\{\hat{s}^2(n)\}$. So we must minimize the output power, and to do this, we take Equation 6.66, and take the expectation over the left- and right-hand sides to give:

$$E\{s^2(n)\} = E\{y^2(n)\} - 2E\{y(n)\mathbf{w}^T\mathbf{x}(n)\} + E\{(\mathbf{w}^T\mathbf{x}(n))^2\} \tag{6.70}$$

$$= \sigma_y^2 - 2E\{y(n)\mathbf{w}^T\mathbf{x}(n)\} + E\{\mathbf{w}^T\mathbf{x}(n)\mathbf{x}^T(n)\mathbf{w}\}. \tag{6.71}$$

We have a number of vector-vector and scalar-vector products here,[2] so for simplicity, let us define the vector and matrix products:

$$\mathbf{r} = E\{y(n)x(n)\} \tag{6.72}$$

$$\mathbf{R} = E\{\mathbf{x}(n)\mathbf{x}^T(n)\}. \tag{6.73}$$

So then the cost function J that we need to minimize is:

$$J = E\{\hat{s}^2(n)\} \tag{6.74}$$

$$= \sigma_y^2 - 2\mathbf{r}^T\mathbf{w} + \mathbf{w}^T\mathbf{R}\mathbf{w}. \tag{6.75}$$

Using the identities

$$\frac{\partial}{\partial \mathbf{w}}\left(\mathbf{w}^T\mathbf{r}\right) = \mathbf{r} \tag{6.76}$$

$$\frac{\partial}{\partial \mathbf{w}}(\mathbf{w}^T\mathbf{R}\mathbf{w}) = 2\mathbf{R}\mathbf{w}, \tag{6.77}$$

the derivative becomes

$$\frac{\partial J}{\partial \mathbf{w}} = -2\mathbf{r} + 2\mathbf{R}\mathbf{w}. \tag{6.78}$$

Setting this to zero to find the minimum-power solution:

$$\mathbf{w} = \mathbf{R}^{-1}\mathbf{r}. \tag{6.79}$$

[2]Recall that scalars are italic font, column vectors are lower case bold typeface, and matrices bold uppercase.

Thus, we have a matrix solution to the estimation problem. The solution relies, again, on correlations formed as a vector and matrix. We can employ the **reshape()** function in MATLAB to aid in forming the matrices, as shown in Listing 6.2. In this example, we have a noise-free sinusoid which is *not* visible, and a noise-corrupted sinusoid which *is* visible. The noise is also measured separately, although of course it is only an estimate of the noise. Note how the **reshape()** operator is used to form the sequence of $\mathbf{x}(n)$ and $\mathbf{y}(n)$ vectors, as matrices.

Listing 6.2 Testing the Wiener Filter Theory

```
N = 10,000;                            % number of samples in test
x = randn(1, N);                       % noise in system
n = 0:N - 1;
yc = 2.0 * sin(2*pi*n/(N - 1)*5);      % clean signal (not visible)
% noise coupling-unknown transfer function
b = [1 0.8 -0.4 0.1];
v = filter(b, 1, x);
y = yc + v;
L = 4; % coefficient length
% Wiener optimal solution
% reshape signal vectors in order to calculate covariances
ym = reshape(y, L, N/L);    % the output (y) vector
xm = reshape(x, L, N/L);    % the input (x) vector
xv = xm(1 ,:);             % y vector at intervals corresponding
                            % to the start of the xm's above
R = (xm * xm')/(N/L);       % E(X X^T)
r = (ym * xv')/(N/L);       % E(y X)
% optimal weights
wopt = inv(R) * r;
% noise estimate
vest = filter(wopt, 1, x);
% error = signal - noise estimate
e = y - vest;
wopt =
          1.0405
          0.7406
          -0.3747
          0.0843
```

The result is that `wopt` forms an estimate of `b`. This is shown for a block of noise-corrupted data in Figure 6.10. The correlations are formed for the block of input data samples.

The process of finding the optimal solution can be imagined in two dimensions as shown in Figure 6.11 as searching for the bottom of a bowl-shaped surface, where the two orthogonal axes form the weight vectors, and the cost function forms the third axis.

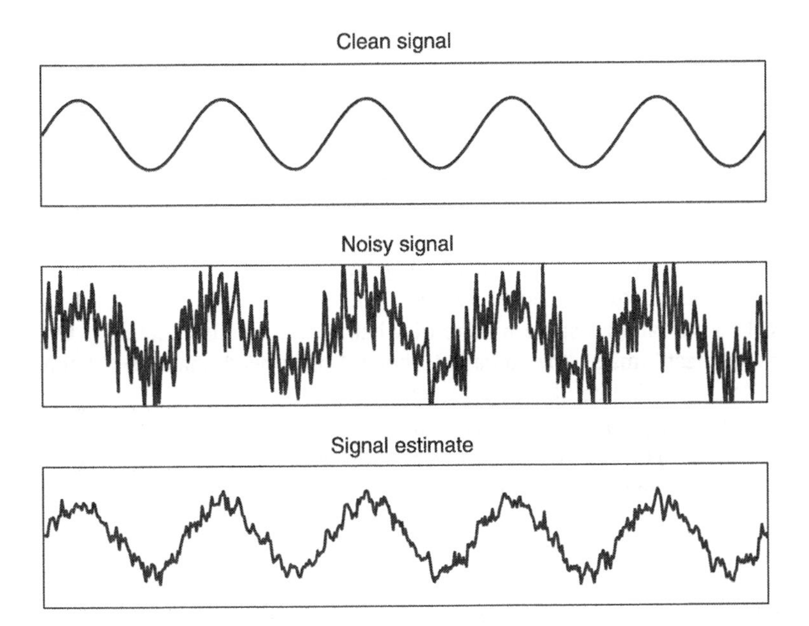

FIGURE 6.10 Wiener filtering example. The clean signal is what we desire, but it is not available to us. The noisy signal is available, and after the optimal filter is applied, we have the signal estimate.

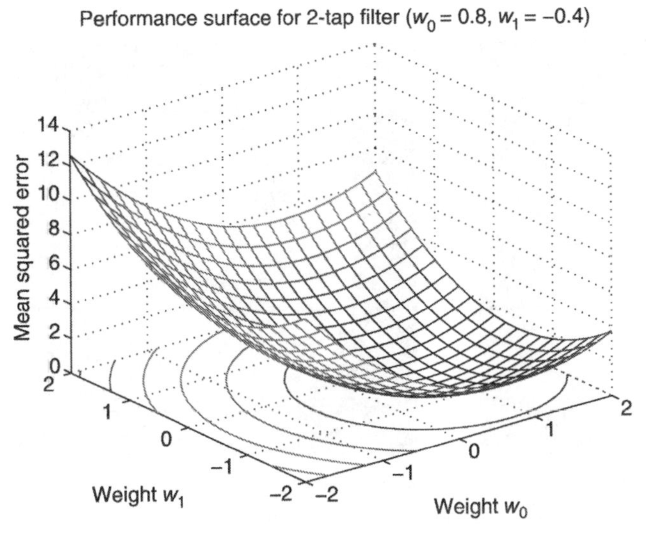

FIGURE 6.11 The performance surface in terms of the filter coefficients. For two coefficients, we can imagine the error surface as a bowl-shape with the cost function as the vertical axis. The goal is to find the minimal cost, and hence the corresponding filter weights.

6.6 TOMOGRAPHY

The temporal correlation of sampled data is clearly important in many signal processing applications, such as filtering and system identification as discussed in the previous sections. We now turn to the concept of spatial correlation, and in particular, we investigate the concept of tomography, which enables us to "see inside" an object with only external measurements. This has enabled a great many medical and other applications in recent decades. We cover the basic principles, and introduce the class of algorithm for reconstruction called *backprojection*.

Figure 6.12 gives an indicative picture of what we are trying to achieve. On the left we have a "test" image, the so-called Shepp–Logan head phantom (Shepp and Logan 1974), which is intended to represent the relative densities of tissue in the human head. On the right is a reconstructed image, formed using only projections through the test image. The relative brightness of the various ellipses is intended to be proportional to the density encountered during passage through the object's components.

In a real application, the various projections are taken by a measurement technique such as X-radiation of a cross-section. The projections are illustrated in Figure 6.13. Clearly, many cross-sections are required to build up a reconstruction in two dimensions of the interior densities of the object.

The following sections first introduce the physical concepts, and then the signal processing approach to solving the problem.

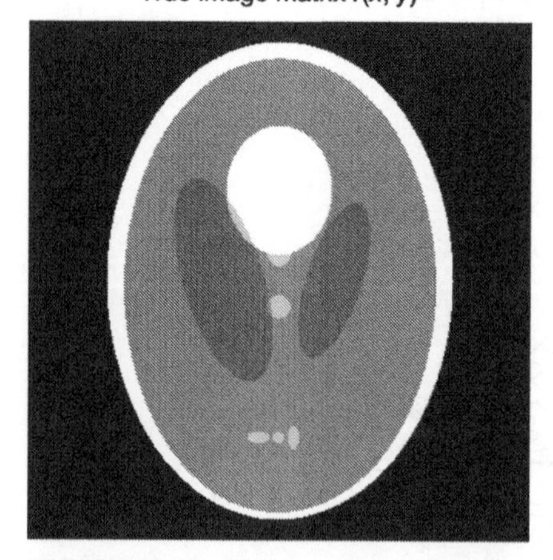

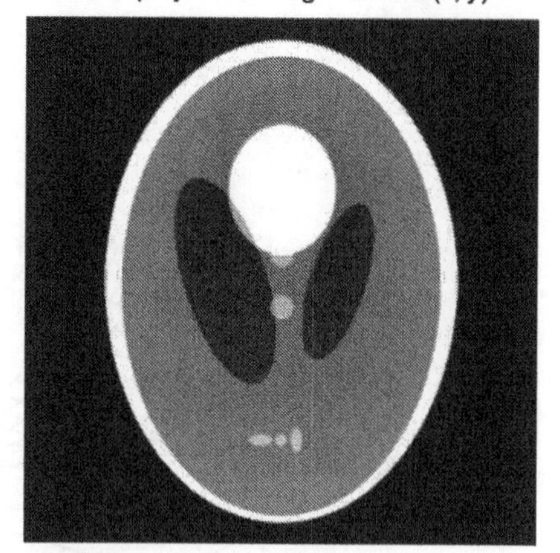

FIGURE 6.12 The tomography problem and one possible approach using the backprojection algorithm. The so-called head phantom (Shepp and Logan 1974) consists of ellipses of varying densities, intended to represent a cross-section of human tissue. This gives a basis on which to compare algorithms. The reconstructed interior projection is shown on the right.

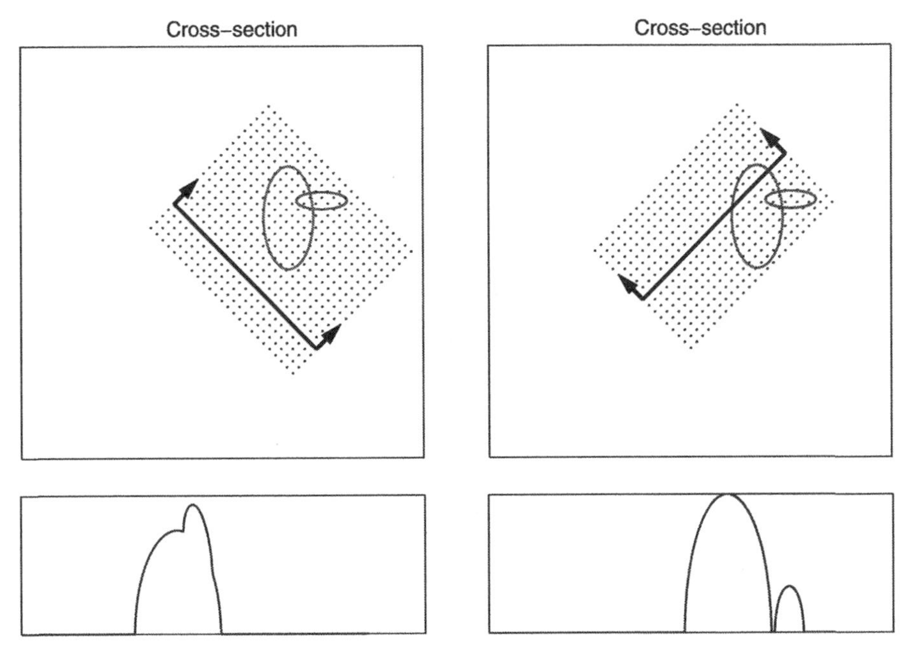

FIGURE 6.13 Projection through an object composed of two ellipses. The upper plots are cross-sections viewed from above, showing all the projections in the direction of the arrows. The lower plots show the projected value, which is proportional to both the amount of time a ray spends in the particular ellipse, as well as the density of that ellipse.

6.6.1 Radon Transform

Each of the cross-sections taken through the object to be imaged may be thought of as a line integral, with the relative strength of the output signal proportional to the density, as illustrated in Figure 6.13. Obviously this is a simplification of the underlying physics, but is sufficient for the purposes of developing the reconstruction algorithms. Further details on the physics of the measurement process may be found in many sources, such as Kak and Slaney (1988), Rosenfeld and Kak (1982) and Jain (1989).

The Radon transformation was developed some time ago, as a purely mathematical concept (see the translation in Radon 1986). The projection through an object is, in effect, an accumulation over each parallel ray passing through the object.[3] This projection results in a density cross-section, as shown. Mathematically, the Radon transform provides a function $P_\theta(s)$, where s is the perpendicular

[3]There are other implementations such as fan-beam projections.

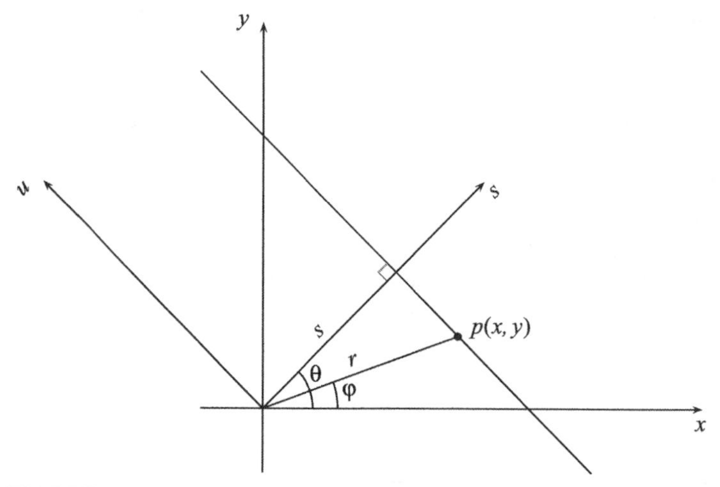

FIGURE 6.14 Axes for the Radon transformation, and the ray path along which measurements are accumulated. Point p moves along the ray as indicated, tracing through regions of various internal densities along the way.

displacement, and θ is the angle of the source plane. The value of $P_\theta(s)$ for a particular projection is dependent on the density encountered by each ray, and is modeled with a point density $f(x, y)$ at point (x, y) within the object. The Radon transform is the line integral over densities f, along the ray inclined at θ and distance s, as shown Figure 6.14.

The Radon transform gives us an approach to model how $P_\theta(s)$ is formed from the internal densities $f(x, y)$. We can further detail the geometry of the situation as in Figure 6.14, and write the equations describing the projection ray. In Figure 6.14, the point p represents a point traveling along the ray, and for a given projection, the value of angle θ is a function of the external geometry. The ray itself is a known distance s from the origin; again, this is a function of the geometry of the projecting device.

Defining u as the distance in the **u** axis direction from p to the intersection with the **s** axis line in Figure 6.14, we can drop perpendicular lines from $p(x, y)$, and use similar triangles for angle θ to find that:

$$x = s\cos\theta - u\sin\theta \tag{6.80}$$

$$y = s\sin\theta + u\cos\theta, \tag{6.81}$$

or, in matrix form,

$$\begin{pmatrix} x \\ y \end{pmatrix} = \begin{pmatrix} \cos\theta & -\sin\theta \\ \sin\theta & \cos\theta \end{pmatrix} \begin{pmatrix} s \\ u \end{pmatrix}. \tag{6.82}$$

This is seen to be a rotation of the coordinate axes. We can invert this equation and solve for (s, u)

$$s = x\cos\theta + y\sin\theta \tag{6.83}$$

$$u = -x\sin\theta + y\cos\theta, \tag{6.84}$$

or

$$\begin{pmatrix} s \\ u \end{pmatrix} = \begin{pmatrix} \cos\theta & \sin\theta \\ -\sin\theta & \cos\theta \end{pmatrix} \begin{pmatrix} x \\ y \end{pmatrix}. \tag{6.85}$$

With this geometry[4] understood, and the projection measurements $P_\theta(s)$, we can write the Radon line integral or "ray-sum" of $f(x, y)$ along s at angle θ as:

$$P_\theta(s) = \int f(s\cos\theta - u\sin\theta, s\sin\theta + u\cos\theta)du \quad -\infty < s < \infty \quad 0 \le \theta \le \pi. \tag{6.86}$$

This describes the formation of the projection. To solve the problem, we need the reverse: we want the internal densities $f(x, y)$, which are computed indirectly from the projection measurements.

This gives us an insight into how the projections are generated in a mathematical sense. In order to determine the shape and density of internal objects that are hidden, we need to take additional projections. This is illustrated in Figures 6.15 and 6.16, where the source plane has been rotated 90°. With multiple measurements, we ought to be able to generate an internal projection of the object, which corresponds mathematically to the values of $f(x, y)$. The situation is akin to looking at a building with the windows open, or trees in a park, and walking around the perimeter taking snapshot views at each angle, and attempting to determine a two-dimensional interior map from one-dimensional views.

6.6.2 Backprojection

Now that the generation of the projections is understood, we need to derive an algorithm for determining the internal values from these external projection measurements. There are several possible approaches to this problem, including the solution of a set of linear equations, the approximate solution to a set of similarly formed equations, the use of the Fourier transform, and an accumulation algorithm called *backprojection*. Although all but the first are workable in practice, the backprojection algorithm is often employed and has the advantage of not requiring an iterative solution. Of course, it is not without shortcomings, too, as will be seen.

The development of the concepts for the backprojection approach are shown in Figure 6.17. In this case, we have two measurement views, which in practice correspond to two rotational positions of the X-ray emission and measurement apparatus. The two views at differing angles θ give us two perspectives at offset s,

[4]In passing, note that in the literature, different conventions exist; in particular, the rotated axes s and u are also termed t and s, respectively, and the projection $P_\theta(s)$ is also denoted $g(s, \theta)$. In this text, we use axes s and u, and denote the projection as $P_\theta(s)$ so as to avoid possible confusion with t being time. The use of the subscript also indicates that θ is a discrete quantity.

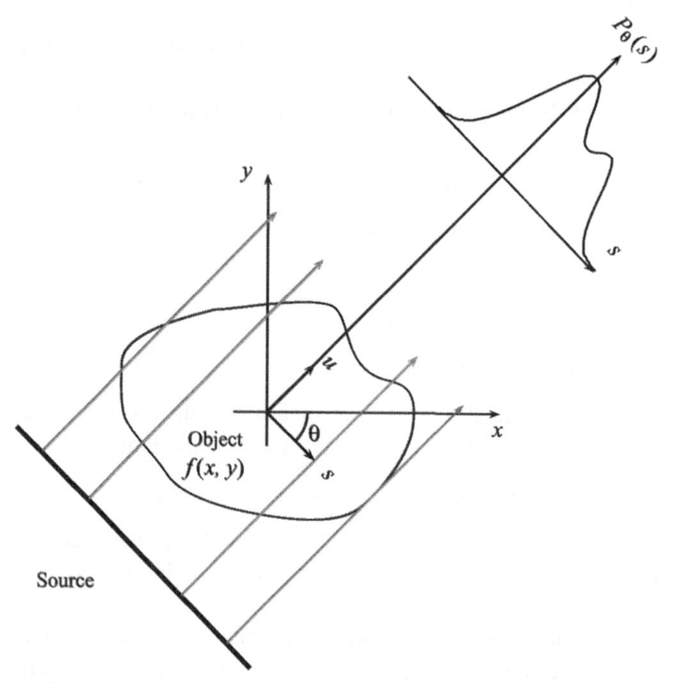

FIGURE 6.15 The cross-sectional projection through an object. The source consists of parallel emitters, or one emitter traveling along the indicated path. The impediment along each ray from source to measurement forms the resultant density projection $P_\theta(s)$ for that particular angle θ and parallel displacement from the origin s.

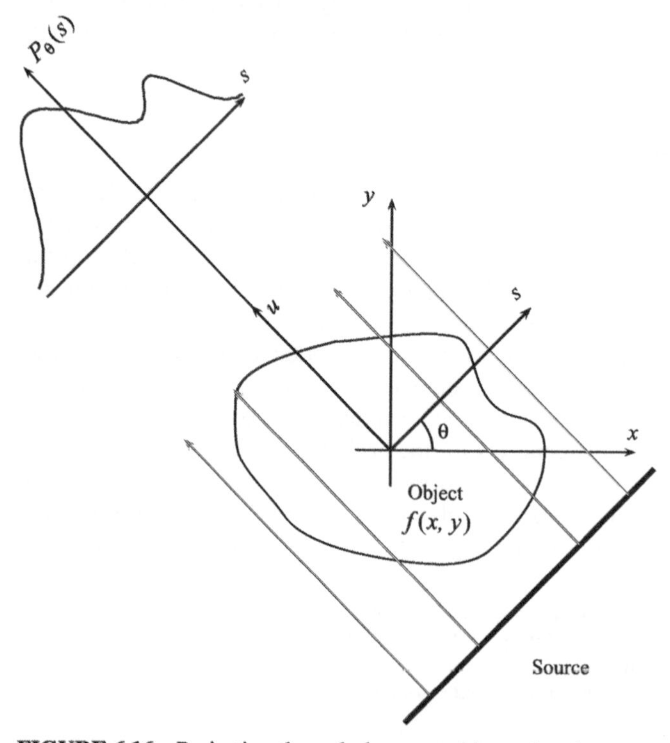

FIGURE 6.16 Projection through the same object, after the measurement apparatus has been rotated through 90°. A different projection is generated, and this gives us a second cross-section with which to build up a map of the object's interior density.

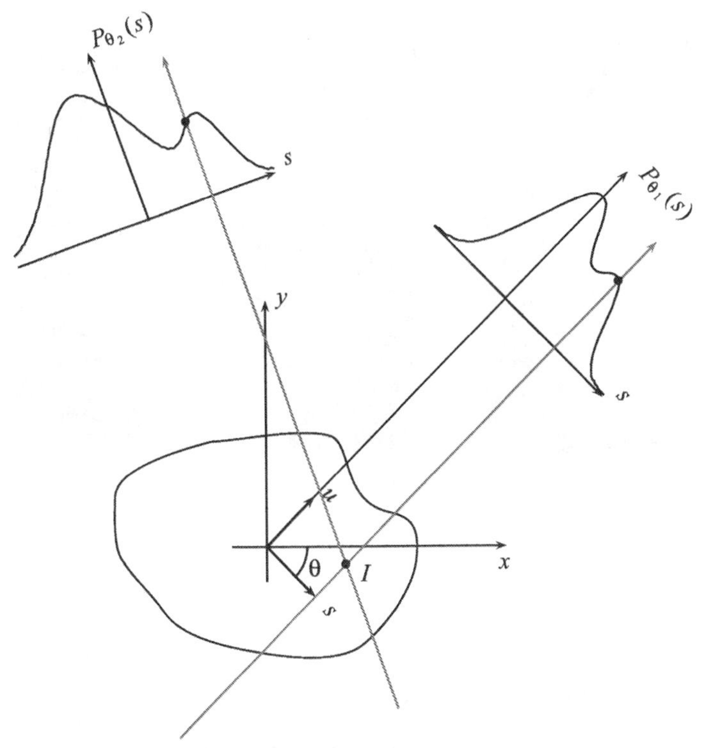

FIGURE 6.17 Interpretation of the backprojection algorithm for reconstructing the internal density of an object. Point I is the intersection of the two projections illustrated, and its intensity is the sum of the two projection values.

and projecting these rays back, we find their intersection point I. Clearly, the intensity $f(x, y)$ at point I must have contributions from both measurements $P_{\theta_1}(s)$ and $P_{\theta_2}(s)$. To obtain a complete set of contributions, we need measurements over 180° in order to accumulate the value of density f at point I.

Mathematically, the backprojection may then be defined as

$$b(x,y) = \int_0^{\pi} P_\theta(x\cos\theta + y\sin\theta)d\theta. \tag{6.87}$$

This takes into account the value of each projection $P_\theta(s)$ at angle θ and offset s, and incorporates the axis rotations as developed earlier. The integration is performed over all angles θ for a particular value of s. One may imagine this as pivoting the projection apparatus around a semicircle, with each discrete angular measurement k resulting in a projection $P_{\theta_k}(s)$ contributing to the estimate $\hat{f}(x, y)$.

In practice, of course, we do not undertake a complete rotation for each point, but rather take a complete "slice" of measurements to yield one set of projection values, and then rotate the projection system. Thus, all values of $P_{\theta_k}(s)$ are stored, the measurement apparatus rotated, and then another set of measurements taken. Finally, the backprojection algorithm is applied to reconstruct the interior map.

True image matrix $f(x, y)$

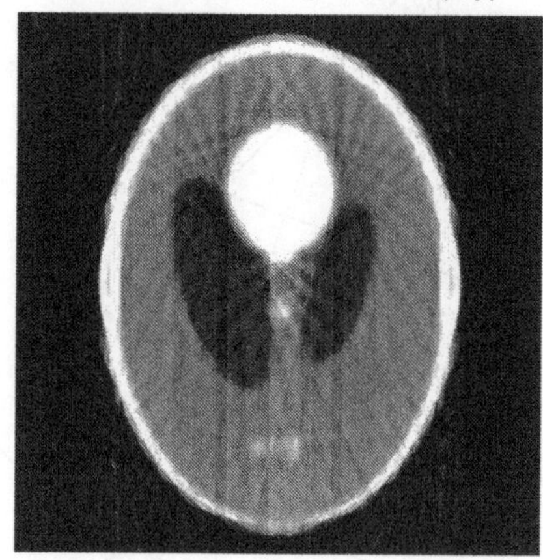

FIGURE 6.18 An example of the backprojection algorithm for tomography. The Shepp–Logan head phantom image is shown on the left, with a low-resolution reconstruction shown on the right. The lower resolution allows the scan lines to be seen, as illustrated in the previous figures.

True image matrix $f(x, y)$

Backprojection image matrix $b(x, y)$

FIGURE 6.19 A backprojection example, using an ellipsoidal-shaped object with constant density. The object is positioned at a known offset from the origin, and rotated by a known angle.

Figure 6.18 shows the reconstructed image, at a lower resolution. This enables the projection lines to be seen in the reconstruction. Note that no additional alteration of the image was performed to see this—the lines are an artifact of the reconstruction. Figure 6.19 shows the backprojection reconstruction of a simple elliptical object with constant density. This shows one of the limitations of the approach. In effect, all projections are smeared across each ray through the object, and thus some points along a projection ray accumulate a value when they ought not to. These erroneous contributions decay with distance, and more projections and/or closer sampling of each projection does not eliminate the problem. Various solutions have been proposed by researchers, with the most widely used being a filtering approach applied to the projections. This high-pass filtering applied does, however, tend to emphasize any noise in the measurements, which is undesirable.

6.6.3 Backprojection Example

The backprojection algorithm as described can be tested using a mathematical model for an object. An ellipse is one such useful shape, and multiple ellipses may be combined as in the Shepp–Logan head phantom encountered earlier.

Recall that the formal definition of an ellipse is that it is the locus of points equidistant from the foci. The equation of an ellipse is:

$$\frac{x^2}{A^2}+\frac{y^2}{B^2}=1. \tag{6.88}$$

It is not difficult to verify that the x intercept on the semi-major axis is $\pm A$, and that the y intercept on the semi-minor axis is $\pm B$.

To be useful in tomography, we need to model the interior of the ellipse as well. This is most easily done by assuming a constant interior density ρ, where the units of ρ are determined by the physics of the object (but which are unimportant from the signal processing perspective). What we require is a set of equations to model the ellipse: specifically, a projection density given the angle θ and offset s for an ellipse specified by parameters A and B.

This situation is depicted in Figure 6.20. To begin with, we utilize an ellipse centered on the origin. We first derive the equations for the tangent at point (x_t, y_t), then for the ray at perpendicular offset s as shown. The quantity $a(\theta)$ is useful to simplify the notation.

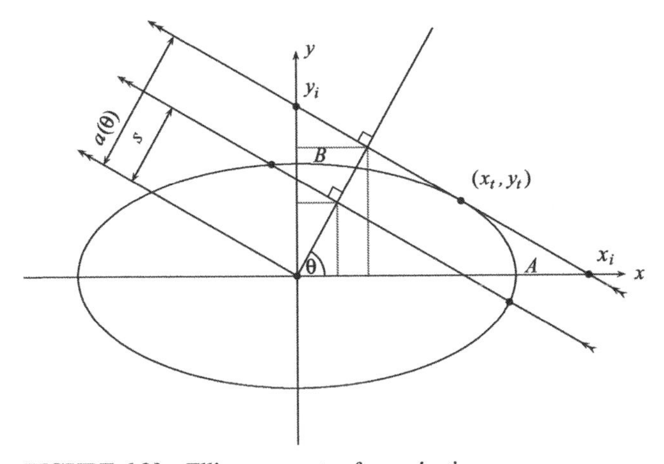

FIGURE 6.20 Ellipse geometry for projection.

After algebraic manipulation of the geometry as shown, we can determine the equation for the projection through the ellipse $P_\theta(s)$ at a given rotational angle θ and offset s for known ellipse geometry specified by A and B. The projection is

$$P_\theta(s) = \frac{2AB}{a^2(\theta)}\sqrt{a^2(\theta) - s^2}, \tag{6.89}$$

$$\text{where } a^2(\theta) = A^2\cos^2\theta + B^2\sin^2\theta. \tag{6.90}$$

So, for an assumed constant density ρ, we can determine the projection quantity depending on whether the ray passes through the ellipse or not. If it passes through the ellipse, the projection is multiplied by the density ρ. If not, the projection is zero, and so $P_\theta(s)$ becomes

$$P_\theta(s) = \begin{cases} \dfrac{2\rho AB}{a^2(\theta)}\sqrt{a^2(\theta) - s^2} & \text{for } |s| \le a(\theta) \\ 0 & \text{for } |s| > a(\theta) \end{cases}. \tag{6.91}$$

Next, we need to position the ellipse at any point in the measurement plane, and with any offset. We can effect this not be rederiving the above, but more easily by compensating for the rotation and translation separately. If we offset the ellipse center of (0, 0) by (x_o, y_o), as illustrated in Figure 6.21, the distance to the center and angle to the center are simply

$$d = \sqrt{x_o^2 + y_o^2} \tag{6.92}$$

$$\gamma = \arctan\left(\frac{y_o}{x_o}\right). \tag{6.93}$$

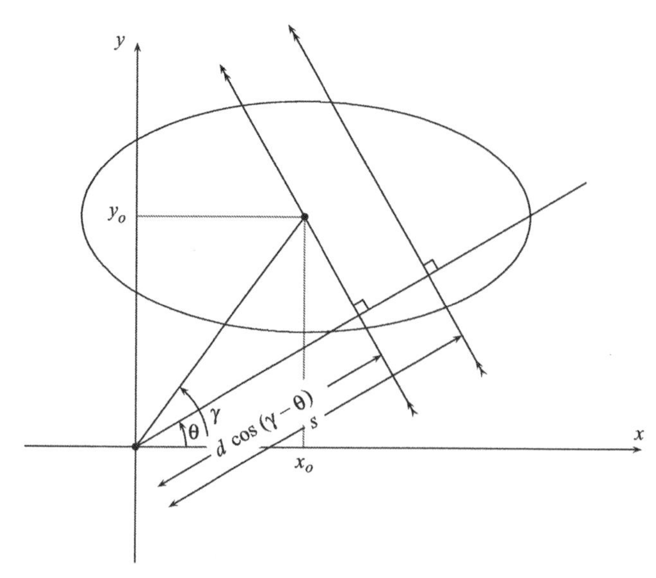

FIGURE 6.21 Ellipse translated by (x_o, y_o).

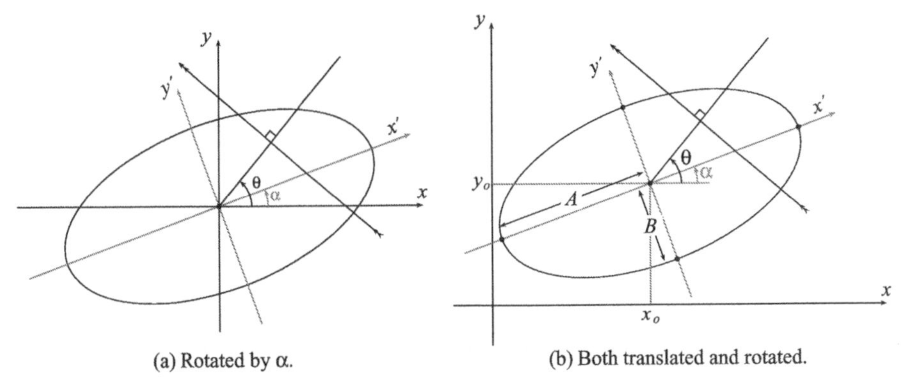

(a) Rotated by α.

(b) Both translated and rotated.

FIGURE 6.22 Ellipse rotation and translation.

Next, consider a rotation by angle α as shown in Figure 6.22a. Clearly, the ray spends a longer time within the ellipse, and hence encounters proportionally more of the internal density ρ. Considering the rotated axes (x', y'), it is seen from the diagram that effectively, this case corresponds to the unrotated ellipse but with angle $\theta - \alpha$.

Combining both translation and rotation, we have the situation depicted in Figure 6.22b. From this diagram, the following substitutions can be derived. We simply have an altered effective angle θ' according to the rotational component, and an effective displacement s' that is reduced after translation by an amount $d \cos(\gamma - \theta)$. So the substitutions we need are

$$\theta' \rightarrow \theta - \alpha \tag{6.94}$$

$$s' \rightarrow s - d\cos(\gamma - \theta), \tag{6.95}$$

with d and γ as given in Equations 6.92 and 6.93. The projection can then be implemented using these equations as follows. We need the ellipse geometry (specified by A and B), the ellipse translation (determined by $[x_o, y_o]$), the ellipse rotation α, and the density within the ellipse ρ. For a given projection angle θ and offset s, we may determine the projection $P_\theta(s)$ using the following function.

```
function [P] = ellipseproj(A, B, rho, theta, s, alpha, xo, yo)
gamma = atan2(yo, xo);
d = sqrt( xo * xo + yo * yo);
thetanew = theta - alpha;
snew = s - d * cos(gamma - theta);
% use translated/rotated values
s = snew;
theta = thetanew;
% find a^2 (theta)
ct = cos(theta);
st = sin(theta);
a2 = A * A * ct * ct + B * B * st * st;
```

```
atheta = sqrt(a2);
% return value if outside ellipse
P = 0;
if( abs(s) <= atheta )
    % inside ellipse
    P = 2 * rho * A * B / a2 * sqrt(a2 - s * s);
end
```

This is the projection of one ray through the ellipse, with all possible parameters accounted for. The size of the ellipse A, B, the ray angle θ, the ellipse translation x_o, y_o, ellipse angle α, and finally the ellipse density ρ, must all be known.

We now need to take projections at discrete angles θ_k and determine the composite result from all projections. Each projection forms a vector, and it is natural to store all projections in a matrix `projmat` with one projection set per row as follows:

```
function [projmat, svals, thetavals] = ...
    ellipseprojmat(A, B, ntheta, ns, srange, rho, alpha, xo, yo)
thetamin = 0;
thetamax = pi;
% each row is a projection at a certain angle
projmat = zeros(ntheta, ns);
smin = -srange;
smax = srange;
dtheta = pi/(ntheta - 1);
ds = (smax - smin)/(ns - 1);
svals = smin:ds:smax;
thetavals = thetamin:dtheta:thetamax;
pn = 1;
for theta = thetavals
    % calculate all points on the projection line
    P = zeros(ns, 1);
    ip = 1;
    for s = svals
        % simple ellipse
        [p] = ellipseproj(A, B, rho, theta, s, alpha, xo, yo);
        P(ip) = p;
        ip = ip + 1;
    end
    % save projection as one row of matrix
    projmat(pn, :) = P';
    pn = pn + 1;
end
```

Finally, we are in a position to implement our backprojection algorithm. The backprojection operation itself was defined in Equation 6.87 as

$$b(x,y)=\int_0^{\pi} P_\theta(x\cos\theta + y\sin\theta)d\theta. \tag{6.96}$$

This example translates the width and height of the reconstructed image into normalized coordinates in the range ±1. That is, real x coordinates from −1 to +1 are translated to indexes 1 to W. The translation equations are then

$$x = 2\left(\frac{x_i - 1}{W-1}\right) - 1,$$

for mapping image indexes to ±1, and

$$x_i = \left(\frac{x+1}{2}\right)(W-1)+1,$$

for mapping from real coordinate x to the image index. For the y coordinates, if we wish to use a range of ±1, we must incorporate the fact that the image matrix dimensions are 1 to H, but in fact from top (1) to bottom (H), whereas a more natural real-number co-ordinate system goes from −1 (bottom) to +1 (top). So $y = -1$ corresponds to index H, and $y = +1$ corresponds to index 1. The index to real mapping is

$$y = 1 - 2\left(\frac{y_i - 1}{H-1}\right),$$

for index to real, and for real to index, it is

$$y_i = \left(\frac{1-y}{2}\right)(H-1)+1.$$

We incorporate all these considerations into the backprojection solution code as shown below.

```
function [b] = bpsolve(W, H, projmat, svals, thetavals)
ntheta = length(thetavals);
ns = length(svals);
srange = svals(ns);
b = zeros(H, W);
for iy = 1:H
    for ix = 1:W
        x = 2 * (ix - 1)/(W - 1) - 1;
        y = 1 - 2 * (iy - 1)/(H - 1);
        % projmat is the P values, each row is P(s) for a given theta
        bsum = 0;
        for itheta = 1:ntheta
            theta = thetavals(itheta);
```

```
                s =x*cos(theta) + y*sin(theta);
                is = (s + srange)/(srange*2)*(ns - 1) + 1;
                is = round(is);
                if(is < 1)
                   is = 1;
                end
                if(is > ns)
                   is = ns;
                end
                Ptheta = projmat(itheta, is);
                bsum = bsum + Ptheta;
           end
           b(iy, ix) = bsum;
     end
end
```

We can now call the m-files, thus developed as follows. Figure 6.19 shows the resulting reconstructed image using this code.

```
% image parameters
W = 400;
H = 400;
% object parameters
rho = 200;
A = 0.4;
B = 0.2;
alpha = pi/10;
xo = 0.3;
yo = 0.5;
% backprojection parameters
ntheta = 100;
srange = 2;
ns = 100;
% generate projections
[projmat, svals, thetavals] = ...
  ellipseprojmat(A, B, ntheta, ns, srange, rho, alpha, xo, yo);
% solve using backprojection
[b] = bpsolve(W, H, projmat, svals, thetavals);
% scale for image display
b = b/max(max(b));
b = b * 255;
bi = uint8(b);
```

```
figure(1);
clf
image(bi);
colormap(gray(256));
box('on');
axis('off');
```

6.7 CHAPTER SUMMARY

The following are the key elements covered in this chapter:

- a review of correlation theory, including auto- and cross-correlation.
- the application of correlation in various signal processing problems and its limitations.
- the key concepts of optimal filtering, as applied to one-step prediction, as well as noise estimation and removal.
- the key concepts of tomography, and the backprojection algorithm used to solve the inverse-Radon approximation.

PROBLEMS

6.1. Using the step-by-step approach in Section 6.3.1, determine whether autocorrelation is symmetrical about the origin. Repeat for cross-correlation.

6.2. Verify that `conv()` may be used to calculate correlation, as described in Section 6.3.1. Does the order of vectors passed to the function matter?

6.3. Derive an expression for the autocorrelation $R_{xx}(\lambda)$ for a sinusoid $x(t) = A \sin \Omega t$.

6.4. A discrete signal sequence is denoted $x(n)$, with a corresponding offset version $x(n+k)$ at integer offset k. Since $\alpha^2 > 0$ for real values of α, it follows that $(x[n] - x[n+k])^2 \geq 0$.

(a) From this expression, take the summation over a block of length N samples and write down the result by expanding the square and taking the summation through.

(b) Assuming that the average signal power remains constant, $\sum x(n)x(n) \approx \sum x(n-k)x(n-k)$.

Take the limit of the summation in (a) as $N \to \infty$, and rewrite the summation terms as correlations (i.e., $\sum x(n)x(n-k) = R(k)$).

(c) From this, prove that the autocorrelation of a sequence at lag k is always less than, or equal to, the autocorrelation at lag zero.

6.5. Extend the second-order derivation in Equation 6.43 to a fourth-order predictor. How many correlation values need to be computed? What is the dimension of the matrix equations to be solved?

6.6. By writing each vector and matrix as its components, prove the results in Equation 6.76 and 6.77 used in the derivation of the Wiener filter.

6.7. The transformation in the example given for orthogonal decomposition uses a loop to transform each vector sample:

```
for k = 1 : N
    v = X( :, k );
    vt = T*v;
    XT( :, k ) = vt;
end
```

What is an equivalent matrix-matrix multiplication that could effect this without a loop?

6.8. Obtain an ECG waveform from the Signal Processing Information Base (SPIB) at http://www.spib.rice.edu/spib.html. Estimate the heart rate using detection of zero-crossings (or crossing of any arbitrary threshold). Then perform autocorrelation on the waveform, and investigate whether the peak of the correlation corresponds to the heart period.

6.9. Consider the backprojection algorithm for tomography. If 100 samples are taken at each slice, at increments of 2 degrees, over the range 0–180 degrees, estimate how much memory is required to store all the projections. If the projection slice is to be imaged using a $1{,}000 \times 1{,}000$ image, what are the computations required for each pixel?

6.10. Using the projection diagram of Figure 6.14, prove Equations 6.80 and 6.81.

6.11. Implement the MATLAB code as developed in Section 6.6.3. Instead of one simple ellipse, use three nonoverlapping ellipses. What happens if the ellipses overlap?

CHAPTER 7

FREQUENCY ANALYSIS OF SIGNALS

7.1 CHAPTER OBJECTIVES

On completion of this chapter, the reader should be able to:

1. define the *Fourier series* and derive equations for a given time series.
2. define the *Fourier transform* and the relationship between its input and output.
3. scale and *interpret* a Fourier transform output.
4. explain the use of frequency *window functions*.
5. explain the derivation of the *fast Fourier transform (FFT)*, and its computational advantages.
6. define the *discrete cosine transform (DCT)* and explain its suitability for data compression applications.

7.2 INTRODUCTION

This chapter introduces techniques for determining the frequency content of signals. This is done primarily via the *Fourier transform*, a fundamental tool in digital signal processing. We also introduce the related but distinct *DCT)*, which finds a great many applications in audio and image processing.

7.3 FOURIER SERIES

The *Fourier series*[1] is an important technique for analyzing the frequency content of a signal. An understanding of the Fourier series is crucial to understanding a number of closely related "transform" techniques. Used in reverse, it may also

[1]Named after the French mathematician, Jean Baptiste Joseph Fourier.

Digital Signal Processing Using MATLAB for Students and Researchers, First Edition. John W. Leis.

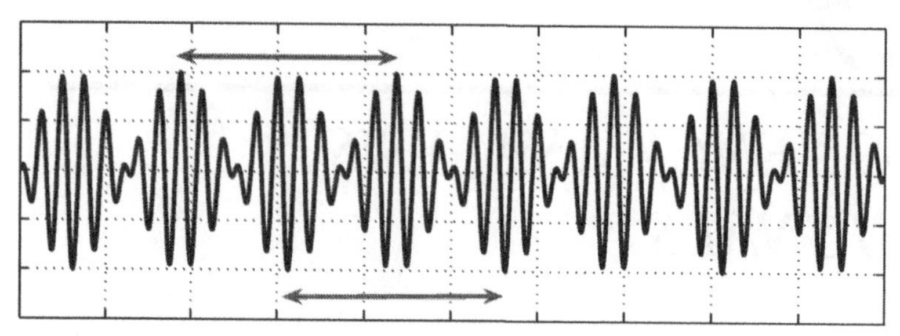

FIGURE 7.1 A simple periodic waveform. In this simple case, it is easy to visually determine the time over which the waveform repeats itself. However, there may be multiple underlying periodic signals present.

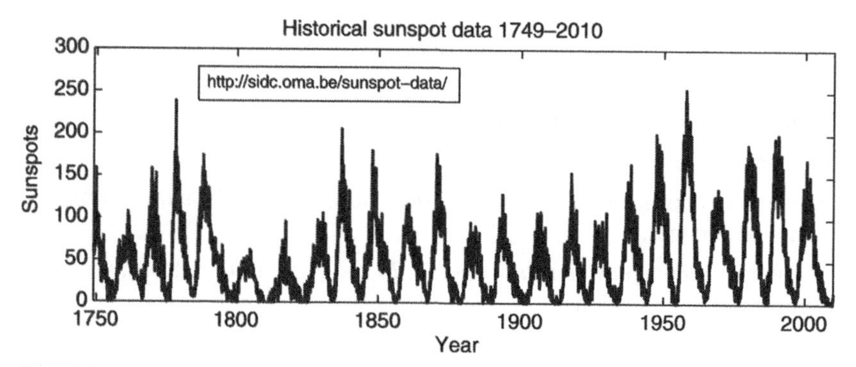

FIGURE 7.2 Historical sunspot count data from SIDC-team (1749–2010). The signal appears to have some underlying periodicity, but it is difficult to be certain using direct observation of the waveform alone.

be used to synthesize arbitrary periodic waveforms—those that repeat themselves over time.

A *periodic* waveform is one which repeats itself over time, as illustrated in Figure 7.1. Many signals contain not one, but multiple periodicities. Figure 7.2 shows some naturally occurring data: the monthly number of sunspots for the years 1749–2010 (SIDC-team 1749–2010). An algorithm ought to be able to determine the underlying periodic frequency component(s) of any such signal.

The sunspot observation signal appears to have some periodicity, but determining the time range of this periodicity is difficult. Furthermore, there may in fact be more than one underlying periodic component. This is difficult, if not impossible, to spot by eye.

Mathematically, a waveform which repeats over some interval τ may be described as $x(t) = x(t + \tau)$. The *period* of the waveform is τ. For a fundamental frequency of Ω_o radians/second or f_o Hz,

$$\Omega_o = \frac{2\pi}{\tau} \tag{7.1}$$

$$\tau = \frac{1}{f_o} \tag{7.2}$$

Fourier's theorem states that *any* periodic function $x(t)$ may be decomposed into an infinite series of sine and cosine functions:

$$\begin{aligned} x(t) = a_0 + & \\ & a_1 \cos\Omega_o t + a_2 \cos 2\Omega_o t + a_3 \cos 3\Omega_o t + \cdots \\ & b_1 \sin\Omega_o t + b_2 \sin 2\Omega_o t + b_3 \sin 3\Omega_o t + \cdots \\ = a_0 + & \sum_{k=1}^{\infty} \left(a_k \cos k\Omega_0 t + b_k \sin k\Omega_0 t \right). \end{aligned} \tag{7.3}$$

The coefficients a_k and b_k are determined by solving the following integral equations, evaluated over one period of the input waveform.

$$a_0 = \frac{1}{\tau}\int_0^{\tau} x(t)\,dt \tag{7.4}$$

$$a_k = \frac{2}{\tau}\int_0^{\tau} x(t)\cos k\Omega_o t\,dt \tag{7.5}$$

$$b_k = \frac{2}{\tau}\int_0^{\tau} x(t)\sin k\Omega_o t\,dt. \tag{7.6}$$

The integration limits could equally be $-\frac{\tau}{2}$ to $+\frac{\tau}{2}$. This would still be over one period of the waveform, but with a different starting point. Using complex numbers, it is also possible to represent the Fourier series expansion more concisely as the series:

$$x(t) = \sum_{k=-\infty}^{k=+\infty} c_k e^{jk\Omega_o t} \tag{7.7}$$

$$c_k = \frac{1}{\tau}\int_0^{\tau} x(t) e^{-jk\Omega_o t}\,dt. \tag{7.8}$$

7.3.1 Fourier Series Example

To illustrate the application of the Fourier series, consider a square wave with period $\tau = 1$ second and peak amplitude ± 1 ($A = 1$) as shown in Figure 7.3. The waveform is composed of a value of $+A$ for $t = 0$ to $t = \tau/2$, followed by a value of $-A$ for $t = \tau/2$ to $t = \tau$.

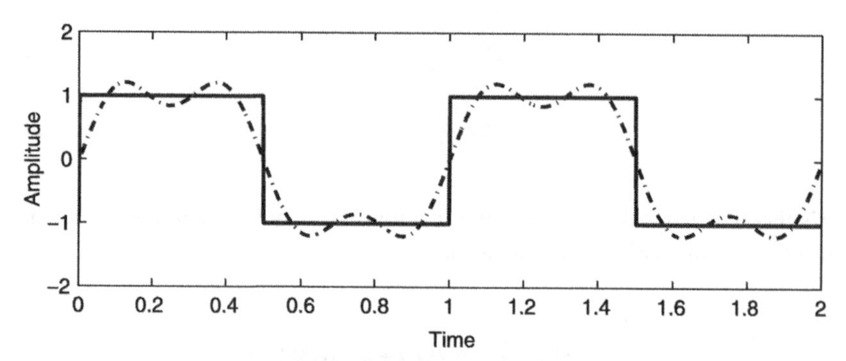

FIGURE 7.3 Approximating a square waveform with a Fourier series. The Fourier series approximation to the true waveform is shown. The Fourier series has a limited number of components, hence is not a perfect approximation.

The coefficient a_0 is found from Equation 7.4 as:

$$\begin{aligned}
a_0 &= \frac{1}{\tau}\int_0^{\tau} x(t)\,dt \\
&= \frac{1}{\tau}\int_0^{\frac{\tau}{2}} A\,dt + \frac{1}{\tau}\int_{\frac{\tau}{2}}^{\tau} (-A)\,dt \\
&= \frac{A}{\tau}t\Big|_{t=0}^{t=\frac{\tau}{2}} + \frac{-A}{\tau}t\Big|_{t=\frac{\tau}{2}}^{\tau} \\
&= \frac{A}{\tau}\left(\frac{\tau}{2}-0\right) - \frac{A}{\tau}\left(\tau-\frac{\tau}{2}\right) \\
&= 0.
\end{aligned} \tag{7.9}$$

The coefficients a_k are found from Equation 7.5:

$$\begin{aligned}
a_k &= \frac{2}{\tau}\int_0^{\tau} x(t)\cos k\Omega_o t\,dt \\
&= \frac{2}{\tau}\int_0^{\frac{\tau}{2}} A\cos k\Omega_o t\,dt + \tfrac{2}{\tau}\int_{\frac{\tau}{2}}^{\tau} (-A)\cos k\Omega_o t\,dt \\
&= \frac{2A}{\tau}\frac{1}{k\Omega_o}\sin k\Omega_o t\Big|_{t=0}^{t=\frac{\tau}{2}} - \frac{2A}{\tau}\frac{1}{k\Omega_o}\sin k\Omega_o t\Big|_{t=\frac{\tau}{2}}^{t=\tau} \\
&= \frac{2A}{\tau}\frac{1}{k}\frac{\tau}{2\pi}\left(\sin k\frac{2\pi}{\tau}\frac{\tau}{2} - \sin 0\right) \\
&\quad - \frac{2A}{\tau}\frac{1}{k}\frac{\tau}{2\pi}\left(\sin k\frac{2\pi}{\tau}\tau - \sin k\frac{2\pi}{\tau}\frac{\tau}{2}\right) \\
&= \frac{A}{k\pi}\sin k\pi - \frac{A}{k\pi}(\sin 2k\pi - \sin k\pi) \\
&= 0.
\end{aligned} \tag{7.10}$$

In a similar way, the coefficients b_k are found from Equation 7.6:

$$b_k = \frac{2}{\tau}\int_0^{\tau} x(t)\sin k\Omega_o t\, dt,$$

and may be shown to equal

$$b_k = \frac{2A}{k\pi}(1-\cos k\pi). \tag{7.11}$$

When the integer k is an odd number (1, 3, 5, …). the value $\cos k\pi = -1$, and hence this reduces to:

$$b_k = \frac{4A}{k\pi} \quad k = 1, 3, 5, \ldots, \tag{7.12}$$

when k is an even number (2, 4, 6, …), the value $\cos k\pi = 1$, and hence the equation for b_k reduces to 0. The completed Fourier series representation obtained by substituting the specific coefficients for this waveform (Equation 7.11) into the generic Fourier series Equation 7.3, giving

$$x(t) = \frac{4A}{\pi}\left(\overbrace{1\sin 1\frac{2\pi}{\tau}t}^{k=1} + \overbrace{\frac{1}{3}\sin 3\frac{2\pi}{\tau}t}^{k=3} + \ldots\right). \tag{7.13}$$

The *components* of the Fourier series in this case are shown in Figure 7.4. These are the sine and cosine waveforms of each harmonic frequency, weighted by the coefficients as determined mathematically. Added together, these form the Fourier series approximation. The magnitudes of these components are identical to the coefficients of the sin(·) and cos(·) components derived mathematically. Figure 7.4 shows that the cos(·) components are all zero in this case. Note that this will not happen for all waveforms; it depends on the symmetry of the waveform, as will be demonstrated in the next section.

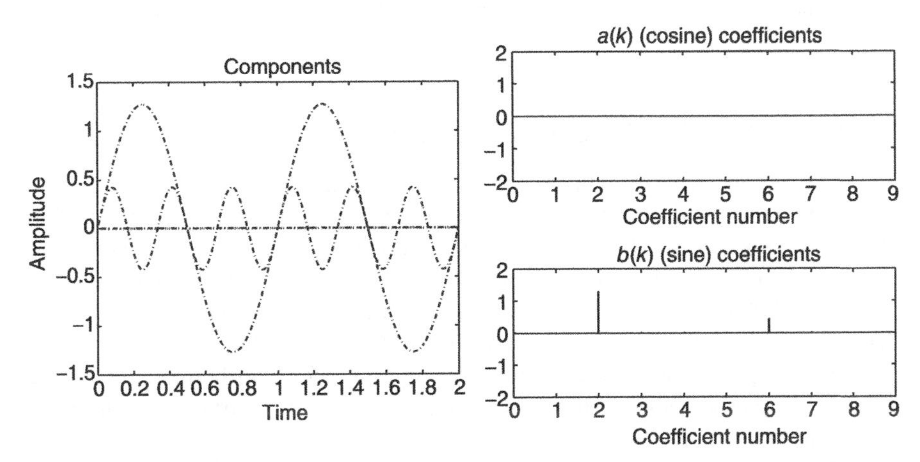

FIGURE 7.4 Approximating a square waveform with a Fourier series. The original waveform and its approximation was shown in Figure 7.3. The component sine and cosine waves are shown at the left, and their respective magnitudes are shown on the right.

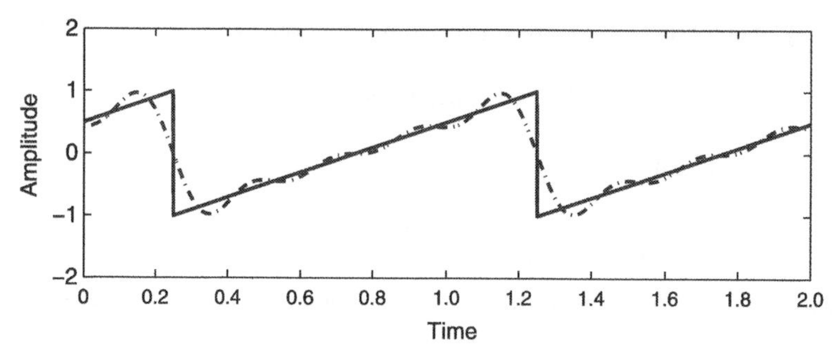

FIGURE 7.5 Approximating a shifted sawtooth waveform with a Fourier series. The Fourier series approximation to the true waveform is shown. The Fourier series only has a limited number of components, hence is not a perfect approximation. The components are shown in Figure 7.6.

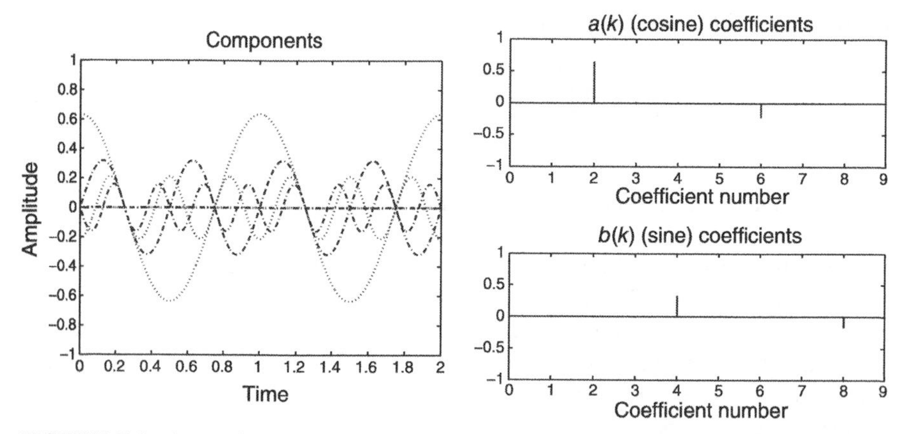

FIGURE 7.6 Approximating a shifted sawtooth waveform with a Fourier series. The component sine and cosine waves are shown at the left, and their respective magnitudes are shown on the right.

7.3.2 Another Fourier Series Example

To investigate the general applicability of the technique, consider a second example waveform as shown in Figure 7.5. This is a so-called *sawtooth* waveform, in this case with a phase shift. To see that it has a phase shift, imagine extrapolating the waveform backwards for negative time. The extrapolated waveform would not be symmetrical about the $t = 0$ axis. Again, it is seen that the sine plus cosine approximation is valid, as the approximation in Figure 7.5 shows. The components which make up this particular waveform are shown in Figure 7.6. Furthermore, as Figure 7.6 shows, we need both sine and cosine components in varying proportions to make up the approximation.

The code shown in Listing 7.1 implements the equations to determine the Fourier series coefficients, and plots the resulting Fourier approximation for 10 terms

Listing 7.1 Generating a Fourier Series Approximation

```
dt = 0.01;
T = 1;
t = [0:dt:T]';
omega0 = 2*pi/T;
N = length(t);
N2 = round(N/2);
x = ones(N, 1);
x (N2 + 1:N) = -1*ones(N - N2, 1);
a(1) = 1/T*(sum(x)*dt);
xfs = a(1)*ones(size(x));
for k = 1:10
    ck = cos(k*omega0*t);                    % cosine component
    a(k + 1) = 2/T*(sum(x.*ck)*dt);
    sk = sin(k*omega0*t);                    % sine component
    b(k + 1) = 2/T*(sum(x.*sk)*dt);
    % Fourier series approximation
    xfs = xfs+a(k+1)*cos(k*omega0*t)+b(k+1)*sin(k*omega0*t);
    plot(t, x, '-', t, xfs, ':');
    legend('desired', 'approximated');
    drawnow; pause(1);
end
```

in the expansion. A square waveform is shown, but other types of waveform can be set up by changing the preamble where the sample values in the x vector are initialized. The Fourier series calculation component need not be changed in order to experiment with different wave types.

7.4 HOW DO THE FOURIER SERIES COEFFICIENT EQUATIONS COME ABOUT?

We have so far applied the Fourier series equations to determine the coefficients for a given waveform. These equations were stated earlier without proof. It is instructive, however, to see how these equations are derived.

For convenience, we restate the basic Fourier waveform approximation, which is

$$x(t) = a_0 + \sum_{k=1}^{\infty} (a_k \cos k\Omega_0 t + b_k \sin k\Omega_0 t)$$

with coefficients a_k and b_k, for $k = 1, 2, \ldots$. To derive the coefficients, we take the integral of both sides over one period of the waveform.

$$\int_0^\tau x(t)\,dt = \int_0^\tau a_0\,dt = \sum_{k=1}^{\infty}\int_0^\tau a_k \cos k\Omega_0 t\,dt + \sum_{k=1}^{\infty}\int_0^\tau b_k \cos k\Omega_0 t\,dt. \tag{7.14}$$

The sine and cosine integrals over one period are zero, hence:

$$\begin{aligned}\int_0^\tau x(t)\,dt &= a_0\tau + 0 + 0\\ \therefore\quad a_0 &= \frac{1}{\tau}\int_0^\tau x(t)\,dt.\end{aligned} \tag{7.15}$$

This yields a method of computing a_0. Next, multiply both sides by $\cos n\Omega_o t$ (where n is an integer) and integrate from 0 to τ,

$$\begin{aligned}\int_0^\tau x(t)\cos n\Omega_o t\,dt &= \int_0^\tau a_0 \cos n\Omega_o t\,dt\\ &\quad + \sum_{k=1}^{\infty}\int_0^\tau a_k \cos k\Omega_o t \cos n\Omega_o t\,dt\\ &\quad + \sum_{k=1}^{\infty}\int_0^\tau b_k \sin k\Omega_o t \cos n\Omega_o t\,dt\\ \int_0^\tau x(t)\cos n\Omega_o t\,dt &= 0 + \frac{\tau}{2}a_n + 0\\ \therefore\quad a_n &= \frac{2}{\tau}\int_0^\tau x(t)\cos n\Omega_o t\,dt.\end{aligned} \tag{7.16}$$

Note that several intermediate steps have been omitted for clarity—terms such as $\int_0^\tau a_k \cos k\Omega_0 t \cos n\Omega_0 t\,dt$ end up canceling out for $n \neq k$. The algebra is straightforward, although a little tedious.

Similarly, by multiplying both sides by $\sin n\Omega_o t$ and integrating from 0 to τ, we obtain

$$\begin{aligned}\int_0^\tau x(t)\sin n\Omega_o t\,dt &= \sum_{k=1}^{\infty}\int_0^\tau a_0 \sin n\Omega_o t\,dt\\ &\quad + \sum_{k=1}^{\infty}\int_0^\tau a_k \cos k\Omega_o t \sin n\Omega_o t\,dt\\ &\quad + \sum_{k=1}^{\infty}\int_0^\tau b_k \sin k\Omega_o t \sin n\Omega_o t\,dt\\ \int_0^\tau x(t)\sin n\Omega_o t\,dt &= 0 + 0 + \frac{\tau}{2}b_n\\ \therefore\quad b_0 &= \frac{2}{\tau}\int_0^\tau x(t)\sin n\Omega_o t\,dt.\end{aligned} \tag{7.17}$$

Replacing n with k in the above equations completes the derivation.

7.5 PHASE-SHIFTED WAVEFORMS

The sine and cosine terms in the Fourier series may be combined into a single sinusoid with a phase shift using the relation:

$$a_k \cos k\Omega_o t + b_k \sin k\Omega_o t = A\sin(k\Omega_o t + \varphi). \tag{7.18}$$

To prove this and find expressions for A and φ, let

$$a\cos\theta + b\sin\theta = A\sin(\theta + \varphi).$$

Expanding the right-hand side:

$$\begin{aligned} a\cos\theta + b\sin\theta &= A\sin\theta\cos\varphi + A\cos\theta\sin\varphi \\ &= A\cos\varphi\underline{\sin\theta} + A\sin\varphi\underline{\cos\theta}. \end{aligned}$$

Equating the coefficients of cos θ and sin θ in turn,

$$\begin{aligned} a &= A\sin\varphi \\ b &= A\cos\varphi. \end{aligned}$$

Dividing, we get an expression for the phase shift:

$$\begin{aligned} \frac{a}{b} &= \frac{A\sin\varphi}{A\cos\varphi} \\ &= \tan\varphi \\ \therefore \quad \varphi &= \tan^{-1}\frac{a}{b}. \end{aligned}$$

The magnitude is found by squaring and adding terms,

$$\begin{aligned} a^2 + b^2 &= A^2\sin^2\varphi + A^2\cos^2\varphi \\ &= A^2\left(\sin^2\varphi + \cos^2\varphi\right) \\ &= A^2 \\ \therefore \quad A &= \sqrt{a^2 + b^2}. \end{aligned}$$

To illustrate these equations involving phase shift, consider the following MATLAB code. The result, shown in Figure 7.7, illustrates that a summation of sine and cosine can indeed be composed of a single phase-shifted sinusoid.

```
t = 0:0.01:1;
a = 3;
b = 2;
omega = (2*pi)*4;
y1c = a*cos(omega*t);
y1s = b*sin(omega*t);
% sine + cosine directly
y1 = a*cos(omega*t) + b*sin(omega*t);
```

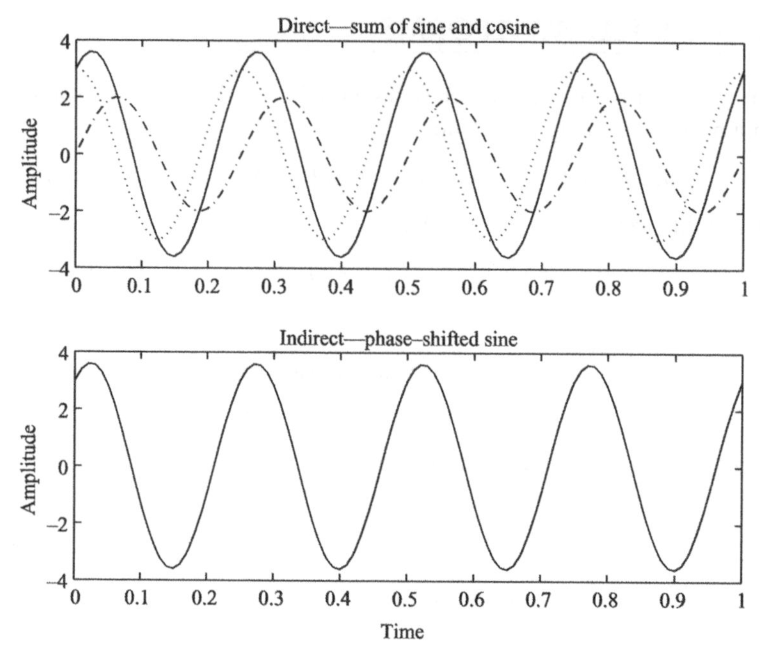

FIGURE 7.7 Converting a sine plus cosine waveform to a sine with phase shift.

```
figure(1);
plot(t, y1);
% sine with phase shift
phi = atan2(a, b);
A = sqrt(a*a + b*b);
y2 = A*sin(omega*t + phi);
figure(2);
plot(t, y2);
```

7.6 THE FOURIER TRANSFORM

The Fourier transform is closely related to the Fourier series described in the previous section. The fundamental difference is that the requirement to have a periodic signal is now relaxed. The Fourier transform is one of the most fundamental algorithms—if not *the* fundamental algorithm—in digital signal processing.

7.6.1 Continuous Fourier Transform

The continuous-time Fourier transform allows us to convert a signal $x(t)$ in the time domain into its frequency domain counterpart $X(\Omega)$, where Ω is the true frequency in radians per second. Note that *both* signals are continuous. The Fourier transform, or (more correctly) the continuous-time/continuous-frequency Fourier transform, is defined as

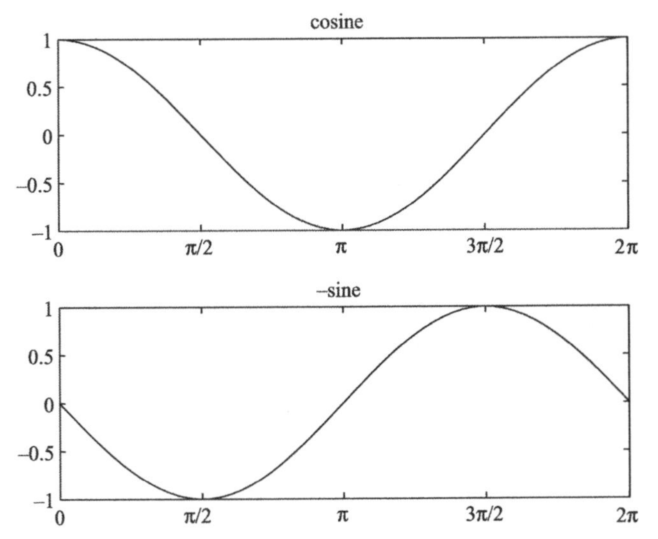

FIGURE 7.8 Sinusoidal basis functions for Fourier analysis: cosine and negative sine.

$$X(\Omega) = \int_{-\infty}^{\infty} x(t)e^{-j\Omega t}\, dt. \tag{7.19}$$

The signal $x(t)$ is multiplied by a complex exponential $e^{-j\Omega t}$. This is really just a concise way of representing sine and cosine, since Euler's formula for the complex exponential gives us $e^{-j\Omega t} = \cos \Omega t - j\sin \Omega t$.

The functions represented by the $e^{-j\Omega t}$ term are sometimes called the *basis functions*, as depicted in Figure 7.8. In effect, for a given frequency Ω, the signal $x(t)$ is multiplied by the basis functions at that frequency, with the result integrated ("added up") to yield an effective "weighting" of that frequency component. This is repeated for all frequencies (all values of Ω)—or at least, all frequencies of interest. The result is the *spectrum* of the time signal. As will be seen later with some examples, the basis functions may be interpreted on the complex plane as shown in Figure 7.9.

The inverse operation—to go from frequency $X(\Omega)$ to time $x(t)$—is:

$$x(t) = \frac{1}{2\pi}\int_{-\infty}^{\infty} X(\Omega)e^{j\Omega t}\, d\Omega. \tag{7.20}$$

Note that the difference is in the exponent of the basis function, and the division by 2π. The latter is related to the use of radian frequency, rather than Hertz (cycles per second).

These two operations—Fourier transform and inverse Fourier transform—are completely reversible. That is, taking any $x(t)$, computing its Fourier transform, and then computing the inverse Fourier transform of the result, returns the original signal.

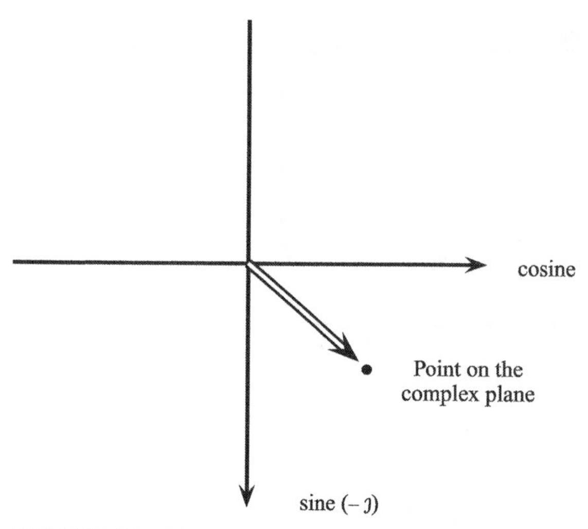

FIGURE 7.9 The sine/cosine basis functions may be viewed as a vector on the complex plane.

To summarize the fundamental result from this section, we have:

Fourier Transform:
Continuous time → Continuous frequency

7.6.2 Discrete-Time Fourier Transform

The next stage is to consider a sampled signal. When a signal $x(t)$ is sampled at time instants $t = nT$, resulting in the discrete sampled $x(n)$, what happens to the spectrum of the signal? The Fourier transform of a continuous signal is:

$$X(\Omega) = \int_{-\infty}^{+\infty} x(t)e^{-\jmath\Omega t}\,dt.$$

The sampling impulses are impulse or "delta" functions $\delta(t)$ spaced at intervals of T:

$$r(t) = \sum_{n=-\infty}^{+\infty} \delta(t - nT).$$

The sampled function is the product of the signal at the sampling instants, and the sampling impulses:

$$\begin{aligned} x_s(t) &= x(t)r(t) \\ &= \sum_{n=-\infty}^{+\infty} x(t)\delta(t - nT). \end{aligned} \tag{7.21}$$

Since at the sample instants $t = nT$, this becomes:

$$x_s(t) = \sum_{n=-\infty}^{+\infty} x(nT)\delta(t - nT).$$

The Fourier transform of the sampled signal is thus:

$$\begin{aligned} X(\Omega) &= \int_{-\infty}^{+\infty} x_s(t)e^{-j\Omega t}\,dt \\ &= \int_{-\infty}^{+\infty} \sum_{n=-\infty}^{+\infty} x(nT)\delta(t - nT)e^{-j\Omega t}\,dt \\ &= \sum_{n=-\infty}^{+\infty} \int_{-\infty}^{+\infty} x(nT)\delta(t - nT)e^{-j\Omega t}\,dt \\ &= \sum_{n=-\infty}^{+\infty} x(nT)e^{-jn\Omega T}. \end{aligned} \tag{7.22}$$

Note that the delta function has been used here, and the property that:

$$\int_{-\infty}^{+\infty} \delta(t_k) f(t)\,dt = f(t_k). \tag{7.23}$$

Since $t = nT$ (time) and $\omega = \Omega T$ (frequency), this yields the Fourier transform of a *sampled* signal as

$$X(\omega) = \sum_{n=-\infty}^{+\infty} x(n)e^{-jn\omega}. \tag{7.24}$$

with

$$t \text{ seconds} = n \text{ samples} \times T \frac{\text{seconds}}{\text{sample}} \tag{7.25}$$

and

$$\omega \frac{\text{radians}}{\text{sample}} = \Omega \frac{\text{radians}}{\text{second}} \times T \frac{\text{seconds}}{\text{sample}} \tag{7.26}$$

The (continuous) Fourier transform of a sampled signal as derived above shows that since the complex exponential function is periodic, the spectrum will be periodic—it repeats cyclically, about intervals centered on the sample frequency. Furthermore, the frequency range $f_s/2$ to f_s is a mirror image of the range 0 to $f_s/2$. The discrete time Fourier transform (DTFT) may be visualized directly as the multiplication by basis function as illustrated in Figure 7.10. The difference is that the basis functions of sine and cosine are multiplied by the sample values $x(n)$. The frequency ω is, however, free to take on any value (it is *not* sampled).

Since the frequency spectrum is still continuous, the inverse of the DTFT above is

$$x(n) = \frac{1}{2\pi} \int_{-\pi}^{\pi} X(\omega)e^{jn\omega}\,d\omega \tag{7.27}$$

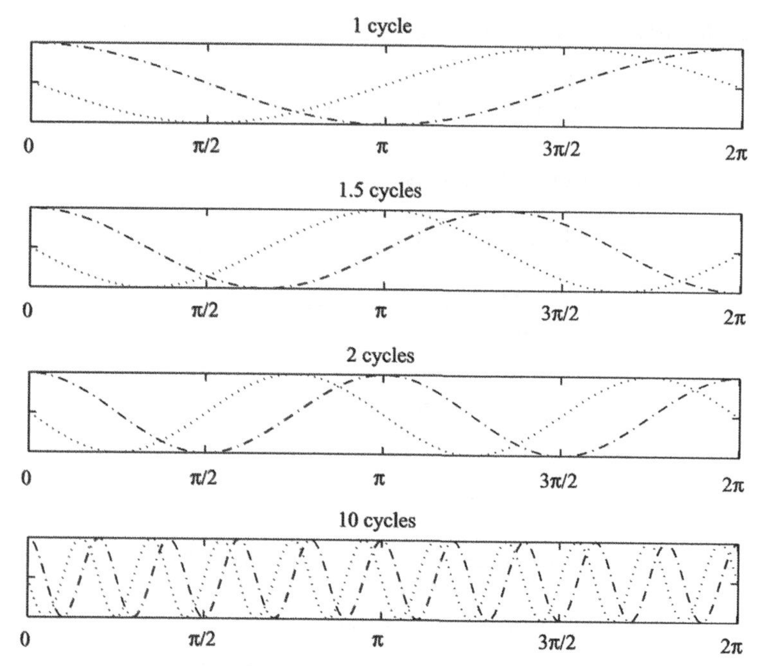

FIGURE 7.10 Basis functions for the DFT.

To summarize the fundamental result from this section, we have

Discrete Time Fourier Transform: *Sampled time* → *Continuous frequency*

7.6.3 The Discrete Fourier Transform (DFT)

The discrete-time/discrete-frequency Fourier transform—usually known as just the discrete Fourier transform or DFT—is the final stage in the conversion to an all-sampled system. As well as sampling the time waveform, only discrete frequency points are calculated. The DFT equation is:

$$X(k) = \sum_{n=0}^{N-1} x(n) e^{-jn\omega_k}, \tag{7.28}$$

where

$$\omega_k = \frac{2\pi k}{N} \tag{7.29}$$

$= \text{frequency of the } k^{\text{th}} \text{ sinusoid.}$

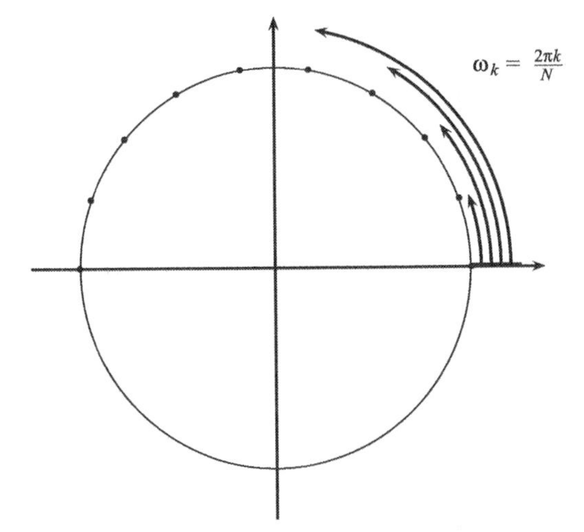

FIGURE 7.11 DFT basis functions as points on the unit circle. This is described by a complex exponential $e^{j\omega_k}$.

The discrete basis functions are now at points on the complex plane, or $1 \cdot e^{-jn\omega_k}$, as illustrated in Figure 7.11. This gives rise to the result that the DFT is now cyclic, and in effect, only the frequencies up to $\omega = \pi$ are of interest. A consequence of the frequency points going all the way around the unit circle is that when we take the DFT, the upper half of the resulting components are a mirror image of the lower half. This will be illustrated later using some examples.

Recall that the z transform was defined in Section 5.4 as the expansion:

$$\begin{aligned} F(z) &= f(0)z^0 + f(1)z^{-1} + f(2)z^{-2} + \cdots \\ &= \sum_{n=0}^{\infty} f(n)z^{-n}. \end{aligned} \tag{7.30}$$

Comparing this with the definition of the DFT in Equation 7.28, it may be seen that the DFT is really a special case of the z transform with the substitution

$$z = e^{j\omega_k}. \tag{7.31}$$

That is, the DFT is the z transform evaluated at the discrete points around the unit circle defined by the frequency ω. Understanding this result is crucial to understanding the applicability and use of Fourier transforms in sampled-data systems.

As in previous cases, there is an inverse operation to convert from the frequency domain back to the time domain samples. In this case, it is the inverse DFT to convert frequency samples $X(k)$ back to time samples $x(n)$, and is given by:

$$x(n) = \frac{1}{N}\sum_{k=0}^{N-1} X(k)e^{jn\omega_k}. \tag{7.32}$$

Note that the forward and inverse operations are quite similar: the difference is in the sign of the exponent and the division by N.

To summarize the fundamental result from this section, we have:

Discrete Fourier Transform:
Sampled time $\rightarrow$ *Sampled frequency*

7.6.4 The Fourier Series and the DFT

Clearly, the Fourier series and Fourier transform are related, and we now wish to explore this idea. Recall that the Fourier series for a continuous function $x(t)$ was shown in Section 7.3 to be:

$$\begin{aligned} x(t) = a_0 + & \\ & a_1 \cos \Omega_o t + a_2 \cos 2\Omega_o t + a_3 \cos 3\Omega_o t + \cdots \\ & b_1 \sin \Omega_o t + b_2 \sin 2\Omega_o t + b_3 \sin 3\Omega_o t + \cdots \\ = a_0 + & \sum_{k=1}^{\infty} (a_k \cos k\Omega_0 t + b_k \sin k\Omega_0 t) \end{aligned} \tag{7.33}$$

with coefficients a_k and b_k:

$$a_0 = \frac{1}{\tau} \int_0^{\tau} x(t)\, dt \tag{7.34}$$

$$a_k = \frac{2}{\tau} \int_0^{\tau} x(t) \cos k\Omega_o t\, dt \tag{7.35}$$

$$b_k = \frac{2}{\tau} \int_0^{\tau} x(t) \sin k\Omega_o t\, dt. \tag{7.36}$$

As before, the integration limits could equally be $-\tau/2$ to $+\tau/2$. Also, the complex form is:

$$x(t) = \sum_{k=-\infty}^{k=+\infty} c_k e^{jk\Omega_o t}. \tag{7.37}$$

with coefficients:

$$c_k = \frac{1}{\tau} \int_0^{\tau} x(t) e^{-jk\Omega_o t}\, dt. \tag{7.38}$$

The discrete Fourier transform was shown to be

$$X(k) = \sum_{n=0}^{N-1} x(n)e^{-jn\omega_k}, \tag{7.39}$$

with $\omega_k = \dfrac{2\pi k}{N}$, which is just the frequency of the kth sinusoid.

Comparing the c_k terms in the Fourier series equation with the $X(k)$ terms in the DFT equation, it may be seen that they are essentially the same except for division by the total time in the Fourier series. The DFT is really a discretized version of the Fourier series, with N samples spaced dt apart. In the limit as $T \to 0$, the DFT becomes a continuous Fourier transform.

By expanding the sine-cosine Fourier series equations and the complex-number form, it may be observed that there is a correspondence between the terms:

$$\begin{aligned} a_0 &= c_0 \\ a_k &= +2\mathcal{Re}\{c_k\} \\ b_k &= -2\mathcal{Im}\{c_k\}. \end{aligned}$$

The division by τ (total waveform period) in the Fourier series equations becomes division by N in the Fourier transform equations. So, the Fourier transform outputs may be interpreted as scaled Fourier series coefficients:

$$\begin{aligned} a_0 &= \frac{1}{N} X(0) \\ a_k &= +\frac{2}{N} \mathcal{Re}\{X(k)\} \\ b_k &= -\frac{2}{N} \mathcal{Im}\{X(k)\}. \end{aligned}$$

Another way to see the reason for the factor of two is to consider the "folding" from 0 to $N/2-1$ and $N/2$ to $N-1$. Half the "energy" of the terms is in the lower half of the output, the other half is in the upper half of the output. An easy solution is to take the first half of the transform coefficients and to double them. The first coefficient $X(0)$ is *not* doubled, as its counterpart is the very first coefficient *after* the output sequence, or the Nth coefficient. This will be illustrated shortly with some examples.

Note that it is important not to misunderstand the relationship between the Fourier series and the DFT. The Fourier series builds upon harmonics (integer multiples) of a fundamental frequency. The DFT dispenses with the integer harmonic constraint, and the limit of frequency resolution is dictated by the number of sample points.

7.6.5 Coding the DFT Equations

Some examples will now be given on the use and behavior of the DFT. First, we need a MATLAB implementation of the DFT equations:

```
% function [XX] = dft(x)
%
% compute the DFT directly
% x = column vector of time samples (real)
% XX = column vector of complex-valued
%     frequency-domain samples
function [XX] = dft(x)
N = length(x);
XX = zeros(N, 1);
for k = 0:N - 1
    wk = 2*pi*k/N;
    for n = 0:N - 1
        XX(k + 1) = XX(k + 1) + x(n + 1)*exp(-j*n*wk);
    end
end
```

For the sake of efficiency, the loops may also be coded in MATLAB using vectorized operations as follows (vectorizing code was covered in Chapter 2—the numerical result does not change, but the calculation time is reduced).

```
% MATLAB vectorized version
function [XX] = dftv(x)
x = x(:);
N = length(x);
XX = zeros(N, 1);
n = [0:N - 1]';
for k = 0:N - 1
    wk = 2*pi*k/N;
    XX(k + 1) = sum(x.*exp(-j*n*wk));
end
```

7.6.6 DFT Implementation and Use: Examples in MATLAB

So, what about the relationship between the DFT component index and the actual frequency? The DFT spans the relative frequency range from 0 to 2π radians per sample, which is equivalent to a real frequency range of $0 - f_s$. There are $N + 1$ components in this range, hence the spacing in frequency of each is f_s/N. The "zeroth" component is the zero-frequency or "average" value. The first component is one N^{th} of the way to f_s. This is reasonable, since a waveform of exactly one sinusoidal cycle over N components has a frequency of one N^{th} of f_s.

The scaling in time and frequency of the DFT components will now be illustrated by way of some MATLAB examples. The following examples use the built-in MATLAB function `fft()`, or fast Fourier transform. The details of this algorithm will be explained in Section 7.12, but for now it is sufficient to know that it is identical to the DFT, *provided the number of input samples is a power of 2.*

To begin with, recall some basic sampling terminology and equations from Chapter 5. A sinusoidal waveform was mathematically described as:

$$x(t) = A \sin \Omega t.$$

Sampled at $t = nT$, this became the samples $x(n)$ at instant n:

$$\begin{aligned} x(n) &= A \sin \Omega nT \\ &= A \sin n\Omega T. \end{aligned}$$

Defining ω as

$$\omega = \Omega T,$$

we have:

$$x(n) = A \sin n\omega, \tag{7.40}$$

where ω is measured in radians per sample.

For one complete cycle of the waveform in N samples, the corresponding angle ω swept out in one sample period will be just $\omega = 2\pi/N$.

7.6.6.1 DFT Example 1: Sine Waveform To begin with a simple signal, we use MATLAB to generate one cycle of a sinusoidal waveform, and analyze it using the DFT. Since there is only one sinusoidal component, the resulting DFT array $X(k)$ should have only one component. Sixty-four samples are used for this example:

```
N = 64;
n = 0:N - 1;
w = 2*pi/N;
x = sin(n*w);
stem(x);
XX = fft(x);
```

The zero-frequency or DC coefficient is zero as expected. The value is not precisely zero, because normal rounding errors apply:

```
XX(1)/N
ans =
      8.8524e - 018
```

The scaled coefficients 1, 2, 3 are:

```
XX(2:4)/N*2
ans =
     0.0 - 1.0i   0.0 - 0.0i   0.0 - 0.0i
```

Note[2] that coefficient number 1 (at index 2) is $-1j$. This may be interpreted in terms of the basis functions, since real numbers correspond to cosine components, and imaginary correspond to negative sine. So a positive sine coefficient equates to a negative imaginary value. Coefficients 64, 63, 62 are the complex conjugates of the samples at indexes 2, 3, and 4.

```
XX(N - 2:N)/N*2
ans =
     0.0 + 0.0i   0.0 + 0.0i   0.0 + 1.0i
```

This is due to the fact that the frequency ω_k goes all the way around the unit circle, as shown previously. The range from 0 to π corresponds to frequencies 0 to $f_s/2$ (half the sampling rate). The range from π to 2π is a "mirror image" of the lower frequency components. Plotting the DFT magnitudes and phases is accomplished with:

```
m = abs(XX);
p = angle(XX);
iz = find(abs(m < 1));
p(iz) = zeros(size(iz));
subplot(2, 1, 1); stem(m);
subplot(2, 1, 2); stem(p);
```

Note that the phase angle of components with a small magnitude was set to zero for clarity in lines 3 and 4.

7.6.6.2 *DFT Example 2: Waveform with a Constant Offset* Again, we generate a 64-sample sine waveform, but this time with a zero-frequency or DC component. The code to test this is shown below, along with the results for the zero-frequency (DC) component, and the first frequency component. These have values of 5 and $-1j$, as expected.

[2]Remember that MATLAB uses i rather than j to denote the complex quantity $j = \sqrt{-1}$.

```
N = 64;
n = 0:N - 1;
w = 2*pi/N;
x = sin(n*w) + 5;
stem(x);
XX = fft(x);
% DC component
XX(1)/N
ans =
      5
% first two frequency components
XX(2:4)/N*2
ans =
      0.0 - 1.0i   0   0.0 - 0.0i
```

Note that, as always, index 1 in the vector is the subscripted component $X(0)$. This result is as we would expect—we have a zero-frequency or DC component, together with a $-1j$ for the sinusoidal component.

7.6.6.3 DFT Example 3: Cosine Waveform The previous example used a sine component. In this example, we generate a 64-sample cosine waveform with a DC component:

```
N = 64;
n = 0:N - 1;
w = 2*pi/N;
x = cos(n*w) + 5;
stem(x);
XX = fft(x);
XX(1)/N
ans =
      5
XX(2:4)/N*2
ans =
      1.0 - 0.0i   0   0.0 + 0.0i
```

As with the earlier discussion on basis functions, the value of $1 + j0$ indicates the presence of a cosine component. The magnitude of the value also tells us the magnitude of this component.

7.6.6.4 DFT Example 4: Waveform with Two Components This example illustrates the process of resolving the underlying components of a waveform. We take a 256-sample waveform with two additive components, and examine the DFT:

```
N = 256;
n = 0:N - 1;
w = 2*pi/N;
x = 7*cos(3*n*w) + 13*sin(6*n*w);
plot(x);
XX = fft(x);
XX(1)/N
ans =
        -2.47e - 016
XX(2:10)/N*2
ans =
        Columns 1 through 4
        0.0 + 0.0i   0.0 + 0.0i    7.0 - 0.0i  0.0 + 0.0i
        Columns 5 through 8
        0.0 + 0.0i   0.0 - 13.0i  0.0 - 0.0i  0.0 - 0.0i
        Column 9
        0.0 - 0.0i
```

Note that the DFT is able to resolve the frequency components present. The frequency resolution may be seen from the fact that there are N points spanning up to f_s, and so the resolution of each component is f_s/N.

7.6.6.5 DFT Example 5: Limits in Frequency To look more closely at the frequency scaling of the DFT, we generate an alternating sequence of samples having values +1, −1, +1, −1, ... This is effectively a waveform at half the sampling frequency.

```
N = 256;
x = zeros(N, 1);
x(1) = 1;
for n = 2:N
     x(n) = -1*x(n - 1);
end
f = fft(x);
stem(abs(f));
[v i] = max(abs(f))
v =
     256
i =
     129
```

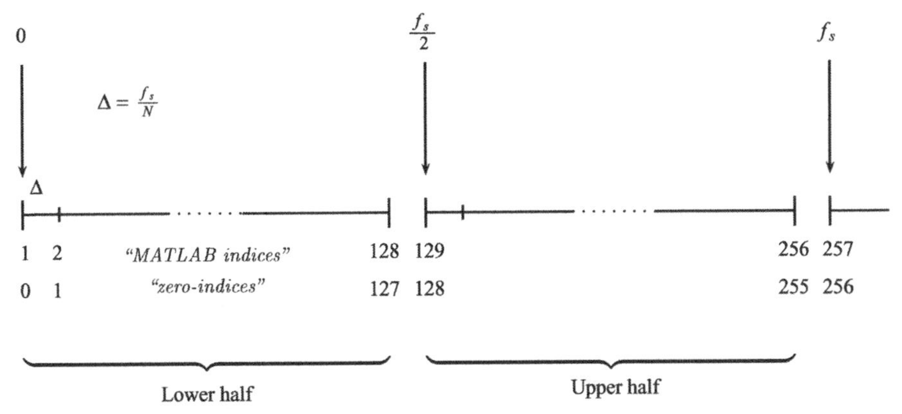

FIGURE 7.12 Illustrating DFT components for an $N = 256$ transform. MATLAB indices start at 1, whereas the DFT equations assume zero-based indexing. The set of upper half returned values is a mirror image of the lower half, with the "pivot point" being the $f_s/2$ sample.

In MATLAB, vector indices for a 256-point vector are numbered from 1 to 256, as illustrated in Figure 7.12. Starting at 0, this corresponds to the range 0–255. Component number 129 contains the only frequency component. Now, the first half of the frequency components are numbered 1–128, so component 129 is the first component in the second half—and this corresponds to the input frequency, namely $f = f_s/2$. The frequency $f = f_s$ is, in fact, the first component *after* the last, or the 257th component (not generated here). Thus the frequency interval is f_s/N. The first component, number 1, is actually the zero-frequency or "DC" component. Hence, it is easier to imagine the indices from 0 to $N - 1$. Using the MATLAB indices of $k = 1, \ldots, N$, we have:

The "true" frequency of component 1 is $0 \times \frac{f_s}{N} = 0$.

The true frequency of component 2 is $1 \times \frac{f_s}{N}$.

The true frequency of component 3 is $2 \times \frac{f_s}{N}$.

The true frequency of component k is $(k-1) \times \frac{f_s}{N}$.

In the example,

The true frequency of component 257 (not generated) is $256 \times \frac{f_s}{256} = f_s$.

The true frequency of component 129 is $128 \times \frac{f_s}{256} = \frac{f_s}{2}$.

Figure 7.12 illustrates these concepts.

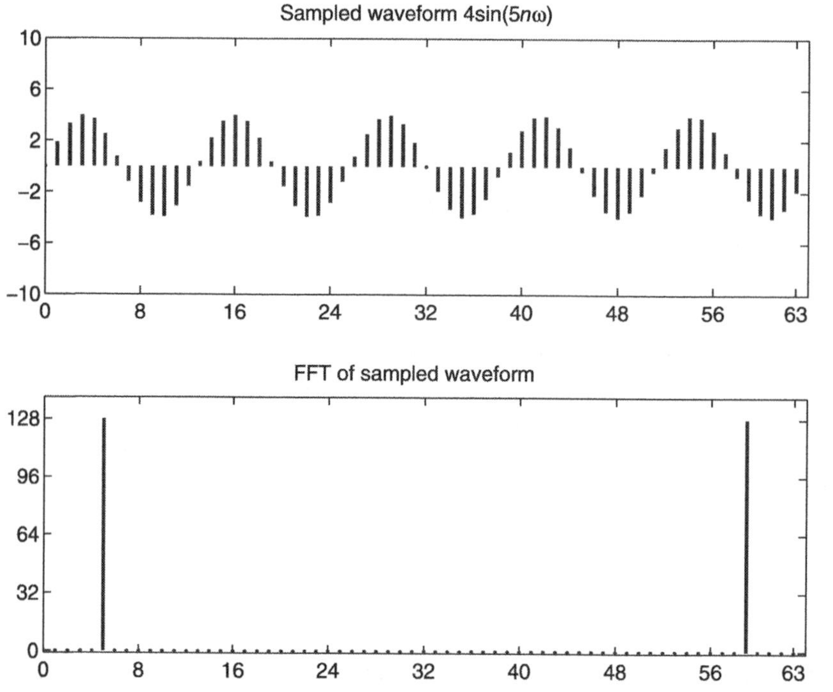

FIGURE 7.13 DFT example: 5 cycles of amplitude 4 in the time and frequency domains.

7.6.6.6 DFT Example 6: Amplitude and Frequency If we now alter both the amplitude and frequency of the input signal, the DFT result ought to reflect this. The waveform with five cycles of a simple sinusoidal waveform of amplitude 4 is calculated as:

$$\begin{aligned} N &= 64 \\ n &= 0, 1, \ldots, N-1 \\ \omega &= \frac{2\pi}{N} \\ x(n) &= 4\sin(5n\omega). \end{aligned} \tag{7.41}$$

The waveform and its DFT are shown in Figure 7.13. Note in particular:

1. The index of the lower frequency peak and its corresponding frequency.
2. The location of the higher frequency peak (the mirror image).
3. The height of the frequency peaks.

7.6.6.7 DFT Example 7: Resolving Additive Waveforms Extending the previous example, we now take the DFT of two additive sinusoidal waveforms, each with a different amplitude and frequency:

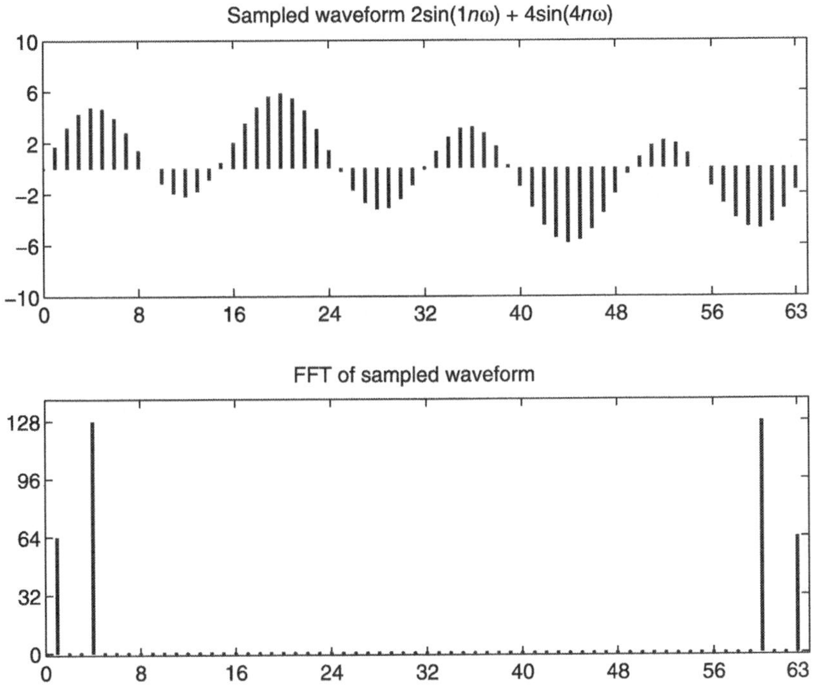

FIGURE 7.14 DFT example: two sinusoidal waveforms in the time and frequency domains.

$$
\begin{aligned}
N &= 64 \\
n &= 0, 1, \ldots, N-1 \\
\omega &= \frac{2\pi}{N} \\
x(n) &= 2\sin(1n\omega) + 4\sin(4n\omega).
\end{aligned}
\tag{7.42}
$$

The waveform and its DFT are shown in Figure 7.14. Examination of the time domain waveform alone does not indicate the frequencies which are present. However, the Fourier transform clearly shows the frequencies and their amplitudes. Combining the sine/cosine components in this way eliminates the phase information, however. That is, it is not possible to determine from the figures whether the underlying waveforms are sine or cosine. However, the complex components themselves yield this information, as we have seen in earlier examples.

7.6.6.8 DFT Example 8: Spectral Leakage What happens when the period of the underlying waveform does not match that of the sampling window? That is, we do not have an integral number of cycles of the waveform to analyze. The waveform parameters are similar to those encountered previously:

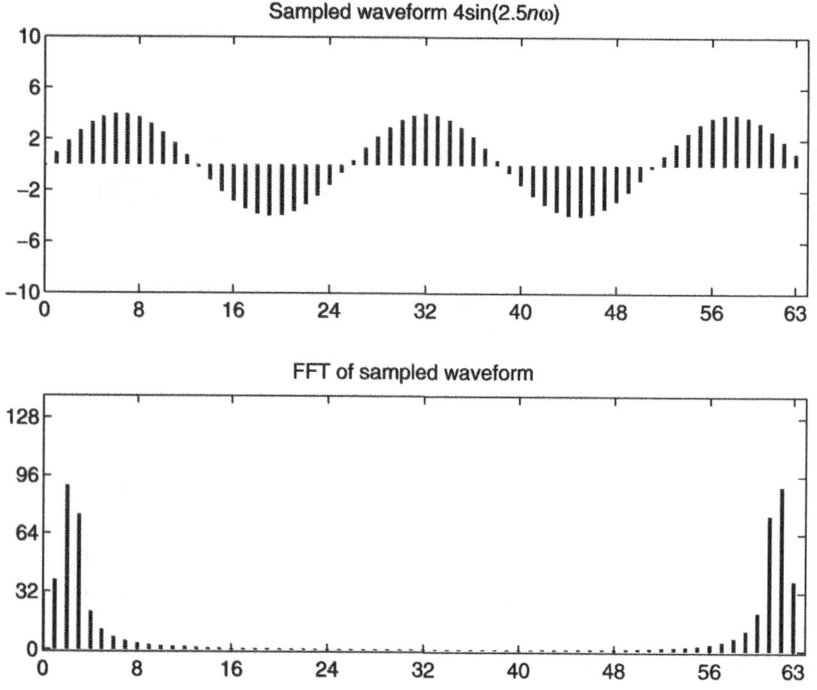

FIGURE 7.15 DFT example: a nonintegral number of cycles of a sinusoidal waveform. In this case, the DFT does not resolve the waveform into a single component as might be expected.

$$\begin{aligned} N &= 64 \\ n &= 0, 1, \ldots, N-1 \\ w &= \frac{2\pi}{N} \\ x(n) &= 4\sin(2.5\, n\omega). \end{aligned} \tag{7.43}$$

The sampled waveform is shown in Figure 7.15—note that there are $2\frac{1}{2}$ cycles in the analysis window, as expected. The product of some of the basis-vector waveforms and the input waveform will be nonzero, resulting in some nonzero sine and cosine components around the frequency of the input.

The DFT of this waveform is shown in Figure 7.15. Note the "spillover" in the frequency components, due to the fact that a noninteger number of cycles was taken for analysis. This is a fundamental problem in applying the DFT, and motivates the question: if we do not know in advance the expected frequencies contained in the waveform, how long do we need to sample for, in order for this problem not to occur? We have already determined that the resolution in the DFT is governed by the number of samples in the time domain, but now we have the further problem as illustrated in this example. To answer this question fully requires some further analysis and consideration of the "sampling window" in Section 7.9. But clearly, we must be careful in interpreting the results produced by the DFT.

7.6.6.9 DFT Example 9: Additive Noise Now consider the case of nonperiodic noise added to sinusoidal components. This type of scenario is typical of communications and speech processing systems, where one or more periodic signals have noise superimposed.

$$\begin{aligned} N &= 64 \\ n &= 0, 1, \ldots, N-1 \\ w &= \frac{2\pi}{N} \\ x(n) &= 2\sin(1\,n\omega) + 4\sin(4\,n\omega) + \alpha v(n). \end{aligned} \tag{7.44}$$

Here, $v(n)$ represents the additive noise signal, with the amount controlled by parameter α. The waveform and its DFT are shown in Figure 7.16. From these plots, it is evident that the periodic components still feature strongly in the frequency domain. The DFT step makes the underlying components clearer than in the time domain waveform.

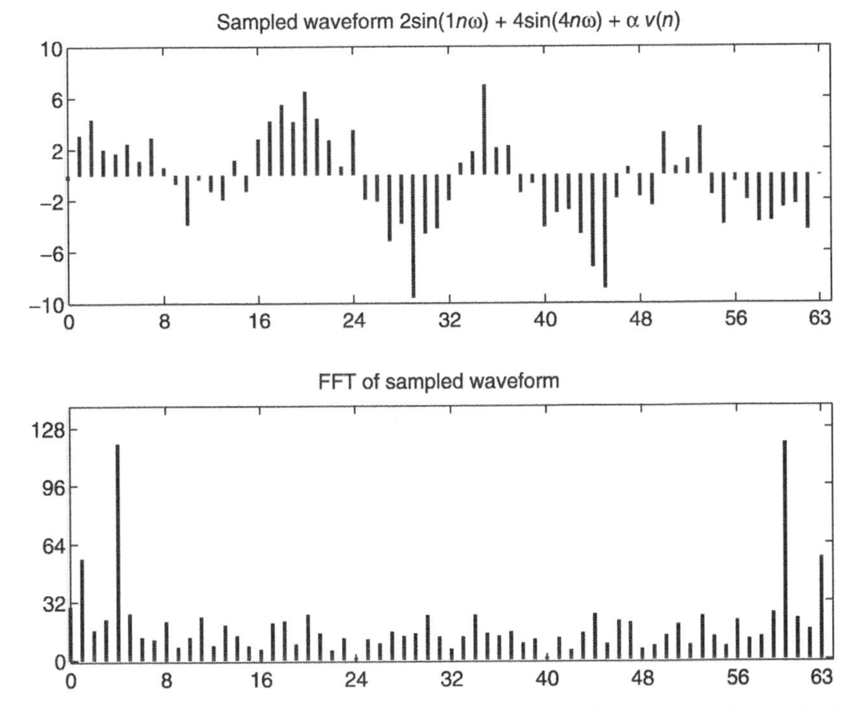

FIGURE 7.16 DFT example: a complex sinusoidal waveform with additive noise in the time and frequency domains.

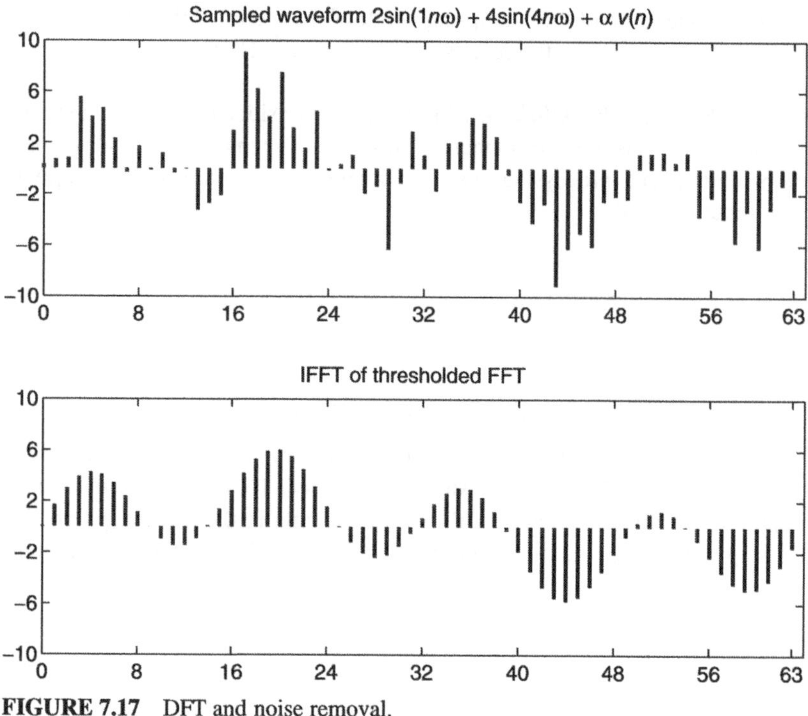

FIGURE 7.17 DFT and noise removal.

7.6.6.10 DFT Example 10: Noisy Waveform Restoration This example considers a waveform with additive noise. Thresholding is done in the frequency domain (note the MATLAB vectorized method of doing this). The real part of the inverse FFT, **`ifft()`**, is taken, because although the imaginary part should be zero in the time domain, it may have a very small imaginary component due to rounding and the effects of finite numerical precision.

The result, shown in Figure 7.17, shows that the operation of taking the frequency components, thresholding, and taking the inverse DFT is able to restore the signal. In practice, however, taking an arbitrary threshold may not be desirable, since the actual signal as presented may vary from the test signal shown. Thus, a more advanced approach may be warranted—for example, if more than one frame of data signal is available, we may be able to average over multiple frames of signal rather than base the results on one single frame.

```
N = 64;
n = 0:N - 1;
w = 2*pi/N;
x = 2*sin(1*n*w) + 4*sin(4*n*w) + 1*randn(1,N);
```

```
subplot(2,1,1);
plot(n,x);
f = fft(x);
i = find(abs(f) < 40);
f(i) = zeros(size(i));
xr = ifft(f);
subplot(2,1,2);
plot(n,xr);
```

7.7 ALIASING IN DISCRETE-TIME SAMPLING

The problem of choosing the sampling rate for a given signal was discussed in Section 3.8. The Fourier transform now provides us with a more analytical way to look this problem, and the related problem of aliasing when we sample a signal.

In Section 3.5, sampling was introduced as a multiplication of the continuous-time analog signal with the sampling pulses (the "railing" function). The Fourier series expansion of the sampling impulses of frequency Ω_s can be shown to be:

$$\begin{aligned} r(t) &= \frac{1}{T} + \frac{2}{T}\cos\Omega_s t + \frac{2}{T}\cos 2\Omega_s t + \frac{2}{T}\cos 3\Omega_s t + \cdots \\ &= \frac{2}{T}\left(\frac{1}{2} + \cos\Omega_s t + \cos 2\Omega_s t + \cos 3\Omega_s t + \cdots\right). \end{aligned} \tag{7.45}$$

We know that sampling is equivalent to multiplying the signal $x(t)$ by the railing function $r(t)$ to obtain the sampled signal $x_s(t)$. To simplify notation, we replace the cosine function with its complex exponential equivalent:

$$\begin{aligned} r(t) &= \frac{2}{T}\left(\frac{1}{2} + \cos\Omega_s t + \cos 2\Omega_s t + \cos 3\Omega_s t + \cdots\right) \\ &= \frac{1}{T}\left(1 + 2\cos\Omega_s t + 2\cos 2\Omega_s t + 2\cos 3\Omega_s t + \cdots\right) \\ &= \frac{1}{T}\left(e^{j0} + \left(e^{j\Omega_s t} + e^{-j\Omega_s t}\right) + \left(e^{2j\Omega_s t} + e^{-2j\Omega_s t}\right) + \left(e^{3j\Omega_s t} + e^{-3j\Omega_s t}\right) + \cdots\right) \\ &= \frac{1}{T}\sum_{k=-\infty}^{+\infty} e^{jk\Omega_s t}. \end{aligned} \tag{7.46}$$

The sampled signal is $x_s(t) = x(t)r(t)$, and we need to determine the effect of the sampling in the frequency domain. As before, the Fourier transform of the continuous signal is:

$$X(\Omega) = \int_{-\infty}^{+\infty} x(t)e^{-j\Omega t}\,dt.$$

Using these results, we can derive the Fourier transform of the sampled signal as follows.

$$
\begin{aligned}
X_s(\Omega) &= \int_{-\infty}^{+\infty} x_s(t) e^{-j\Omega t}\, dt \\
&= \int_{-\infty}^{+\infty} x(t) r(t) e^{-j\Omega t}\, dt \\
&= \int_{-\infty}^{+\infty} x(t) \overbrace{\left(\frac{1}{T} \sum_{-\infty}^{+\infty} e^{jk\Omega_s t} \right)}^{r(t)} e^{-j\Omega t}\, dt \\
&= \frac{1}{T} \sum_{k=-\infty}^{+\infty} \left(\int_{-\infty}^{+\infty} x(t) e^{jk\Omega_s t} e^{-j\Omega t}\, dt \right) \\
&= \frac{1}{T} \sum_{k=-\infty}^{+\infty} \left(\int_{-\infty}^{+\infty} x(t) e^{-j(\Omega - k\Omega_s)t}\, dt \right) \\
&= \frac{1}{T} \sum_{k=0}^{+\infty} \left(\int_{-\infty}^{+\infty} x(t) e^{-j(\Omega + k\Omega_s)t}\, dt \right).
\end{aligned}
\tag{7.47}
$$

The integral (in brackets) is $X(\Omega \pm k\Omega_s)$, which is the spectrum $X(\Omega)$ shifted by $\pm k\Omega_s$. So the spectrum of the sampled waveform is:

$$
X_s(\Omega) = \frac{1}{T} \sum_{k=0}^{\infty} X(\Omega \pm k\Omega_s) \tag{7.48}
$$

The above equation demonstrates why aliasing occurs, and may be expressed in words as follows:

> The spectrum of the signal $X(\Omega)$ is "mirrored" at
> $\pm\Omega_s, \pm 2\Omega_s, \pm 3\Omega_s, \ldots$

We can understand this further by returning to the first introduction of aliasing in Chapter 3. Figure 7.18 shows the problem in the time domain, and now we are able to interpret this in terms of frequency. Referring to Figure 7.18, we first sample below the input frequency. Clearly, we cannot expect to capture the information about the waveform if we sample at a frequency lower than the waveform itself, so this is just a hypothetical case. The waveform *appears* to have a frequency of $|f - f_s| = |4 - 3| = 1\,\text{Hz}$. Next, we sample at 5 Hz, which is a little above the waveform frequency. This is still inadequate, as shown. The frequency is effectively "folded" down to $|f - f_s| = |4 - 5| = 1\,\text{Hz}$. Finally, we sample at greater than twice the frequency ($f_s > 2f$, or $10\,\text{Hz} > 2 \times 4\,\text{Hz}$). The sampled waveform does not produce a lower "folded-down" frequency.

Figure 7.19 illustrates this approach, from the frequency domain perspective. It shows that as the sampling frequency is reduced relative to the frequency spectrum of the signal being sampled, a point will be reached where the baseband spectrum and the "ghost" spectrum overlap. Once this occurs, the signal cannot be inverted back into the time domain without aliasing occurring.

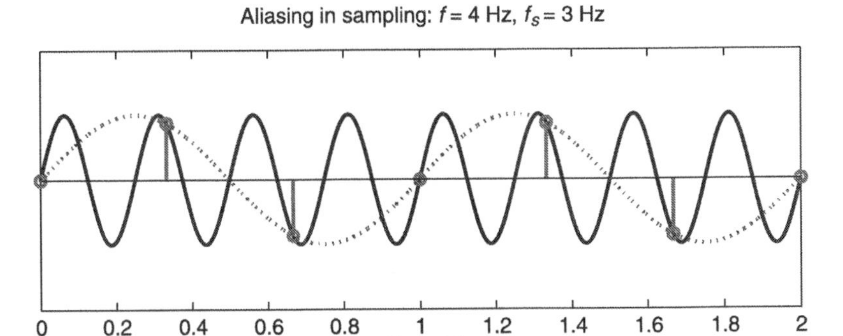

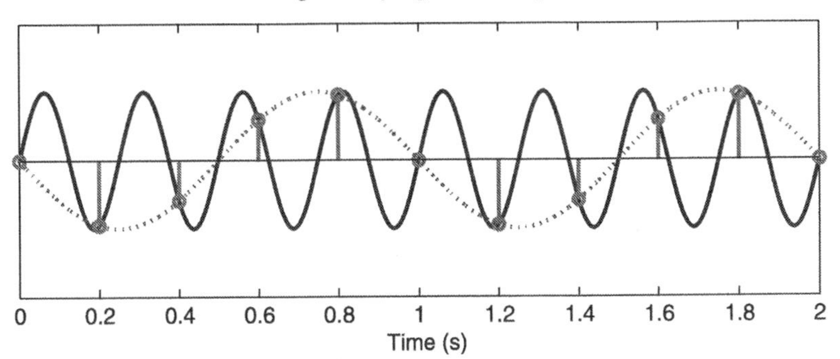

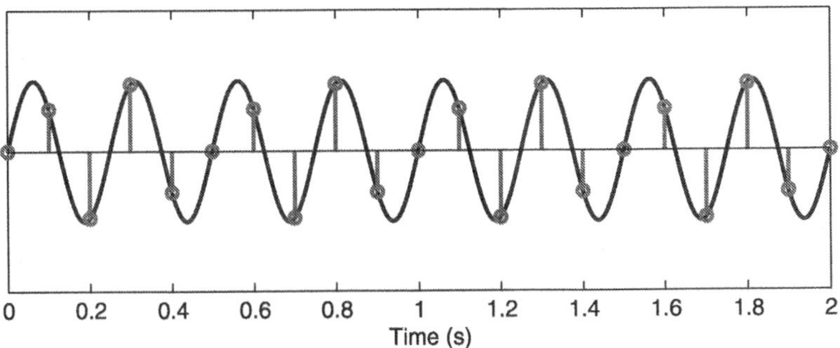

FIGURE 7.18 How aliasing arises in the time domain. A 4-Hz sinewave is sampled, so the Nyquist rate is 8 Hz, and we should sample above that rate.

7.8 THE FFT AS A SAMPLE INTERPOLATOR

As an interesting insight, consider the generation of missing samples in a sequence. This is related to the problem of interpolating the signal value between known samples, as introduced in Section 3.9.1.

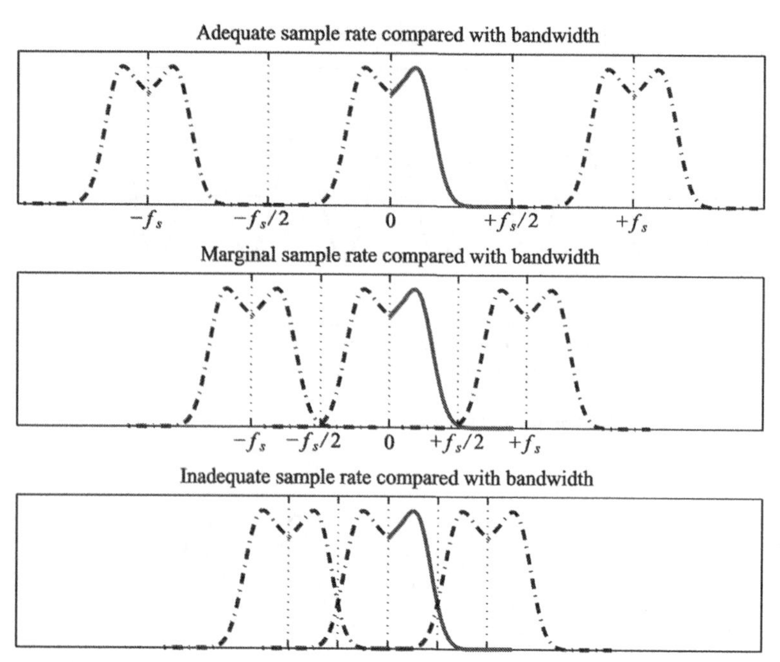

FIGURE 7.19 Sampling rate and "folding frequency". The true spectrum is that shown over the range 0 to $f_s/2$. The secondary spectral images are artifacts of the sampling process. If the spectra overlap, aliasing occurs.

To begin with, consider Figure 7.20. This figure shows zeros inserted into the sample sequence between every known sample, so as to generate a new sample sequence. This is about as simple an operation as we could imagine, and effectively doubles the sample rate (even though the new zero-valued samples obviously bear no relation to the true samples).

But what is the effect in the frequency domain? If we take the zero-inserted sequence and examine its FFT, we see, as in Figure 7.21, that spurious components have been introduced. Considering the frequency-domain interpretation, we may say that these correspond to the high-frequency components introduced by the zero values at every second sample.

If we then remove these high-frequency components and take the inverse FFT, we see that the original signal has been "smoothed" out, or interpolated, as shown in Figure 7.22. The following MATLAB code shows how to use the `reshape()` function to perform zero-sample interpolation:

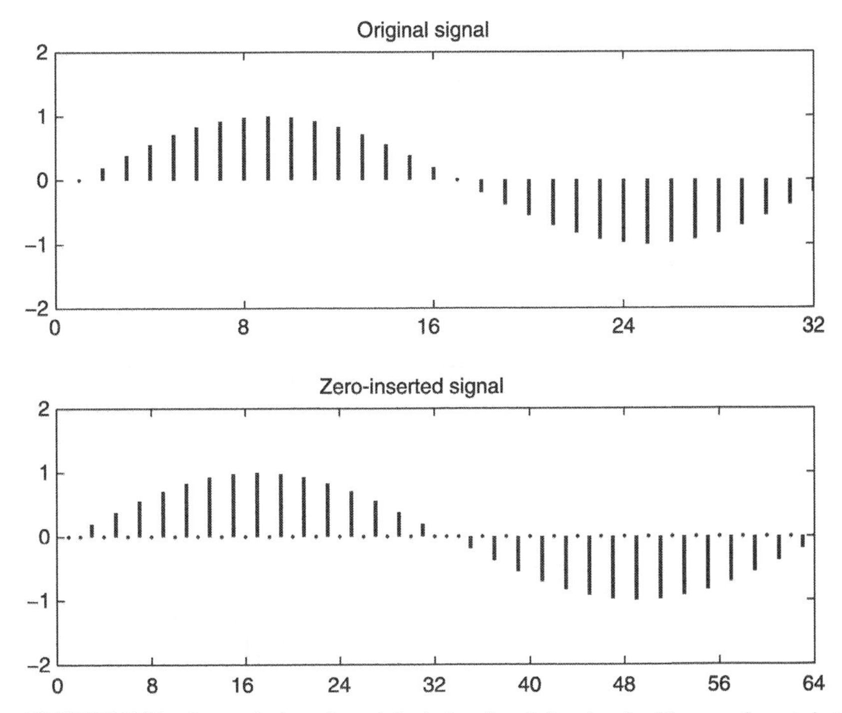

FIGURE 7.20 Interpolation: the original signal and the signal with zeros inserted at every other sample.

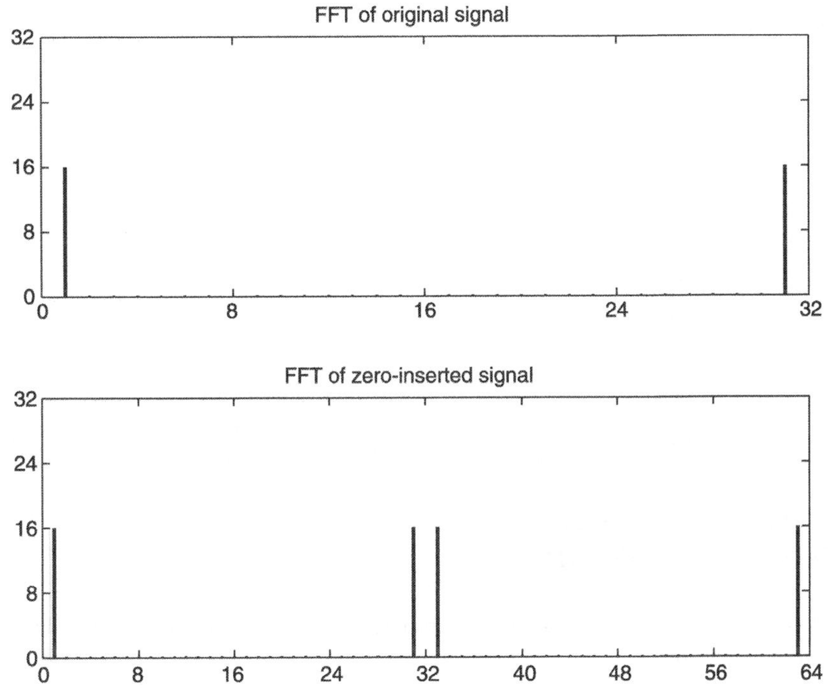

FIGURE 7.21 Interpolation: the DFT of the original signal and the DFT of the signal with zeros inserted at every other sample.

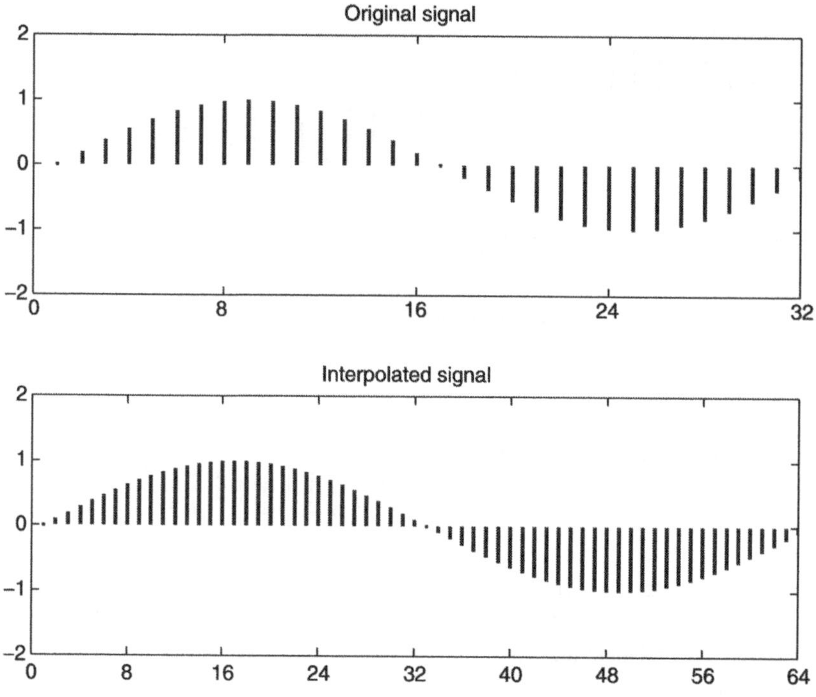

FIGURE 7.22 Interpolation after removing the high-frequency components in the zero-inserted sequence.

```
N = 32;
n = 0:N - 1;
w = 2*pi/N;
x = sin(n*w);
f = fft(x);
xi = [x; zeros(1, N)];
xi = reshape(xi, 2*N, 1);
fi = fft(xi);
fi(32) = 0;
fi(34) = 0;
xr = 2*real(ifft(fi));
```

7.9 SAMPLING A SIGNAL OVER A FINITE TIME WINDOW

Considering the preceding discussion, two problems are apparent with discrete-data spectrum estimation:

1. The sampling rate limits the maximum frequency we can observe, and may introduce aliasing if proper filtering is not present at the input.

2. We can only sample data over a finite time record. The continuous-time Fourier equations used infinite time limits, which is of course impractical.

Furthermore, what happens if the frequency content of the signal changes over the sampling time? We will examine this related issue in Section 7.10. So the question we must examine theoretically is this: since we can only sample for a finite amount of time, what is the effect on the spectrum? As we have seen, the Fourier Transform of a continuous signal is

$$X(\Omega)=\int_{-\infty}^{+\infty} x(t)e^{-j\Omega t}\,dt.$$

If this is truncated to a finite length (because we sample the waveform for a finite time), say $\pm\tau$, then the transform becomes

$$X(\Omega)=\int_{-\tau}^{+\tau} x(t)e^{-j\Omega t}\,dt. \tag{7.49}$$

Suppose $x(t)$ is a sine wave of frequency Ω_o radians/second:

$$x(t)=\sin\Omega_o t. \tag{7.50}$$

In exponential form,

$$x(t)=\frac{1}{2j}\left(e^{j\Omega_o t}-e^{-j\Omega_o t}\right). \tag{7.51}$$

The windowed spectrum becomes:

$$X(\Omega)=\int_{-\tau}^{+\tau}\frac{1}{2j}\left(e^{j\Omega_o t}-e^{-j\Omega_o t}\right)e^{-j\Omega t}\,dt. \tag{7.52}$$

Performing the integration, the magnitude becomes a pair of sinc functions centered at $\pm\Omega_o$,

$$|X(\Omega)|=\tau|\text{sinc}(\Omega_o-\Omega)\tau|+\tau|\text{sinc}(\Omega_o+\Omega)\tau|, \tag{7.53}$$

where the "sinc" function is as previously defined, of the form[3]

$$\text{sinc}\theta=\frac{\sin\theta}{\theta}. \tag{7.54}$$

Noting that the sinc function is a $\sin\theta$ component multiplied by a $1/\theta$ component, we can deduce that the shape is sinusoidal in form, multiplied by an envelope that decays as $1/\theta$. Here, the θ is actually $\Omega\tau$, with Ω offset by the fixed frequency Ω_o. Thus, the resulting spectrum is a pair of sinc functions, centered at $\pm\Omega_o$. The sinc function $(\sin\theta)/\theta$ has zero crossings at $\pm\pi$, $\pm 2\pi$, … , and thus the zero crossings Ω_c of the sinc function are at:

$$(\Omega_o\pm\Omega_c)\tau=\pm k\pi \tag{7.55}$$

$$\therefore\Omega_c=\pm\Omega_o\pm\frac{k\pi}{\tau} \tag{7.56}$$

[3]Some texts define $\text{sinc}\,t=(\sin\pi t)/\pi t$.

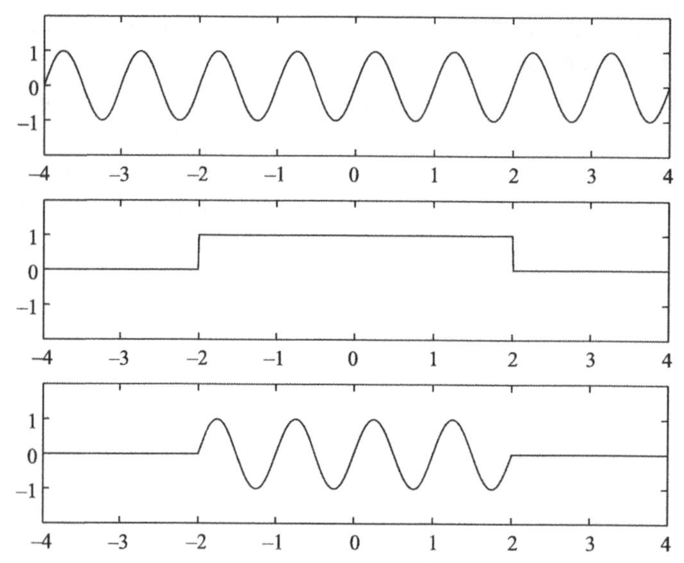

FIGURE 7.23 Windowing: an infinite sine wave (upper) is multiplied by a rectangular window, to yield the windowed waveform (bottom). Note that if we do sample over a finite time limit, we have effectively windowed the waveform. Ambiguity arises because our window (and hence the sampling) is not synchronized to the underlying signal.

That is, as the length of the window increases and $\tau \rightarrow \infty$, the zero crossings tend to be closer and closer to the "true" frequency. In other words, the sinc function tends to an impulse at the frequency of the input sine wave. A *rectangular* window over the time interval $-2 \leq t \leq 2$ is shown in Figure 7.23. Note that the window is "implicit," in that we are taking only a fixed number of samples.

The type of spectral estimate resulting from a rectangular window is shown in the successive plots in Figures 7.24 and 7.25. Each point in the lower panel is one point on the spectrum. This is derived by multiplying the input waveform by each of the sine and cosine basis functions, adding or integrating over the entire time period, and taking the magnitude of the resulting complex number. Figure 7.24 shows the situation when the analysis frequency is relatively low, and Figure 7.25 shows the situation when the analysis frequency is just higher than the input frequency.

This may also be understood by the fundamental theorem of convolution: multiplication in the time domain equals convolution in the frequency domain. Thus, multiplication of a signal which is infinite in extent, by a finite rectangular window in the time domain, results in a spectrum which is the convolution of the true spectrum (an impulse in the case of a sine wave) with the frequency response of the rectangular window. The latter was shown to be a sinc function. Thus, the true frequency estimates are "smeared" by the sinc function.

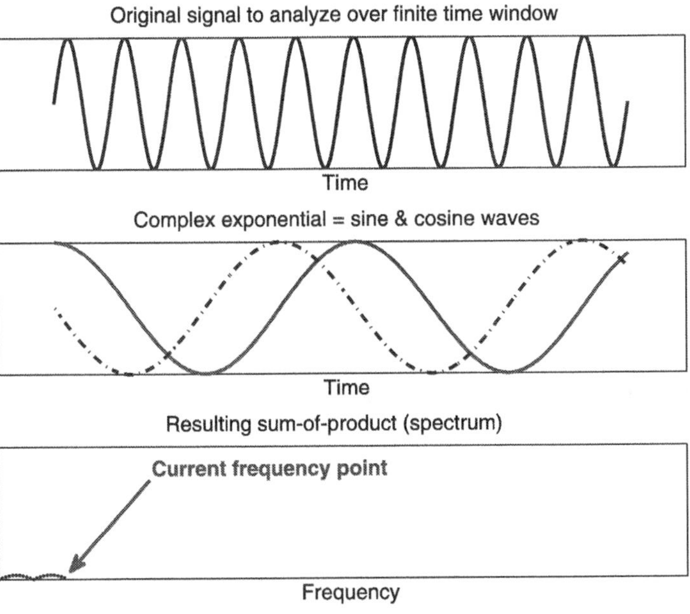

FIGURE 7.24 The effect of a window on the spectrum. The upper panel shows the signal to be analyzed using the sine and cosine basis functions (middle). The resulting spectrum is built up in the lower plot; at this time instant, there is little correlation and so the amplitude is small.

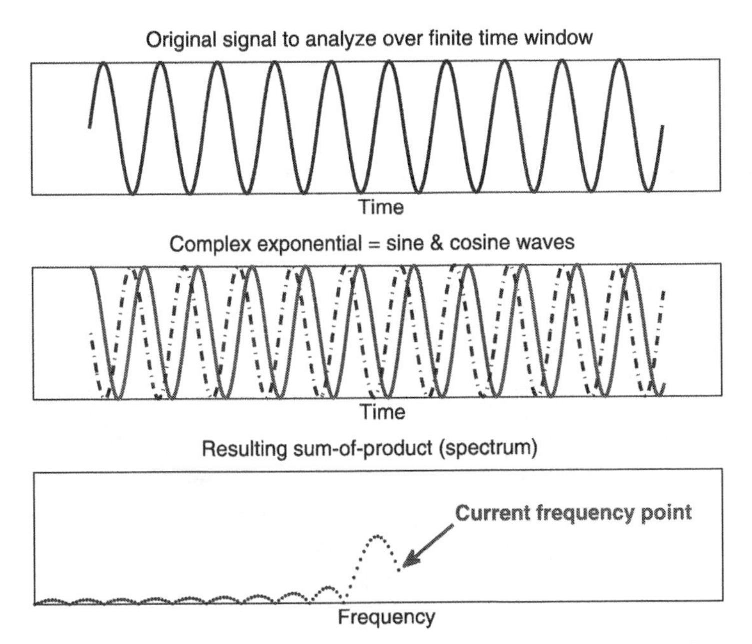

FIGURE 7.25 As the analysis frequency is increased, the signals align to a greater degree and so the spectral magnitude is larger. Clearly, the resultant spectrum is not a single value at the input frequency, as might be expected, but is spread over nearby frequencies. This gives rise to ambiguity in determining the true frequency content of the input signal.

7.10 TIME-FREQUENCY DISTRIBUTIONS

In many applications, we are interested in not simply the spectral content of a signal, but *how* that spectral content *evolves* over time. Probably the best example is the speech signal—the frequency content of speech obviously changes over time. For a simpler real-world example, think of a vehicle with a siren moving toward an observer. Because of the Doppler effect, the frequency increases as the vehicle moves toward the observer, and decreases as it moves away.

An obvious way to approach this problem is to segment the samples into smaller frames or blocks. The Fourier analysis techniques can then be applied to each frame in turn. This is termed a *spectrogram.*

The following generates a "chirp" signal, which is one whose frequency increases over time:

```
Fs = 8000;
dt = 1/Fs;
N = 2^13;     % ensure a power of 2 for FFT
t = 0:dt:(N - 1)*dt;
f = 400;
fi = (1:length(t))/length(t)*f;
y = sin(2*pi*t.*fi);
plot(y(1:2000));
sound(y, Fs);
```

Plotting the FFT magnitude shows only the average power in the signal over *all* time. Using the spectrogram, we obtain a time-frequency plot as shown in Figure 7.26. This shows higher energy with a darker color.

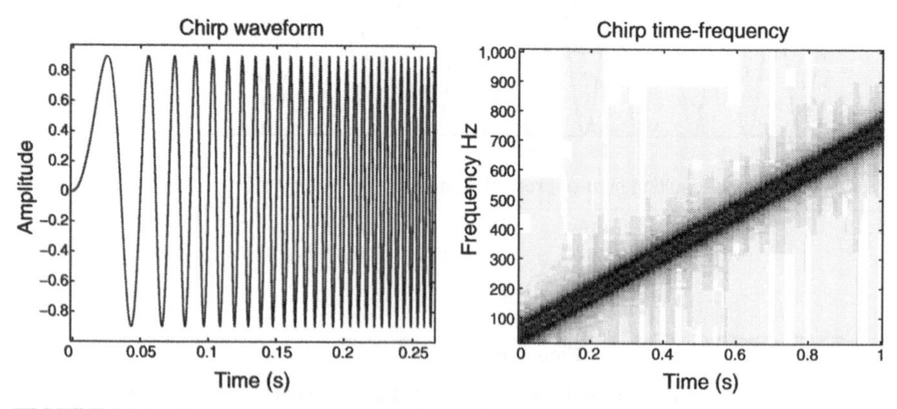

FIGURE 7.26 The evolution of a "chirp" signal, whose frequency increases over time. The darker sections correspond to greater energy in the signal at a particular time and frequency. On the left is a portion of the time domain waveform. On the right is the time-frequency plot.

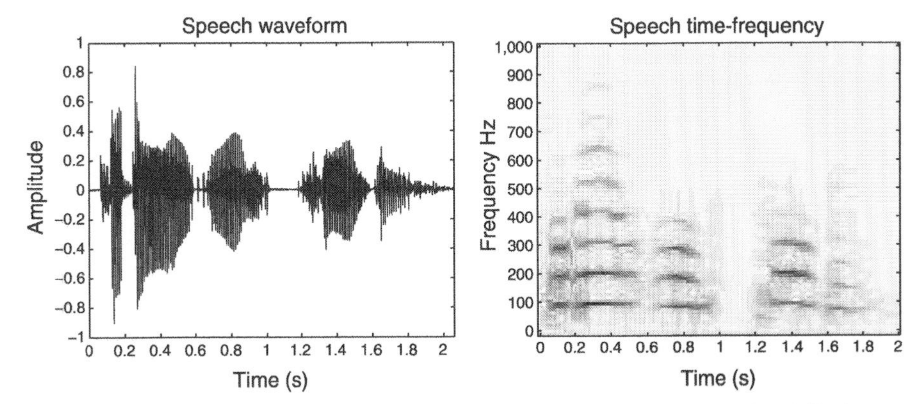

FIGURE 7.27 Spectrogram of voice waveform signal. The time domain plot (left) shows us the source waveform itself, but the time-frequency plot shows the frequency energy distribution over time.

As an illustration of a more complicated signal, Figure 7.27 shows the spectrograms of a speech utterance, the sentence "today is Tuesday." To attain greater accuracy in the frequency domain, a longer sample record is required. However, this means less accuracy in the time domain, because the frequency content of the signal changes relatively quickly.

7.11 BUFFERING AND WINDOWING

The buffering required to implement a spectrogram is shown in Figure 7.28. The samples are split into consecutive frames, and the DFT of each frame is taken. Ideally, a longer frame for the Fourier analysis gives better frequency resolution, since (as seen earlier) the resolution is f_s/N. There is, however, a trade-off required in determining the buffer length: a longer buffer gives better frequency resolution, but since there are fewer buffers, this gives poorer resolution in time. Conversely, smaller buffers give better time resolution, but poorer frequency resolution.

Because we are often trying to capture a "quasi-stationary" signal such as the human voice, the direct buffering method can be improved upon in some situations. Figure 7.29 shows the use of overlapped frames, with a 25% overlap. Each new frame incorporates a percentage of samples from the previous frame. Typical overlap may be from 10 to 25% or more.

To smooth the overlap, a number of so-called "window functions" have been developed. As described above, we have until now been using a rectangular window of the form:

$$w_n = \begin{cases} 1.0 & : \quad 0 \le n \le M \\ 0 & : \quad \text{otherwise} \end{cases}. \tag{7.57}$$

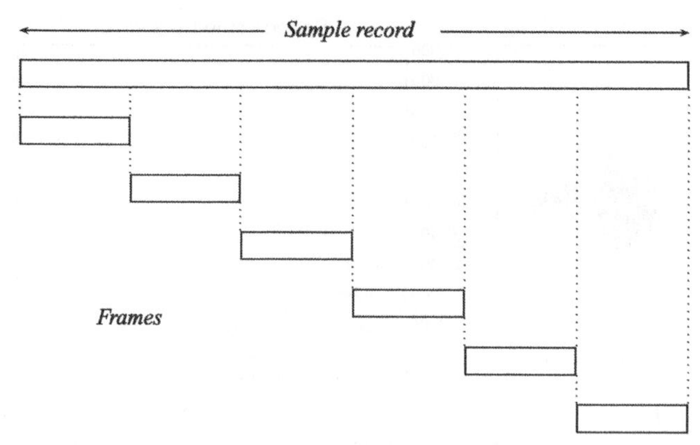

FIGURE 7.28 A signal record split into nonoverlapping frames.

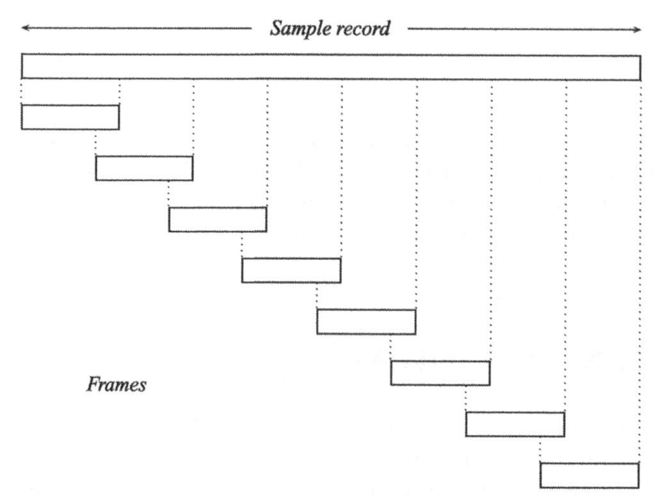

FIGURE 7.29 A signal record split into overlapping frames.

To improve upon this, we need to taper the edges of the blocks so that there is a smoother transition from one block to another. That is, the portions where the blocks overlap ought to have a contribution from the previous block, together with a contribution from the next block. The conventional approach is to use a "window" function to perform this tapering. A straightforward window function is a triangular-shaped window:

$$w_n = \begin{cases} \dfrac{2n}{M} & : \quad 0 \le n \le \dfrac{M}{2} \\ 2 - \dfrac{2n}{M} & : \quad \dfrac{M}{2} \le n \le M. \\ 0 & : \quad \text{otherwise} \end{cases} \tag{7.58}$$

Two other types having a smoother shape are the Hanning window—for even-valued M, defined as

$$w_n = \begin{cases} 0.5 - 0.5\cos\dfrac{2n\pi}{M} & : \ 0 \le n \le M \\ 0 & : \ \text{otherwise} \end{cases}, \tag{7.59}$$

and the Hamming window—for even-valued M, defined as:

$$w_n = \begin{cases} 0.54 - 0.46\cos\dfrac{2n\pi}{M} & : \ 0 \le n \le M \\ 0 & : \ \text{otherwise} \end{cases}. \tag{7.60}$$

The Hamming window is often used, and it is easily implemented in MATLAB, as follows.

```
M = 32;
n = 0:M;
w = 0.54 - 0.46*cos(2*pi*n/M);
stem(w);
```

A comparison of these window shapes is shown in Figure 7.30. Different window shapes produce differing frequency component overflows (termed "side-lobes") in the frequency domain when applied to a rectangular window of sampled data. Windows will be encountered again in Chapter 8, where the Hamming window is again applied, but in the context of designing a digital filter, in order to adjust its frequency response.

7.12 THE FFT

The direct DFT calculation is computationally quite "expensive"—meaning that the time taken to compute the result is significant when compared with the sample period. Not only is a faster method desirable, in real-time applications, it is essential. This section describes the so-called FFT, which substantially reduces the computation required to produce exactly the same result as the DFT.

The FFT is a key algorithm in many signal processing areas today. This is because its use extends far beyond simple frequency analysis—it may be used in a number of "fast" algorithms for filtering and other transformations. As such, a solid understanding of the FFT is well worth the intellectual effort.

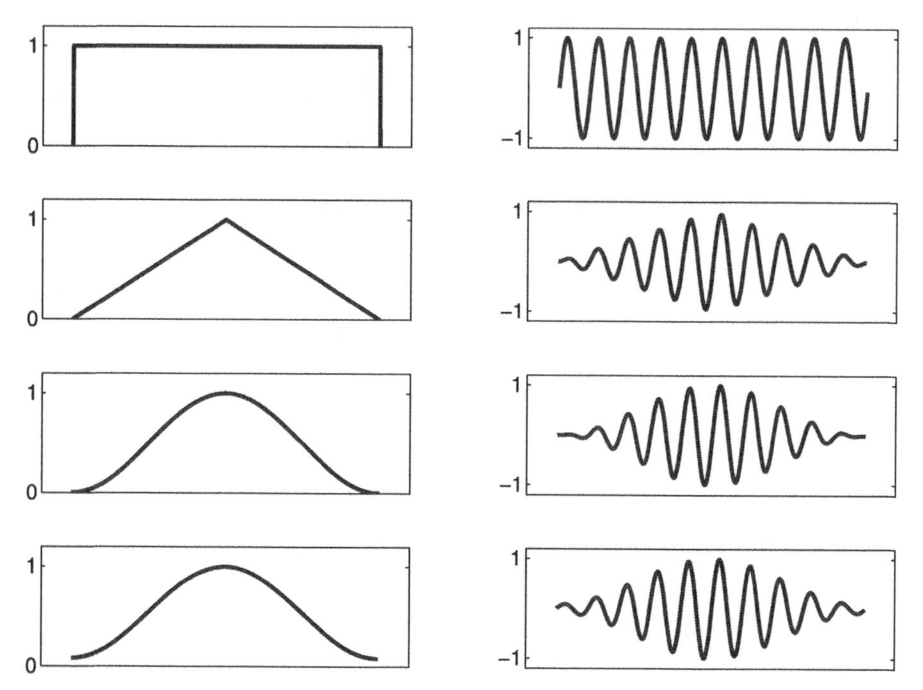

FIGURE 7.30 Window functions for smoothing frames. Top to bottom are Rectangular, Triangular, Hanning, and Hamming. The right-hand side shows the effect on a sine wave block.

7.12.1 Manual Fourier Transform Calculations

To recap, the discrete-time Fourier transform (DFT) is defined by:

$$X(k) = \sum_{n=0}^{N-1} x(n) e^{-jn\omega_k}$$

$$\omega_k = \frac{2k\pi}{N} \quad k = 0, 1, \ldots, N-1.$$

To grasp the basic idea of the number of arithmetic operations required, consider a four-point ($N = 4$) DFT calculation:

$$\begin{array}{lllll}
X(k) & n=0 & n=1 & n=2 & n=3 \\
X(0) = & x(0)e^{-j0\frac{2\pi}{4}0} & + x(1)e^{-j1\frac{2\pi}{4}0} & + x(2)e^{-j2\frac{2\pi}{4}0} & + x(3)e^{-j3\frac{2\pi}{4}0} \\
X(1) = & x(0)e^{-j0\frac{2\pi}{4}1} & + x(1)e^{-j1\frac{2\pi}{4}1} & + x(2)e^{-j2\frac{2\pi}{4}1} & + x(3)e^{-j3\frac{2\pi}{4}1} \\
X(2) = & x(0)e^{-j0\frac{2\pi}{4}2} & + x(1)e^{-j1\frac{2\pi}{4}2} & + x(2)e^{-j2\frac{2\pi}{4}2} & + x(3)e^{-j3\frac{2\pi}{4}2} \\
X(3) = & x(0)e^{-j0\frac{2\pi}{4}3} & + x(1)e^{-j1\frac{2\pi}{4}3} & + x(2)e^{-j2\frac{2\pi}{4}3} & + x(3)e^{-j3\frac{2\pi}{4}3}.
\end{array}$$

This is a small example, at about the limit for manual calculation. In practice, typically $N = 256$ to 1024 samples are required for reasonable resolution (depending on the application, and of course the sampling rate), and it is not unusual to require many more than this. From the above manual calculation, we can infer:

1. The number of complex additions $= 4 \times 3 \approx N^2$.
2. The number of complex multiplications $= 4 \times 4 = N^2$.

Note that complex numbers are comprised of real and imaginary components, so there are in fact more operations—a complex number multiplication really requires four simple multiplications and two arithmetic additions:

$$\begin{aligned}(a + \jmath b)(c + \jmath d) &= ac + \jmath ad + \jmath bc - bd \\ &= (ac - bd) + \jmath(ad + bc).\end{aligned}$$

If $N \approx 1000$, and each multiply/add takes $T_{\text{calc}} = 1\ \mu\text{s}$, the Fourier transform calculation would take of the order of $N^2 \cdot T_{\text{calc}} = 1000^2 \times 10^{-6} = 1$ second.

Referring to Figure 7.31, we see that the exponents are cyclical, and this leads to the simplification:

$$p \cdot \frac{2\pi}{4} \rightarrow \left(\frac{2\pi}{4}\right)(p \bmod 4),$$

where $N = 4$ and "mod" is the modulo remainder after division:

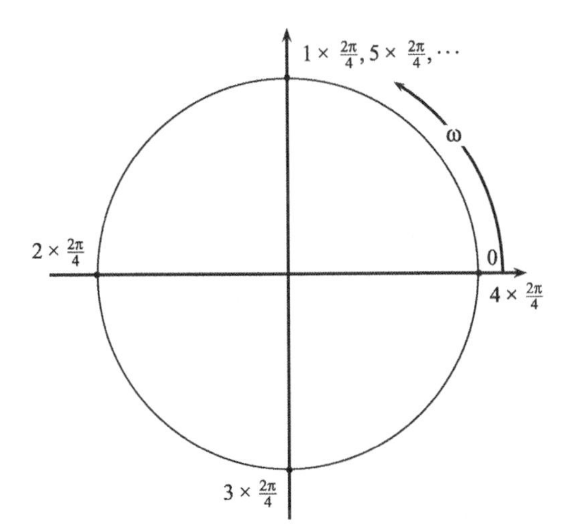

FIGURE 7.31 The DFT exponent as it moves around the unit circle. Clearly, the point at angle $2nk\pi/N$ will be recalculated many times, depending on the values which n and k take on with respect to a given N.

$$
\begin{array}{llll}
0 & \bmod & 4 & \to 0 \\
1 & \bmod & 4 & \to 1 \\
2 & \bmod & 4 & \to 2 \\
3 & \bmod & 4 & \to 3 \\
4 & \bmod & 4 & \to 0 \\
5 & \bmod & 4 & \to 1
\end{array}
$$

Now, grouping the even-numbered and odd-numbered terms, we have:

$$
\begin{aligned}
X(k) &\quad \overbrace{\qquad\qquad\qquad\qquad}^{\text{Events }(0,2)} \quad \overbrace{\qquad\qquad\qquad\qquad}^{\text{Odds }(1,3)} \\
X(0) &= x(0)e^{-j0\frac{2\pi}{4}0} + x(2)e^{-j2\frac{2\pi}{4}0} + x(1)e^{-j1\frac{2\pi}{4}0} + x(3)e^{-j3\frac{2\pi}{4}0} \\
X(1) &= x(0)e^{-j0\frac{2\pi}{4}1} + x(2)e^{-j2\frac{2\pi}{4}1} + x(1)e^{-j1\frac{2\pi}{4}1} + x(3)e^{-j3\frac{2\pi}{4}1} . \\
X(2) &= x(0)e^{-j0\frac{2\pi}{4}2} + x(2)e^{-j2\frac{2\pi}{4}2} + x(1)e^{-j1\frac{2\pi}{4}2} + x(3)e^{-j3\frac{2\pi}{4}2} \\
X(3) &= x(0)e^{-j0\frac{2\pi}{4}3} + x(2)e^{-j2\frac{2\pi}{4}3} + x(1)e^{-j1\frac{2\pi}{4}3} + x(3)e^{-j3\frac{2\pi}{4}3}
\end{aligned}
$$

Note that there are common terms like ($x[0] + x[2]$) in the $k = 0$ and $k = 2$ calculations. Now, a two-point ($N = 2$) DFT is:

$$
\begin{aligned}
& \qquad\qquad\qquad\quad n=0 \qquad\qquad n=1 \\
k=0:&\quad X(0) = x(0)e^{-j0\frac{2\pi}{2}0} + x(1)e^{-j1\frac{2\pi}{2}0} \\
k=1:&\quad X(1) = x(0)e^{-j0\frac{2\pi}{2}1} + x(1)e^{-j1\frac{2\pi}{2}1}.
\end{aligned}
$$

Comparing with the four-point DFT, some terms are seen to be the same if we renumber the four-point algorithm terms $x(0)$ and $x(2)$ to become $x(0)$ and $x(1)$ for the two-point transform. This is shown in Figure 7.32—the four-point DFT can be calculated by combining two two-point DFTs. This suggests a "divide-and-conquer" approach, as will be outlined in the following section.

7.12.2 FFT Derivation via Decimation in Time

The goal of the FFT is to reduce the time required to compute the frequency components of a signal. As the reduction in the amount of computation time is quite significant (several orders of magnitude), many "real-time" applications in signal processing (such as speech analysis) have been developed.[4]

If we write the discrete-time Fourier transform (DFT) as usual:

$$
X(k) = \sum_{n=0}^{N-1} x(n)e^{-jn\omega_k},
$$

[4]The FFT decimation approach is generally attributed to J. W. Cooley and J. W. Tukey, although there are many variations. Surprisingly, although Cooley and Tukey's work was published in 1965, it is thought that Gauss described the technique as long ago as 1805 (Heideman et al. 1984).

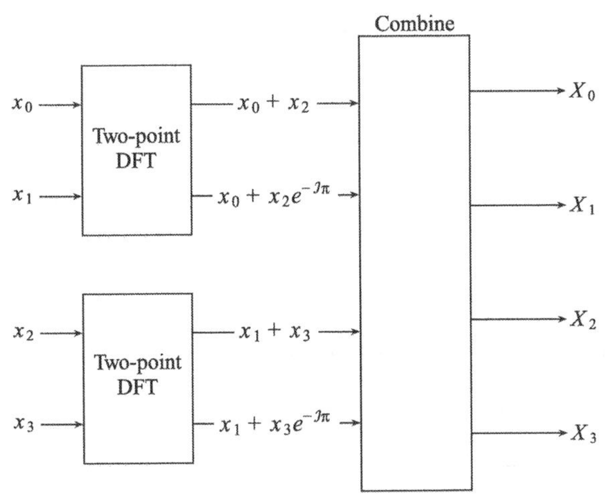

FIGURE 7.32 A four-point FFT decomposition. This idea is used as the basis for the subsequent development of the FFT algorithm.

$$\omega_k = \frac{2k\pi}{N} \quad k = 0, 1, \ldots, N-1,$$

and let $W = e^{-j\frac{2\pi}{N}}$, the DFT becomes

$$X(k) = \sum_{n=0}^{N-1} x(n) W^{nk}. \tag{7.61}$$

We can split the $x(n)$ terms into odd-numbered ones ($n = 1, 3, 5, \ldots$) and even-numbered ones ($n = 0, 2, 4, \ldots$). If N is an integer, then $2N$ is an even integer, and $2N + 1$ is an odd integer. This results in:

$$\begin{aligned}
X(k) &= \overbrace{\sum_{n=0}^{\frac{N}{2}-1} x(2n) W^{(2n)k}}^{\text{Even-numbered terms}} + \overbrace{\sum_{n=0}^{\frac{N}{2}-1} x(2n+1) W^{(2n+1)k}}^{\text{Odd-numbered terms}} \\
&= \sum_{n=0}^{\frac{N}{2}-1} x(2n) W^{2nk} + \sum_{n=0}^{\frac{N}{2}-1} x(2n+1) W^{2nk} W^k \\
&= \sum_{n=0}^{\frac{N}{2}-1} x(2n) W^{2nk} + W^k \sum_{n=0}^{\frac{N}{2}-1} x(2n+1) W^{2nk} \\
&= \sum_{n=0}^{\frac{N}{2}-1} x(2n) \left(W^{nk}\right)^2 + W^k \sum_{n=0}^{\frac{N}{2}-1} x(2n+1) \left(W^{nk}\right)^2.
\end{aligned}$$

Inspecting the above equation, it seems that we need a value for W^2, which can be found as follows:

$$W = e^{-j\frac{2\pi}{N}}$$

$$\therefore W^2 = \left(e^{-j\frac{2\pi}{N}}\right)^2$$

$$= e^{-j\frac{2\pi}{(N/2)}}.$$

The last term is just W for $N/2$ instead of N. So we have:

$$X(k) = \underbrace{\sum_{n=0}^{\frac{N}{2}-1} x(2n)\left(W^2\right)^{nk}}_{\text{DFT of even-indexed } \frac{N}{2} \text{ sequence}} +$$

$$\underbrace{W^k \sum_{n=0}^{\frac{N}{2}-1} x(2n+1)\left(W^2\right)^{nk}}_{\text{DFT of odd-indexed } \frac{N}{2} \text{ sequence}}$$

$$\therefore X(k) = E(k) + W^k F(k).$$

Now if $k \geq N/2$, then the equation for $X(k)$ can be simplified by replacing

$$k \rightarrow \left(k + \frac{N}{2}\right). \tag{7.62}$$

Then,

$$X\left(k+\frac{N}{2}\right) = \sum_{n=0}^{\frac{N}{2}-1} x(2n) W^{2n\left(k+\frac{N}{2}\right)} +$$

$$W^{\left(k+\frac{N}{2}\right)} \sum_{n=0}^{\frac{N}{2}-1} x(2n+1) W^{2n\left(k+\frac{N}{2}\right)}$$

$$= \sum_{n=0}^{\frac{N}{2}-1} x(2n) W^{2nk} W^{nN} +$$

$$W^k W^{\frac{N}{2}} \sum_{n=0}^{\frac{N}{2}-1} x(2n+1) W^{2nk} W^{nN}.$$

As before, $W = e^{-j\frac{2\pi}{N}}$ and hence,

$$\begin{aligned} W^{nN} &= e^{-j\frac{2\pi}{N}nN} \\ &= e^{-j2\pi n} \\ &= 1 \quad \text{for all } n, N \end{aligned},$$

and

$$\begin{aligned} W^{\frac{N}{2}} &= e^{-j\frac{2\pi}{N}\frac{N}{2}} \\ &= e^{-j\pi} \\ &= -1 \quad \text{for all } N. \end{aligned}$$

We previously defined

$$X(k) = E(k) + W^k F(k). \tag{7.63}$$

Hence,

$$X\left(k + \frac{N}{2}\right) = E(k) - W^k F(k). \tag{7.64}$$

Together, these show how the DFT of the original sequence can be calculated via DFTs of half-length sequences. To express this graphically, the "butterfly" diagram of Figure 7.33 is used. This indicates that for inputs p and q, the q value is multiplied by a constant α, and the result summed, to give outputs $p \pm \alpha q$. A 4-to-8 DFT can then be shown as in Figure 7.34.

The complete diagram for an eight-point FFT decomposition is shown in Figure 7.35. The transform calculation begins with eight inputs, and computes four, two-point transforms. Then these are combined into two, four-point transforms, with the final stage resulting in the eight output points. Thus there are three stages for eight input points. In general, for N input points the number of stages will be

$$S = \log_2 N. \tag{7.65}$$

This is where the reduction in complexity comes about: one large computation is reduced to several sequential, smaller computations.

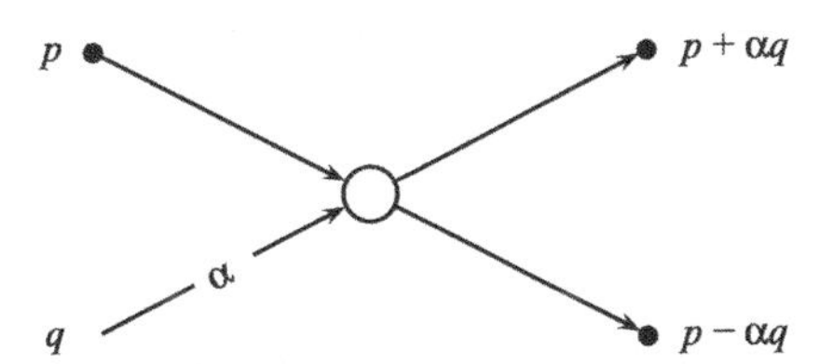

FIGURE 7.33 The flow diagram convention used in the DFT decomposition. The input is the pair of points on the left, and the two outputs are on the right.

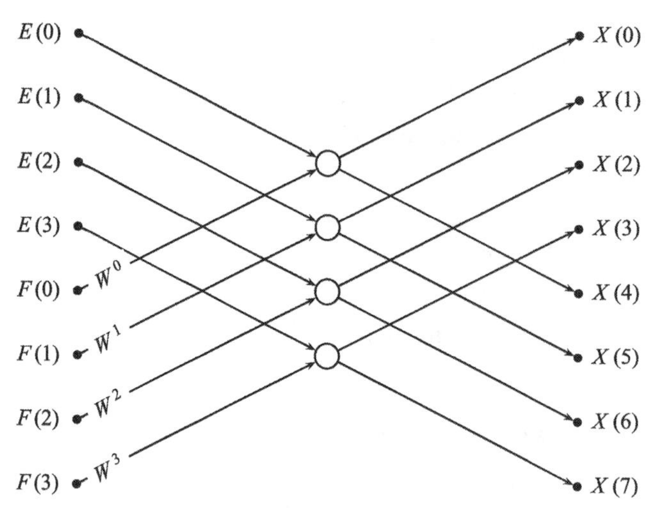

FIGURE 7.34 The calculation of an eight-point DFT from two, 4-point DFTs.

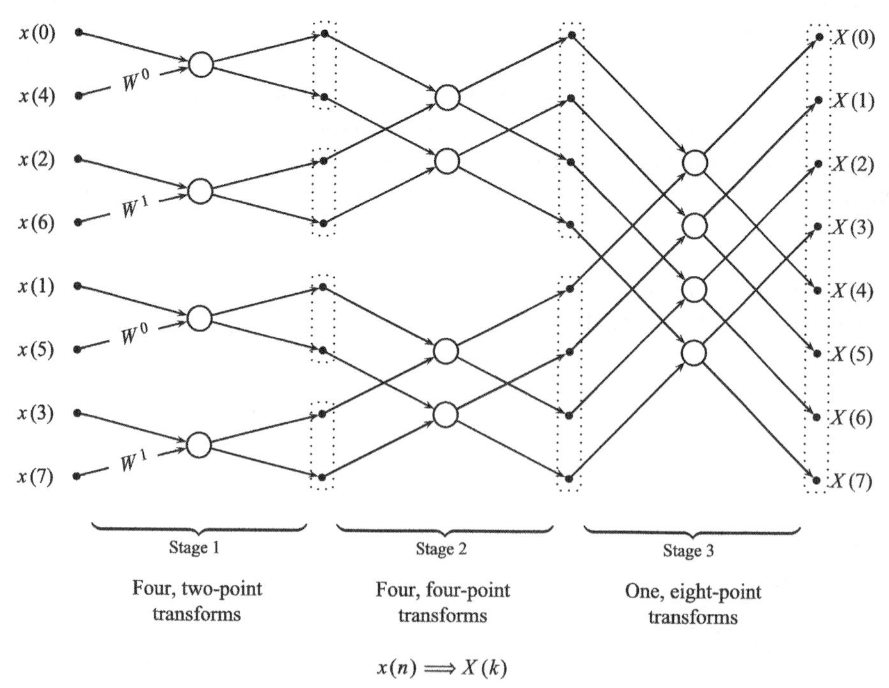

FIGURE 7.35 An eight-point DFT decimation. Although it looks more complicated and employs multiple stages, each stage is recursively computed from the previous stage. Note also that the output is in the correct order (0, 1, 2, $\cdots$) for the given input order.

To summarize, the steps for computing the DFT via decimation are:

1. Shuffle the input order (discussed shortly).
2. Compute $N/2$, two-sample DFTs.
3. Compute $N/4$, four-sample DFTs.
4. Continue until one, N-sample DFT is computed.

Consider stepping through an $N = 8$ point DFT.

$$X(k) = E(k) + W^k F(k)$$
$$X\left(k + \frac{N}{2}\right) = E(k) - W^k F(k)$$
$$W = e^{-j\frac{2\pi}{N}}$$

Stage 1: Four, two-point transforms:

$$X(0) = E(0) + W^0 F(0)$$
$$X(1) = E(0) - W^0 F(0)$$

Stage 2: Two, four-point transforms:

$$X(0) = E(0) + W^0 F(0)$$
$$X(1) = E(1) + W^1 F(1)$$
$$X(2) = E(0) - W^0 F(0)$$
$$X(3) = E(1) - W^1 F(1)$$

Stage 3: One, eight-point transform:

$$X(0) = E(0) + W^0 F(0)$$
$$X(1) = E(1) + W^1 F(1)$$
$$X(2) = E(2) + W^2 F(2)$$
$$X(3) = E(3) + W^3 F(3)$$
$$X(4) = E(0) - W^0 F(0)$$
$$X(5) = E(1) - W^1 F(1)$$
$$X(6) = E(2) - W^2 F(2)$$
$$X(7) = E(3) - W^3 F(3)$$

There are $\log_2 N$ stages, with each stage having $N/2$ complex multiply operations. The complexity is then of the order of $((N/2)\log_2 N)$. This may be compared with a straightforward DFT complexity, which is of the order of N^2.

For example, if $N = 1024$ input samples, a DFT requires approximately 10^6 operations, whereas an FFT requires approximately $500 \times \log_2 1024$ or 5000 operations, which is a 200-fold reduction. To put this in perspective, if a complex arithmetic operation takes (for example) one microsecond, this is equivalent to reducing the total time taken to compute the DFT from from 1 second down to 5 milliseconds, which is quite a substantial reduction.

0	1	2	3	4	5	6	7
000	001	010	011	100	101	110	111

bit 0=0 bit 0=1

0	2	4	6	1	3	5	7
000	010	100	110	001	011	101	111

bit 1=0 bit 1=1 bit 1=0 bit 1=1

0	4	2	6	1	5	3	7
000	100	010	110	001	101	011	111

FIGURE 7.36 DFT decimation indices. At each stage, the lowest-order bit is used to partition the sequence into two.

The only remaining complication is that of the "reordered" input sequence. The original inputs were

$$x(0) \quad x(1) \quad x(2) \quad x(3) \quad x(4) \quad x(5) \quad x(6) \quad x(7)$$

Since each stage involves splitting the sequence into even-numbered and odd-numbered subsequences, it is really equivalent to examining the least-significant bit of the indexes (0 = even, 1 = odd). This is shown in Figure 7.36.

Comparing the bit values of the indexes in the upper row and the lower row of the figure, they are seen to be in reversed bit order:

$$\begin{aligned}
x(000) &\rightarrow x(000)\\
x(001) &\rightarrow x(100)\\
x(010) &\rightarrow x(010)\\
x(011) &\rightarrow x(110)\\
x(100) &\rightarrow x(001)\\
x(101) &\rightarrow x(101)\\
x(110) &\rightarrow x(011)\\
x(111) &\rightarrow x(111)
\end{aligned}$$

For obvious reasons, this is usually termed "bit-reversed" addressing. As a consequence of the successive partitioning of the input sequences by a factor of two, the number of input samples *must* be a power of two—typically 256, 512, 1024, 2048, 4096, or larger.

7.13 THE DCT

The discrete Fourier transform plays a very important—some would say essential—role in digital signal processing. A distinct but related transform, the DCT, has emerged as the method of choice for many digital compression systems. In particular, high definition TV and JPEG still image compression are based on the DCT, as well as important audio coding standards such as MP3. This section introduces the concept of transform-domain coding with a focus on the DCT, and examines the concept of an optimal transform.

7.13.1 Transform Approach to Data Compression

In the preceding sections, we demonstrated that the Fourier transform provides a means to convert a time-sampled waveform into its constituent frequency components. Importantly, the transform is invertible—that is, we can convert back from the frequency components in order to reconstruct a time domain waveform. In the field of data compression, the objective is to minimize the number of bits needed to encode a given source. The key link is that given humans' perception of audio and images, not all frequency components are necessarily important in encoding the source. In the Fourier domain, a simplistic view of this would be to discard the higher-frequency components, and encode only those in the lower bands, which are deemed necessary according to human perceptual criteria. So the system flow is broadly as follows:

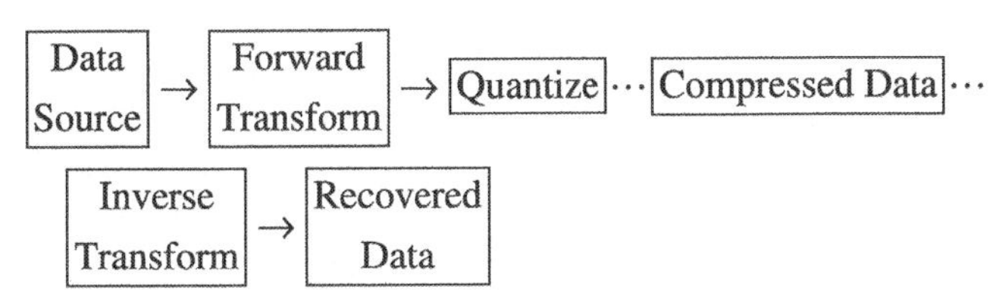

The question then becomes how to determine the best type of transformation to be applied to the source data. Many researchers have investigated the use of the Fourier transform; however, the most widely used transform for image and picture coding/compression is the DCT. In the case of image encoding, a greater degree of redundancy may be removed by processing the image in blocks, rather that as a linear sequence. That is, we process the image in matrix blocks rather than vectors. For computational and practical reasons, a block size of 8×8 is commonly chosen.

7.13.2 Defining the Discrete Cosine Transform

As one might expect, the DCT is not unlike the DFT, in that it is based on trigonometric functions (Ahmed et al. 1974). However, where the DFT utilizes paired sine and cosine functions (condensed into one complex exponential), the DCT uses only cosine components. We will initially investigate the one-dimensional DCT in order to introduce the concepts.

Let the input vector $\mathbf{x}$ define the input samples, taken as a block at some point in time. The output vector $\mathbf{y}$ resulting from the transformation contains the transform coefficients. As will be seen, the process can be written as a matrix multiplication

$$\mathbf{y} = \mathbf{A}\mathbf{x}, \tag{7.66}$$

where the transformation is represented as a transform matrix $\mathbf{A}$. The transformation matrix stores the coefficients, which are normally fixed.

Given this terminology, the DCT in one dimension for a block of N samples is defined as

$$\mathbf{y}(k)=\sqrt{\frac{2}{N}}c_k\sum_{n=0}^{N-1}\mathbf{x}(n)\cos\left(\frac{(2n+1)k\pi}{2N}\right) \quad k=0,1,\cdots,N-1. \tag{7.67}$$

The constant c_k is defined as

$$c_k=\begin{cases}\frac{1}{\sqrt{2}} & : \; k=0 \\ 1 & : \; k\neq 0\end{cases}. \tag{7.68}$$

In order to understand the operation of the DCT, and in particular its application in data compression, consider a simple one-dimensional transform where the block size is two. The transformation matrix is

$$\mathbf{A}_{2\times 2}=\frac{1}{\sqrt{2}}\begin{pmatrix}1 & 1\\ 1 & -1\end{pmatrix}. \tag{7.69}$$

If we apply this transform to two adjacent data points, we see that the transformed output will be the sum and difference of the data points. In effect, we have a low-pass or average output as the first transform coefficient, and a difference or high-pass output as the second transform coefficient. The first coefficient, in effect, provides the broad approximation, and the second coefficient provides the detail. Furthermore, we can define the basis vectors as being formed by the rows of the transform matrix.

The inverse DCT (IDCT) is defined as

$$\mathbf{x}(n)=\sqrt{\frac{2}{N}}\sum_{k=0}^{N-1}c_k\mathbf{y}(k)\cos\left(\frac{(2n+1)k\pi}{2N}\right) \quad n=0,1,\cdots,N-1. \tag{7.70}$$

The DCT is completely invertible. That is, performing the inverse DCT on the DCT of a vector gets back the original vector. This is equivalent to finding the inverse of the transformation matrix **A**.

As with the DFT, the concept of basis vectors helps our understanding of the DCT. Whereas the basis vectors for the DFT were sine and cosine waveforms, the basis vectors for the DCT are

$$\mathbf{b}_k=\sqrt{\frac{2}{N}}c_k\cos\left(\frac{(2n+1)k\pi}{2N}\right) \quad n=0,1,\cdots,N-1 \quad k=0,1,\cdots,N-1. \tag{7.71}$$

The DCT is employed extensively in image encoding. For images, a more natural structure to employ is in two dimensions, since not only are images inherently two-dimensional, there is redundancy in both the vertical and horizontal directions. Thus, we need a two-dimensional DCT. This is an extension in both directions of the one-dimensional DCT. Formally, the two-dimensional DCT is defined as

$$\mathbf{Y}(k,l)=\frac{2}{N}c_k c_l\sum_{m=0}^{N-1}\sum_{n=0}^{N-1}\mathbf{X}(m,n)\cos\left(\frac{(2m+1)k\pi}{2N}\right)\cos\left(\frac{(2n+1)l\pi}{2N}\right) \tag{7.72}$$

$$k,l=0,1,\cdots,N-1.$$

Again, we have an inverse, which is defined as

$$\mathbf{X}(m,n)=\frac{2}{N}\sum_{k=0}^{N-1}\sum_{l=0}^{N-1}c_k c_l\mathbf{Y}(k,l)\cos\left(\frac{(2m+1)k\pi}{2N}\right)\cos\left(\frac{(2n+1)l\pi}{2N}\right) \tag{7.73}$$

$$m,n=0,1,\cdots,N-1.$$

So let us look at the interpretation of the two-dimensional transform. Extracting the basis vectors, a pictorial representation is shown in Figure 7.37. This gives us an intuitive insight into the DCT as a 2D data compressor: the transformed coefficients are, in effect, a linear combination of the basis vectors. As will be observed from the figure, the basis vectors form various patterns according to the cosine functions in each direction. The lower-order coefficients are those in the upper left of the picture, and it may be observed that these map to lower frequency image characteristics. Conversely, the lower-right side of the images shows the higher-frequency basis vectors (or, more correctly, basis matrices).

Since the reconstructed subblock is effectively a linear combination of the basis images, higher frequency components, which correspond to greater perceptual detail, may be omitted by simply not encoding the higher coefficients (those on the

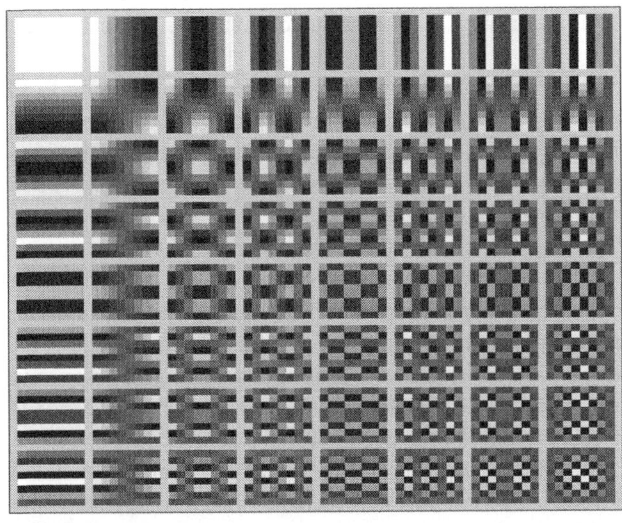

FIGURE 7.37 The DCT basis images for an 8 × 8 transform. Each block represents an 8 × 8 block of source pixels, and each block produces exactly one DCT coefficient. This coefficient represents the "weighting" of the corresponding basis block.

FIGURE 7.38 A test image for DCT compression experiments. The image contains both contours of approximately the same luminance, and background areas of more detail. This allows us to see the effect of bitrate compression on the various aspects of a typical image.

lower-right side of the output matrix). We will see how this is done, and how well it works in practice, in the next section.

7.13.3 Using the Discrete Cosine Transform

We now examine the operation of the DCT as a data compression transform. Figure 7.38 shows a source image to be encoded. The DCT is applied to each 8×8 subblock in the image in turn, and the statistics of the coefficients are examined. The transform coefficients form an array of 8×8 coefficient matrices, one per image subblock.

As discussed, the coefficients in the upper-left of the transformed matrix form the lower-level approximation to the subblock, and the lower-right coefficients add additional detail. Note that coefficient $\mathbf{Y}(0, 0)$ is in effect the average (sometimes called "DC") value of the source subblock, and this is easily seen by putting $k = l = 0$ in the 2D DCT transform specified by Equation 7.72.

The histograms of the first nine coefficients, corresponding to the 3×3 coefficients in the upper left of the transform, are shown in Figure 7.39. Apart from the average coefficient, all others exhibit a highly peaked Laplacian-like distribution, and may be quantized accordingly. Furthermore, since the distribution of the coefficients tends to zero, we reach an important conclusion: that the higher-frequency coefficients require fewer bits, and may in fact require no bits at all. This corresponds to setting that coefficient value to zero at the decompression stage.

So how well does this work in practice? The best way to gage is to examine some test cases; Figure 7.40 shows an image which has been compressed and decompressed using the DCT with the retention of only lower-order DCT transform coefficients. It is obvious that only a limited number of coefficients need to be retained for good quality image reconstruction. The compression attained is

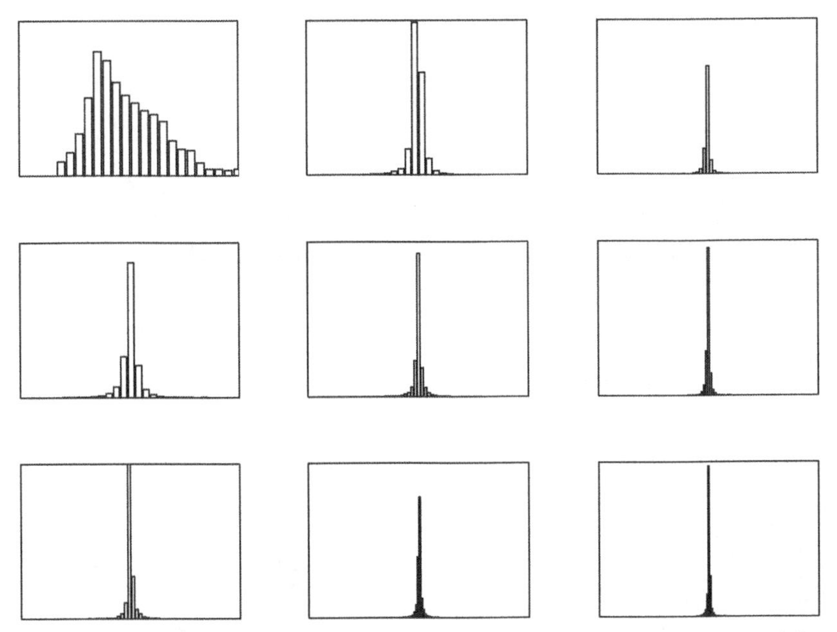

FIGURE 7.39 Histogram of DCT coefficients for a typical image. The upper-left histogram is the average value of the blocks. Each of the other histograms represents detail in each direction, at successively higher frequencies. The scale on the higher-frequency coefficients is identical, and it is seen that the distribution tends towards a value of zero for higher coefficients.

substantial: for example, retaining only 4 coefficients out of 64 is a 16:1 compression ratio, equivalent to 0.5 bits per pixel.

Care should be taken in extrapolating this result: the quality of the reconstructed image depends on many factors, including the image resolution to begin with, the reproduction media, and the image itself. In addition, we have made no effort to quantize the transformed coefficients. The methods discussed in Section 3.6 may be applied, and in practice (in JPEG, for example), more complex bit-assignment codes are employed. However, from this simple example, the potential is clearly demonstrated.

7.13.4 Vectorspace Interpretation of Forward and Inverse Transforms

Let us now look at the transformation and inverse transformation operations themselves, and see how they can be written in terms of vector and matrix operations. The forward transform is a matrix-vector multiplication

$$\mathbf{y} = \mathbf{A}\mathbf{x}, \tag{7.74}$$

FIGURE 7.40 DCT compression—upper left, original image, 8 × 8 sub-blocks. Upper right, 1 × 1 coefficient retained; Lower left, 2 × 2 coefficients retained; Lower right, 3 × 3 coefficients retained.

which takes an input block **x** and yields another (transformed) vector **y**. If we write the transform matrix as row vectors, then we can visualize this product as

$$\mathbf{y} = \begin{pmatrix} - & \mathbf{a}_0 & - \\ - & \mathbf{a}_1 & - \\ & \vdots & \end{pmatrix} \begin{pmatrix} | \\ \mathbf{x} \\ | \end{pmatrix}. \tag{7.75}$$

In this case, we can see that the *basis vectors* are the *rows* of the matrix. Each output coefficient in **y** is nothing more than a vector dot product,

$$y_0 = \begin{pmatrix} | \\ \mathbf{a}_0 \\ | \end{pmatrix} \cdot \begin{pmatrix} | \\ \mathbf{x} \\ | \end{pmatrix}. \tag{7.76}$$

The next output term is $y_1 = \mathbf{a}_1 \cdot \mathbf{x}$, and so forth. In effect, we are multiplying each basis vector by the input vector, to give each output coefficient in turn.

The inverse transform is a multiplication:

$$\mathbf{x} = \mathbf{B}\mathbf{y}. \tag{7.77}$$

Let us write the matrix as column vectors, and the vector y in terms of its components

$$\mathbf{x} = \begin{pmatrix} | & | & | \\ \mathbf{b}_0 & \mathbf{b}_1 & \cdots \\ | & | & | \end{pmatrix} \begin{pmatrix} y_0 \\ y_1 \\ \vdots \end{pmatrix}. \tag{7.78}$$

We have the basis vectors as *columns*. The output may then be written a little differently, as

$$\mathbf{x} = y_0 \begin{pmatrix} | \\ \mathbf{b}_0 \\ | \end{pmatrix} + y_1 \begin{pmatrix} | \\ \mathbf{b}_1 \\ | \end{pmatrix} + \cdots. \tag{7.79}$$

Thus, the output sum **x** is a weighted sum of scalar-vector products. Each scalar weighting is the value taken from the input vector, which is actually a transformed component value. We weight the basis vector by this value, which represents its relative importance in determining the output. Importantly, the weights may be arranged in decreasing order, so as to successively approximate the output vector. This is where the application in transform data compression comes in: less important basis vectors are given less weight, if we have to economize on bandwidth.

7.13.5 Basis Vectors and Orthogonality

To further explore the transform approach, it is necessary to understand the decorrelation properties of the transform.

For the purpose of example, let us return back to the simpler situation of a one-dimensional transform. The transformation is equivalent to multiplying the transform matrix by the data vector. The coefficients that comprise the matrix are what determines the transform (DCT, DFT, or another). Such a matrix-vector product can be visualized as a rotation of a point in space. Consider Figure 7.41, which shows a point defined by vector **x** rotated to **y** using a rotation matrix **A**.

What is the matrix **A** in this case? For the two-dimensional rotation, the result is well known: it is the rotation matrix. To derive this, let the starting point be the vector **x** at angle α:

$$\mathbf{x} = \begin{pmatrix} x_0 \\ x_1 \end{pmatrix} \tag{7.80}$$

$$= \begin{pmatrix} |\mathbf{x}|\cos\alpha \\ |\mathbf{x}|\sin\alpha \end{pmatrix}. \tag{7.81}$$

Similarly, we can write for the rotated point **y**

$$\mathbf{y} = \begin{pmatrix} y_0 \\ y_1 \end{pmatrix} \tag{7.82}$$

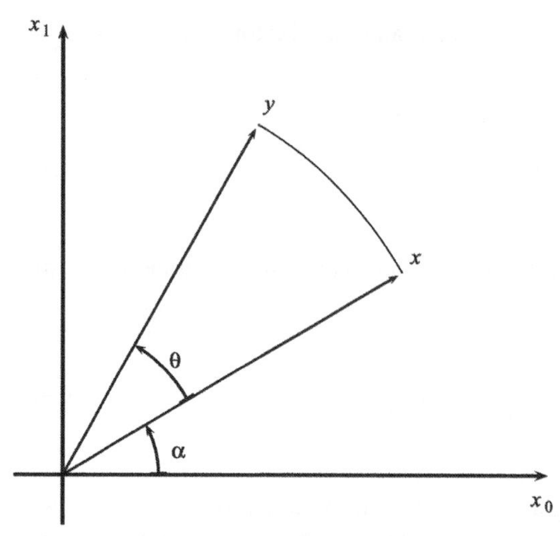

FIGURE 7.41 Transformation as a rotation in two dimensions. Vector **x** at angle α is rotated to vector **y** at angle $(\alpha + \theta)$.

$$= \begin{pmatrix} |\mathbf{x}|\cos(\alpha+\theta) \\ |\mathbf{x}|\sin(\alpha+\theta) \end{pmatrix}. \tag{7.83}$$

We can expand this using standard identities

$$\mathbf{y} = \begin{pmatrix} |\mathbf{x}|(\cos\alpha\cos\theta - \sin\alpha\sin\theta) \\ |\mathbf{x}|(\sin\alpha\cos\theta + \cos\alpha\sin\theta) \end{pmatrix}. \tag{7.84}$$

Using the components from Equation 7.81, we have:

$$\begin{pmatrix} y_0 \\ y_1 \end{pmatrix} = \begin{pmatrix} x_0\cos\theta - x_1\sin\theta \\ x_0\sin\theta + x_1\cos\theta \end{pmatrix}. \tag{7.85}$$

This can now be written in matrix form as:

$$\begin{pmatrix} y_0 \\ y_1 \end{pmatrix} = \begin{pmatrix} \cos\theta & -\sin\theta \\ \sin\theta & \cos\theta \end{pmatrix} \begin{pmatrix} x_0 \\ x_1 \end{pmatrix}. \tag{7.86}$$

So the matrix **A** is in effect:

$$\mathbf{A} = \begin{pmatrix} \cos\theta & -\sin\theta \\ \sin\theta & \cos\theta \end{pmatrix}. \tag{7.87}$$

The rows of this transformation matrix form the *basis vectors*, which define the axes for the new vector **y**. Let us denote the rows of **A** by:

$$\mathbf{b}_0 = \begin{pmatrix} \cos\theta \\ -\sin\theta \end{pmatrix}$$

$$\mathbf{b}_1 = \begin{pmatrix} \sin\theta \\ \cos\theta \end{pmatrix}.$$

These basis vectors are *orthogonal*, which means that the inner or vector dot product is zero

$$\mathbf{b}_k \cdot \mathbf{b}_j = \begin{cases} \text{constant} & : \quad k = j \\ 0 & : \quad \text{otherwise} \end{cases}. \tag{7.88}$$

Equivalently, since the basis vectors form the columns of the inverse transformation matrix **B**, the product of the transpose and the matrix itself equals the identity matrix (possibly scaled by a constant c),

$$\mathbf{B}^T\mathbf{B} = c\mathbf{I}. \tag{7.89}$$

To place this on a practical footing, consider Figure 7.42. Suppose we plot pairs of pixels in an image as a dot, with the horizontal and vertical (x_0 and x_1) axes being the value of the first and second pixel in any given pair. Because of the strong correlation between pixels, we would expect the scatter plot to contain points mostly within the ellipse as shown. To encode a pair of pixels, we need the values along the x_0 and x_1 axes. However, if we use a set of rotated axes b_0 and b_1 as shown, we can encode the relative position of a point in a different way. The distance along the b_0 axis might be relatively large, but the distance along the b_1 axis would be comparatively smaller. Thus, the smaller average value requires less accuracy in encoding its value. The encoded data is formed using the rotated axes, and at the decoder, it is a simple matter to revert to the original axes, since the axes are simply rotated through a certain angle. Note that the **b** axes are still perpendicular to each other, and their directions could be changed by 180° and still be perpendicular.

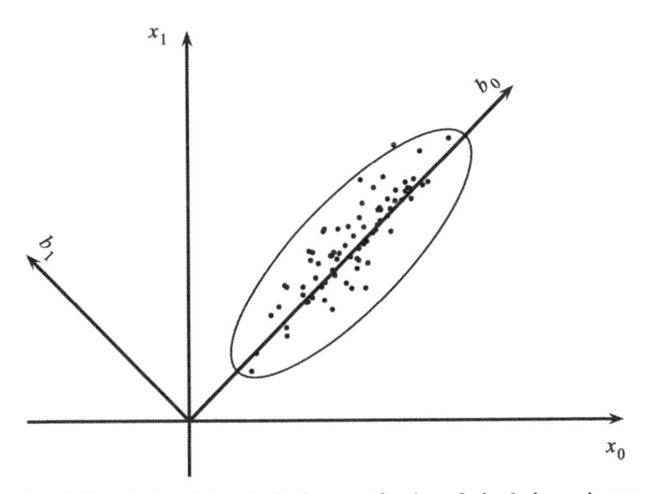

FIGURE 7.42 Hypothetical case of pairs of pixels in an image. If plotted against the x_0, x_1 axes, we expect a concentration as shown by the ellipse. The points are better represented by the rotated axes **b**, and will have a larger component on the b_0 axis and much smaller component on the b_1 axis.

Understanding the orthogonality of the basis vectors helps in understanding the operation of transformation in general, and in particular why an optimal transformation exists. This is the subject of the next section.

7.13.6 Optimal Transform for Data Compression

In practice, the DCT works extremely well, and this is no doubt why it has found its way into many useful applications. A natural question then arises—what is the optimal transform? Investigating this question gives a great deal of insight into the transformation process. Unfortunately, the optimal transform in a mathematical sense is also of little practical use in real applications.

The interdependency of the samples may be examined by means of the correlation matrix, which can be written as:

$$\mathbf{R} = \begin{pmatrix} E\{x_0x_0\} & E\{x_0x_1\} & \cdots & R\{x_0x_{D-1}\} \\ E\{x_1x_0\} & E\{x_1x_1\} & \cdots & E\{x_1x_{D-1}\} \\ \cdots & \vdots & \ddots & \vdots \\ E\{x_{D-1}x_0\} & E\{x_{D-1}x_1\} & \cdots & E\{x_{D-1}x_{D-1}\} \end{pmatrix}. \tag{7.90}$$

If we take the samples as being of dimension D (e.g., $D = 2$ considers two pixels at a time), and create a concatenation of N such vectors, we form a matrix $\mathbf{X}$ of dimension $D \times N$. Then we can easily form the $D \times D$ correlation matrix using

$$\mathbf{R} = \frac{1}{N}\mathbf{X}\mathbf{X}^T. \tag{7.91}$$

So what should the optimal transform do, and how could we derive such a transform? Recall that the *eigenvectors* $\mathbf{e}$ and *eigenvalues* λ of some matrix $\mathbf{M}$ satisfy

$$\mathbf{M}\mathbf{e} = \lambda\mathbf{e} \tag{7.92}$$

$$|\mathbf{M} - \lambda\mathbf{I}| = 0. \tag{7.93}$$

An important property of the eigenvectors is that they are orthogonal,

$$\mathbf{e}_j \cdot \mathbf{e}_k = \begin{cases} 1 & : \quad j = k \\ 0 & : \quad j \neq k \end{cases}. \tag{7.94}$$

Let the new data set $\{\mathbf{y}\}$ be formed from a linear transformation of the original data set $\{\mathbf{x}\}$. If the transformation matrix (as yet unknown) is $\mathbf{A}$, then we transform each vector using

$$\mathbf{y} = \mathbf{A}\mathbf{x} \tag{7.95}$$

as before. The correlation matrix of $\mathbf{x}$ may be formed using

$$\mathbf{R}_x = E\{\mathbf{x}\mathbf{x}^T\}. \tag{7.96}$$

Similarly, the correlation matrix of $\mathbf{y}$ is

$$\mathbf{R}_y = E\{\mathbf{y}\mathbf{y}^T\}. \tag{7.97}$$

So looking at the components of the correlation matrix and how it is formed in Equation 7.90, we can say that we want the correlation matrix $\mathbf{R}_y$ to be a diagonal matrix. That is, the correlation of every component with every other component ought to be zero (or at least, as close to zero as possible). In this case, the (transformed) data is said to be uncorrelated. Of course, the correlation of a component with itself cannot be zero, and these two facts together mean that the correlation matrix of the transformed data must form a diagonal matrix. So the correlation matrix of the new data is

$$\begin{aligned}\mathbf{R}_y &= E\{\mathbf{y}\mathbf{y}^T\} \\ &= E\{\mathbf{A}\mathbf{x}(\mathbf{A}\mathbf{x})^T\} \\ &= E\{\mathbf{A}\mathbf{x}\mathbf{x}^T\mathbf{A}^T\} \\ &= \mathbf{A}\,E\{\mathbf{x}\mathbf{x}^T\}\mathbf{A}^T \\ &= \mathbf{A}\,\mathbf{R}_x\mathbf{A}^T.\end{aligned} \tag{7.98}$$

In deriving this, we have used the general identity $(\mathbf{AB})^T = \mathbf{B}^T\mathbf{A}^T$, and also the fact that $\mathbf{A}$ is a constant and is thus independent of the expectation operator.

The key to the transformation operation lies in the properties of the eigenvalues and eigenvectors of the autocorrelation matrix $\mathbf{R}_x$. So applying Equation 7.92 to the autocorrelation matrix, and then premultiplying by the eigenvector,

$$\mathbf{R}_x\mathbf{e}_k = \lambda_k\mathbf{e}_k \tag{7.99}$$

$$\therefore \mathbf{e}_j^T\mathbf{R}_x\mathbf{e}_k = \mathbf{e}_j^T\lambda_k\mathbf{e}_k \tag{7.100}$$

$$= \lambda_k\mathbf{e}_j^T\mathbf{e}_k. \tag{7.101}$$

Since the eigenvectors are orthogonal, the right-hand side reduces to λ_k if $j = k$, and zero otherwise. This is what we want: the correlation of the transformed data to be zero. If we compare Equation 7.98 with 7.101, we can see that the transform matrix $\mathbf{A}$ would have each row being an eigenvector. So we have a way of defining the optimal transformation matrix $\mathbf{A}$—it ought to have its rows equal to the eigenvectors of the correlation matrix.

$$\begin{pmatrix} | \\ \mathbf{y} \\ | \end{pmatrix} = \begin{pmatrix} - & e_0 & - \\ - & e_1 & - \\ & \vdots & \\ - & e_{D-1} & - \end{pmatrix} \begin{pmatrix} | \\ \mathbf{x} \\ | \end{pmatrix}. \tag{7.102}$$

This may be interpreted as a projection of the input vector onto each eigenvector in turn. So we now use the transformed data and calculate the correlation matrix as derived earlier in Equation 7.98

$$\mathbf{R}_y = \mathbf{A}\mathbf{R}_x\mathbf{A}^T, \tag{7.103}$$

with $\mathbf{A}$ composed of the eigenvectors along its rows, and $\mathbf{R}_x$ being a constant, the previous result yields

$$\mathbf{R}_y = \begin{bmatrix} \lambda_0 & 0 & 0 & 0 \\ 0 & \lambda_1 & 0 & 0 \\ 0 & 0 & \ddots & 0 \\ 0 & 0 & 0 & \lambda_{D-1} \end{bmatrix}. \tag{7.104}$$

So we can see that the transformed data has an autocorrelation matrix which is diagonal, and thus the transformed data is uncorrelated.

Finally, if the eigenvalues are ordered in decreasing order, the transformed data may be *approximated* using a weighted combination of the corresponding eigenvectors. The example below illustrates these calculations. We generate random data vectors $\mathbf{x}$, in which the first component is random with a Gaussian distribution, and the second component is correlated with the first. Thus the matrix $\mathbf{X}$ is 2×400.

```
N = 400;
x1 = randn(N, 1);
x2 = randn(N, 1);
X=[(x1) (0.8*x1+0.4*x2)]'; % each column is an observation vector
C = (X*X')/length(X);
[V D] = eig(C);
% sort according to eigenvalue order
EigVals = diag(D);
[EigVals, EigIdx] = sort(EigVals);
EigVals = flipud(EigVals);
EigIdx = flipud(EigIdx);
% transform matrix
T = V(:, EigIdx)'
fprintf(1, 'Transform matrix:\n');
disp(T);
for k = 1:N
    v = X(:, k);
    vt = T*v;
    XT(:, k) = vt;
end
```

We calculate the eigenvectors and eigenvalues of the correlation matrix using the **eig()**; function. We then transform each observation vector in a loop, to yield the XT vector.

Figure 7.43 illustrates these concepts. The scatter plot shows the x_1, x_2 data, with the ellipse showing the concentration of the data. What we require is a set of two new axes as indicated. Using the untransformed data, the variance along both axes is large.

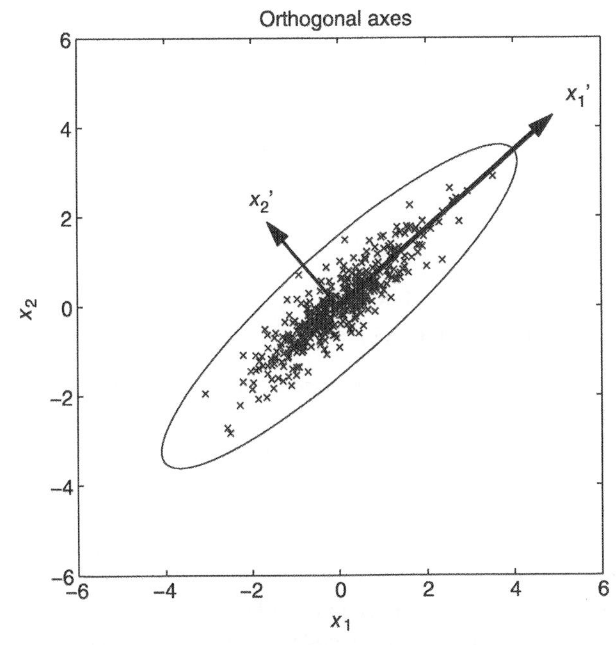

FIGURE 7.43 Orthogonal axes for correlated data are formed by the eigenvectors of the correlation matrix. The rotated coordinate system thus has a larger variance on one axis and smaller variance on the other, corresponding to the major and minor axes of the ellipse in this two-dimensional example.

However, using the rotated axis coordinate system, the variance is large along one axis but small along the other axis. This is the essence of the compression function: we can utilize a smaller number of high-variance components, and coarsely quantize or even not use the low-variance components. The variance corresponds to the major and minor axes of the ellipse, and the axes are determined by the eigenvalues of the correlation matrix.

The above implementation yields a correlation matrix of the raw data of

$$\begin{pmatrix} 1.0766 & 0.8308 \\ 0.8308 & 0.7956 \end{pmatrix}.$$

The eigenvector matrix is:

$$\begin{pmatrix} 0.6455 & -0.7638 \\ -0.7638 & -0.6455 \end{pmatrix}.$$

The transform axes comprise each column vector of this matrix. The transform matrix is formed from the eigenvectors as rows:

$$\mathbf{A} = \begin{pmatrix} -0.7638 & -0.6455 \\ 0.6455 & -0.7638 \end{pmatrix}.$$

The eigenvalues of the transformed data are:

$$\begin{pmatrix} 0.0935 \\ 1.7787 \end{pmatrix},$$

which indicates a high correlation on one axis, and a much smaller correlation on the other.

The scalar dot product of the eigenvectors is zero, as we hoped for. After transformation, the correlation matrix of the transformed data is

$$\begin{pmatrix} 1.7787 & 0.0000 \\ 0.0000 & 0.0935 \end{pmatrix}.$$

This is a diagonal matrix, and shows that one component has significantly higher variance than the other.

So we have derived the optimal transformation to apply to given data in order to compact the input data representation into as few output values as possible. Since the transformation is invertible, we can reconstruct the original data. However, in practice, the eigenvectors are difficult to compute, and of course this computation depends on knowing the autocorrelation matrix, as shown above. The autocorrelation matrix itself is time-consuming to calculate, and requires a large sample of data. In the data compression/transmission problem, the decoder (receiver) also needs to know the transformation matrix to effect the reconstruction of the data. So even if we compute the optimal transform, it must be given somehow to the decoder, and this works against what we are trying to achieve (compressing the data into fewer bits). For these reasons, suboptimal transforms such as the DCT are normally used in practice, since they can approximate the optimal transform for typical sources. Although the compression is not optimum in a mathematical sense, it solves the problem of informing the decoder of the transform coefficients to use, since they are known in advance and determined by the DCT equations.

7.14 CHAPTER SUMMARY

The following are the key elements covered in this chapter:

- the *Fourier series* and *Fourier transform*.
- the *interpretation* of the Fourier transform.
- the development of the *FFT* algorithm.
- the *DCT* algorithm, its use in data compression, and relationship to the optimal decorrelating transform.

PROBLEMS

7.1. Using the MATLAB code shown in Listing 7.1, change the input signal to each of the following signals and examine the a_k and b_k coefficients.

(a) A two-component sine wave: $x(t) = \sin 2\pi t + \sin(4 \times 2\pi t)$

(b) A sine plus cosine waveform: $x(t) = \cos 2\pi t + 2 \sin 2\pi t$

Are they equal to the signal equation amplitudes/frequencies?

7.2. It was stated in Section 7.4 that terms of the form $\int_0^\tau a_k \cos k\Omega_0 t \cos n\Omega_o t \, dt$ equate to zero for $n \neq k$. Prove this statement mathematically by expanding the integration and noting that $\tau = 2\pi/\Omega_0$.

7.3. Use MATLAB to generate an input signal as described below, and use the Fourier series code as given in Listing 7.1 to determine the resulting Fourier series expansion. Examine the a_k and b_k coefficients, and calculate the corresponding A and φ. Does the Fourier series give the expected result in each case?

(a) A phase-shifted sinusoidal wave (positive—shift to left)

$$x(t) = \sin\left(2\pi t + \frac{\pi}{3}\right)$$

(b) A phase-shifted sinusoidal wave (negative—shift to right)

$$x(t) = \sin\left(2\pi t - \frac{\pi}{3}\right)$$

(c) A square wave with a phase shift of $-\pi/2$ radians.

(d) A square wave with a phase shift of $+\pi/2$ radians.

(e) A random (nonperiodic) signal.

(f) A triangular waveform.

(g) A square wave with 10% duty cycle (+ for first 10%, – otherwise).

7.4. Write a MATLAB function to perform the inverse DFT. Check that for any given time waveform, the DFT and IDFT are inverses of one another.

7.5. Section 7.6.5 showed how to implement the DFT equation directly with two nested loops, and subsequently how to perform a conversion to a vectorized version with only one loop and one vector dot product within the loop. It is possible to code the DFT using only one matrix-vector multiplication, where the matrix is composed of fixed coefficient values and the vector is the time-sampled input vector. Show how to implement the DFT in this way, and check your results using the DFT code given.

7.6. Explain the following in terms of frequency:

```
x = [1 -1  1 -1  1 -1  1 -1];
XX = fft(x)
         XX = 0   0   0   0   8   0   0   0
```

7.7. Explain the following:

```
x = [3 5 7 9]
sum(x)
    ans = 24
fft(x)
    ans = 24.0   -4.0+4.0i   -4.0   -4.0-4.0i
```

7.8. An ECG is sampled at 500 Hz, and a data record of length 256 is used for an FFT analysis. What is the frequency resolution of the FFT components? If a resolution of 1 Hz is required, what would the minimum data length be for analysis?

7.9. We wish to investigate the following relationship:

$$\sum_{n=0}^{N-1} |x(n)|^2 = \frac{1}{N} \sum_{k=0}^{N-1} |X(k)|^2.$$

This shows that the sum of the energy in a sampled signal (left-hand side) is proportional to the sum of the energy in the frequency domain (Fourier transformed).

(a) Test this relationship using some data:

```
N = 32;
x = randn(N, 1);
sum(abs(x).^ 2)
X = fft(x);
sum(abs(X).^ 2)/N
```

Verify that this code implements the equations specified. Try for differing N.

(b) Write the energy summation term $\Sigma_n |x(n)|^2$ as $\Sigma_n x(n)x^*(n)$, where the $*$ denotes the complex conjugate.

(c) Substitute the inverse DFT for $x^*(n)$. Reverse the order of the summation and simplify, to give the right-hand side $(1/N)\Sigma_k |X(k)|^2$. This shows that the original relationship holds for all sets of samples $x(n)$. This relationship is a form of what is known as Parseval's theorem. Energy is conserved in going from the time domain to the frequency domain, as we would expect.

7.10. Determine the bits per pixel (bpp) rate for each of the DCT compressed images in Figure 7.40.

7.11. Given a transform matrix $\mathbf{A}$ whose rows are orthogonal vectors, what is the product $\mathbf{A}\mathbf{A}^T$? Given the vectorspace interpretation of transform and inverse transform, is the relationship between forward transform matrix $\mathbf{A}$ and its corresponding inverse $\mathbf{B}$?

7.12. Given a sequence in vector $\mathbf{x}$ which has zero mean and unit variance, and correlation coefficient ρ,

(a) What is the correlation matrix $\mathbf{R}_x$?

(b) Show that the eigenvalues are $1 \pm \rho$.

(c) Show that the eigenvectors are:

$$c\begin{pmatrix}1\\1\end{pmatrix} \text{ and } c\begin{pmatrix}1\\-1\end{pmatrix}.$$

(d) What is the value of c to make these vectors orthonormal (that is, both orthogonal and having unit magnitude)?

(e) What would be the optimal transform in that case? The inverse transform?

7.13. Using the method explained in Section 7.13.2, implement and test a DCT function in MATLAB as follows.

(a) Implement an $N = 8$-point DCT to compute the DCT directly.

(b) For an input of $x(n) = 1$ for all n, verify mathematically that the DCT output $y(k)$ should equal $\sqrt{N}$ for $k = 0$.

(c) Verify that your implementation gives $y(0) = \sqrt{8}$ for the case of $N = 8$ and $x(n) = 1$.

7.14. Fast algorithms have been put forward for computing the DCT. This question is based on using an N-point FFT to compute an N-point DCT (see Narashima and Peterson 1978).

(a) If not done already, implement a DCT in MATLAB as in the previous question.

(b) For an N-point input sequence $x(n)$, create a new sequence $y(n)$, such that $y(n) = x(2n)$ and $y(N - 1 - n) = x(2n + 1)$ for $n = 0, \cdots, N/2$. Using the input sequence $x(n) = [0, 1, 2, 3, 4, 5, 6, 7]$, verify by hand that $y(n) = [0, 2, 4, 6, 7, 5, 3, 1]$. Note that the first half of $y(n)$ is made up of the even-indexed values from $x(n)$, and that the second half of $y(n)$ comprises the odd-indexed values from $x(n)$ in reversed order.

(c) Implement the above reordering in MATLAB, and verify against the same test case.

(d) Compute the transform values $F(k)$ using

$$F(k) = \sqrt{\frac{2}{N}} c_k \mathcal{R}e\left\{e^{-jk\pi/(2N)} \text{FFT}(y(n))\right\},$$

where $c_k = \frac{1}{\sqrt{2}}$ for $k = 0$, and $\mathcal{R}e$ means take the real value (ignoring the imaginary component if present). Be sure to use the new sequence $y(n)$ in computing $F(k)$, and the original sequence $x(n)$ in computing the DCT itself. Verify against the output of the previous question that $F(k)$ produces the same DCT result for $x(n)$ for a random 8-sample input case.

(e) Considering that the above Fast DCT (FDCT) is based on an FFT and a complex-number multiplication for each term, what is the approximate speedup over the direct DCT calculation?

CHAPTER 8

DISCRETE-TIME FILTERS

8.1 CHAPTER OBJECTIVES

On completion of this chapter, the reader should be able to

1. explain the basic types of filters and approaches to *filter design*;
2. design a *finite impulse response* (FIR) filter for a given frequency specification;
3. plot the *frequency response* of a filter;
4. be conversant with filter *windows* and where they are used; and
5. explain and implement *fast filtering algorithms*.

8.2 INTRODUCTION

This chapter introduces digital filters, which are able to remove (filter out) or enhance certain frequency components in a signal. Digital filters are a key technology in many audio processing applications and are fundamental to communications systems. The idea of a digital filter is to either reduce or increase the strength of a signal but only over specific frequency ranges. Digital filters are used in various applications, a few of which are

1. audio recording/playback/enhancement applications,
2. noise removal,
3. sub-band filtering for audio and image storage and/or transmission, and
4. narrow-band filtering for selecting a communications channel.

To give a concrete example, consider Figure 8.1, which shows a sample application—measuring the electrocardiogram (ECG) signals as derived from the beating of the human heart. Such a measurement has a great many clinical and diagnostic applications and is attractive because the signal can be measured at various points on a patient's skin, thus monitoring the heart externally. The problem is that the evoked potentials on the skin are somewhat removed from the heart muscle itself, and hence the signals are tiny (of the order of microvolts to millivolts). Typically, an amplification of the order of 1,000 times is required to adequately quantize such signals. Such

Digital Signal Processing Using MATLAB for Students and Researchers, First Edition. John W. Leis.

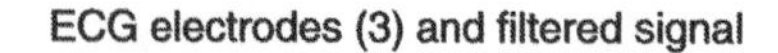

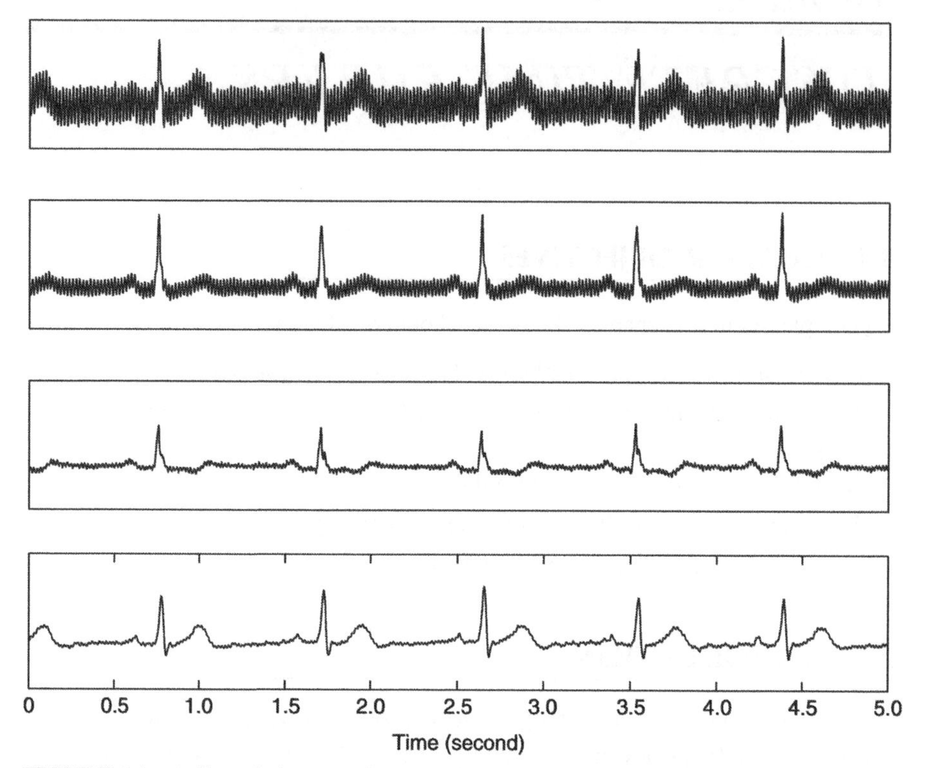

FIGURE 8.1 A filtered electrocardiograph (ECG) signal. Note the significant amount of interference (noise) present in the raw electrode signals and the relatively "clean" signal derived below.

small signals with high amplification are very susceptible to noise from external sources—the body and connecting leads act as antennas which pick up electromagnetic interference from power lines and other sources, and other muscle signals within the body are also present. Thus, a signal processing approach is needed to filter the measured signal and leave only the desired portion. This has to be achieved without removing or altering the desired signal and without introducing any artifacts in the observed waveform.

8.3 WHAT DO WE MEAN BY "FILTERING"?

In order to understand filters, the concepts of the *gain* and *phase* response of a system must be well understood. Figure 8.2 illustrates how we might have a sinusoidal waveform, which is the input to a filter, and the corresponding output. Although most signals encountered in practice are more complex than a simple sinusoid, it may be recalled from the Fourier series in Chapter 7 that any periodic signal may

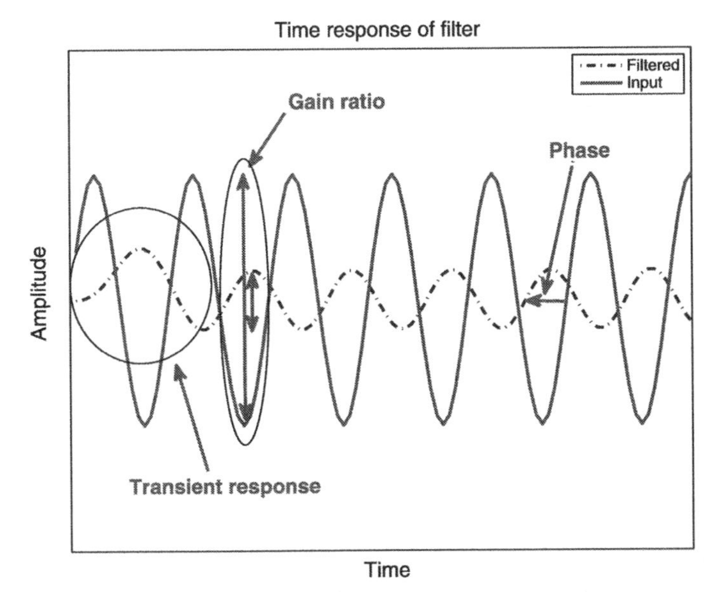

FIGURE 8.2 The output of a filter consists of one or more frequency components. The magnitude of each component will be altered according to the filter's transfer function, and the phase will show a change with respect to the input. Normally, we are interested in the steady-state response only—that is, the time after the transient response.

be broken down into its component sine and cosine waves. Thus, we are quite entitled to consider only a simple sinusoidal waveform in our discussion of digital filter responses, as complex waveforms are made up of many additive sinusoidal components. Put another way, the filter acts on each component of the input waveform independently, with the final output being the summation of the effects on each individual component.

Returning to Figure 8.2, we see that the output of our (as yet hypothetical) filter is another sinusoid, with the following characteristics:

1. The frequency of the output is identical to that of the input.
2. The amplitude may be greater or lesser.
3. The time of occurrence of the output is changed with respect to the input.

The ratio of the output to input amplitudes is termed the gain of the filter. The delay is expressed not in absolute terms but in relative terms of phase with respect to the input. The phase angle is a convenient way to represent this delay, as it inherently incorporates the frequency of the waveform (since phase is a relative quantity).

Our filter may be subjected to not one, but numerous input frequencies. We require a concise method of showing the response of the filter, which conveys the information we require. Such a graphical visualization of the gain and phase at each frequency is illustrated in Figure 8.3. Usually, but not always, we are interested in the gain response. Note that the gain is often quoted in decibels, which is $20\log_{10}|G|$, where $|G|$ is the output/input gain ratio.

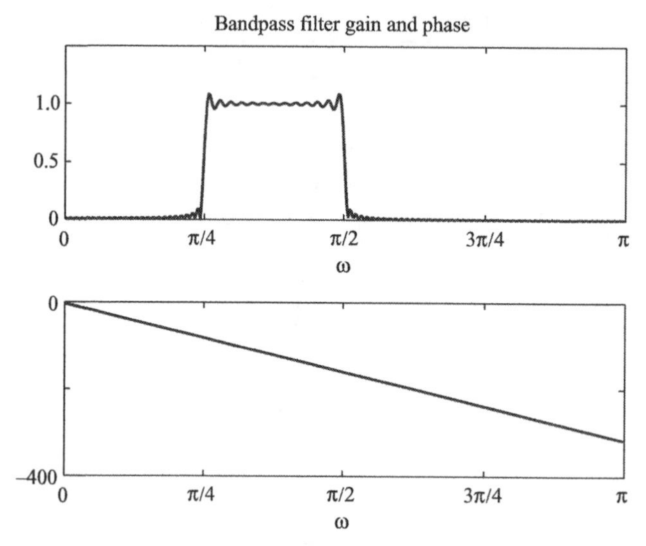

FIGURE 8.3 The gain and phase response curves allow determination of the gain (upper) and phase (lower) of the filter transfer function at any given frequency. The normalized frequency is shown (ω radians per sample). This filter is a finite impulse response (FIR) filter of order $N = 201$. The gain is shown as a ratio, and the phase is shown in radians.

The magnitude of a complex number may be found using the **`abs()`** operator in MATLAB. The phase may be found using **`angle()`**. Since the phase angle is cyclical over 2π, the phase graphs often exhibit a "wrapping" effect. For example, successive phase lags of −350°, −355°, and +5° are better interpreted as −350°, −355°, and −365°. The MATLAB function **`unwrap()`** is able to effect this for a given vector of complex numbers.

8.4 FILTER SPECIFICATION, DESIGN, AND IMPLEMENTATION

Designing and implementing a discrete-time filter requires a three-step process:

1. Specify the frequency response, according to the problem at hand. This requires choosing an appropriate sample rate, f_s, and a response shape.
2. Design the filter itself by selecting an algorithm and by determining the coefficients.
3. Implement the difference equation derived from the design step.

Generally speaking, the design process tends to be iterative since there is no ideal way of designing a filter for all situations. Filtering is normally thought of as an operation in the frequency domain, although the operation of filtering is implemented as a difference equation in the time domain. Recall from Section 3.10 that the general form of a difference equation is

$$y(n) = (b_0 x(n) + b_1 x(n-1) + \cdots + b_N x(n-N)) \\ - (a_1 y(n-1) + a_2 y(n-2) + \cdots + a_M y(n-M)). \tag{8.1}$$

The operation of designing a filter, once the specifications for its response are known, essentially reduces to choosing the required transfer function order, and choosing the difference equation coefficients (all $(N + 1)$, b values, and M, a values). As will be shown in the next section, this may be considered as "placing" the poles and/or zeros of a system on the complex plane.

We may summarize the steps required to design and implement a discrete-time filter as follows:

General Filter Design Steps

1. Specify the response required and determine an appropriate sampling frequency, f_s.
2. Determine the transfer function $G(z)$ from the specification.
3. Let $Y(z)/X(z) = G(z)$. Solve for $Y(z)$.
4. Convert $Y(z)$ into a difference equation, that is, $y(n)$ in terms of $x(n)$, $x(n-1)$, $y(n-1)$, and so on.
5. Code a loop to sample the input $x(n)$, calculate $y(n)$, output $y(n)$, and buffer the samples.

This describes the process in general terms; for specific filter types described in the next sections, the process will vary a little in the details—primarily the calculation of $G(z)$, which is the most technically difficult stage. Since we are examining digital or discrete-time filters, the sampling rate needs to be chosen accordingly, so as to avoid aliasing. This was examined in detail in Section 3.8.

In general, the determination of $G(z)$ is the most difficult step, and there are many ways to approach this. The following sections introduce some possible approaches to solving this problem. First, however, we need to more clearly specify what we mean by *specifying* a filter's response.

8.4.1 Revisiting Poles and Zeros

A filter is simply a transfer function, and we have examined these before. Recall that a transfer function is a ratio of a numerator polynomial in z to a denominator polynomial in z, and that we can factorize each into roots to give zeros and poles, respectively. So we can imagine the design process as one of specifying the poles and zeros in such a way as to give us the desired frequency response.

Of course, the implementation of the filter will be done as a discrete-time difference equation, which is derived from the transfer function. So we are working in several domains:

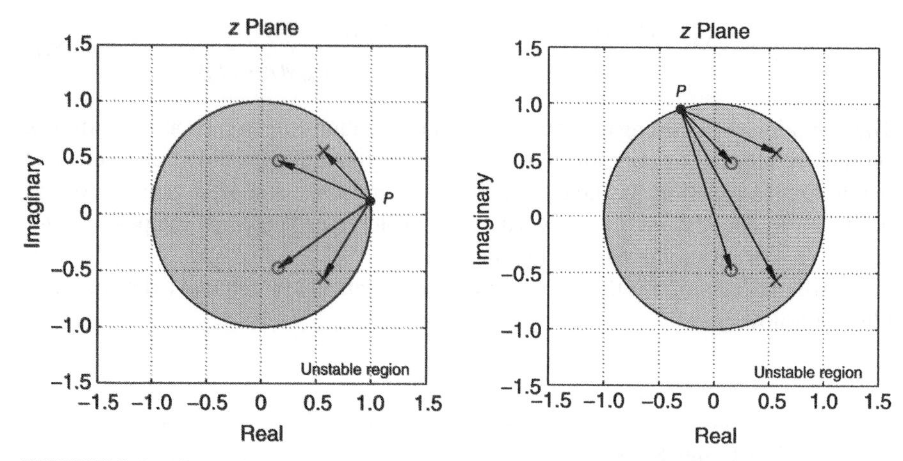

FIGURE 8.4 Illustrating pole–zero vectors on the unit circle. The "real" frequency in hertz relative to the sampling frequency is converted into a corresponding angle. We then draw vectors from the point $P = 1e^{j\omega}$ to the poles (shown as ×) and zeros (shown as O). As we change the input frequency, we change the angle $\omega = \angle P$ from the origin to point P and redraw the vectors.

1. the frequency domain (what we want),
2. the time domain (how we implement what we want), and
3. the transfer function (which describes the system in a mathematical sense).

The key to understanding the linkage between these abstract viewpoints is the poles and zeros themselves. Section 5.8 explained how the poles and zeros may be thought of as a set of vectors, with frequency response conceived graphically as a mapping from the given frequency to a point on the unit circle. Figure 8.4 illustrates this idea, for the case where we have two poles and two zeros. The poles and zeros describe the transfer function, with each factor of the numerator polynomial being of the form "(z − zero)." The denominator is formed in a similar fashion, as a product of "(z − pole)" terms.

The response of the system depends on the frequency of the signal which is applied to that system. This does not mean we are restricted to a single-frequency input—most real-world signals are composed of a multitude of input frequencies. But by the principles of superposition and linearity (refer to Section 3.11), we are able to consider each of these separately.

Figure 8.4 shows a snapshot of two frequencies as points on the unit circle. As introduced in Section 5.7, the response is evaluated for a point, $1e^{j\omega}$, on the complex plane, where ω is the frequency relative to the sampling frequency. Note that the radius of this point is one. Hence, as we vary the frequency from 0 (or "DC") up to $\omega = \pi$ (which corresponds to half the sample frequency), the point moves around the top half of the unit circle. The figure shows the vectors to the poles, and the vectors to the zeros. The frequency response magnitude is then the product of the zero vector lengths divided by the product of the pole vector lengths. Clearly, when the frequency is such that the point on the unit circle is close to the pole, the magnitude will be large. Similarly, when the frequency is such that the point is close to a zero, the magnitude will be small. So, we can imagine that *designing* the filter comes down to a matter of specifying where the poles and/or zeros will be. The only real restriction is that, for stability, the poles must lie inside the unit circle.

8.4.2 Specification of a Filter's Response

In this section, we look at how to *specify* the frequency response of a filter. If we have a broad idea of what frequency response we want to shape our signal into, we must first translate the general specification into something more precise.

The preceding section gave an intuitive insight into filter design—we need to find the appropriate combination of poles and zeros to match the desired frequency response. Before we do that, however, we need to be clear about the frequency response and how we specify that more precisely. In pragmatic terms, ideal, perfectly sharp "brick wall" filters are not able to be physically realized. This can be understood from the pole–zero vector concept—as the frequency point on the unit circle is approaching a pole (or zero), the vector lengths will *gradually* change and will not have the abrupt change which we would like.

Furthermore, since the transfer function is specified in terms of numerator and denominator polynomials, we expect the frequency response (obtained by putting $z = e^{j\omega}$) would be a "smooth" function. We can see this if we start with a second-order polynomial (a quadratic equation), then extend to a third-order (cubic) polynomial, and so forth. As we increase the order, we become increasingly able to approximate any function. However, polynomials are smooth functions, without any discontinuities. If we increase the order of the polynomial, we can get a "sharper" change, but in theory it would require an *infinite* order to get a step change.

To put this in perspective, consider Figure 8.5, which shows a low-pass filter. Such a filter has a *passband* so that lower frequencies are passed through, and frequencies above the cutoff frequency are reduced in magnitude (attenuated). The filter characteristic curve shows the frequency of the input waveform (horizontal) and the desired filter gain (vertical). The hypothetical response plotted is a higher-order polynomial, which was fitted to try to match the ideal response having a passband from 0 to the cutoff frequency f_c. Not only do we see the finite tapering of the response as it crosses the cutoff frequency f_c but we also have a "rippling" response in the passband and also in the stopband. This, of course, is characteristic of a polynomial.

Figure 8.5 then helps us to make the link between what we want and what we can actually obtain in practice. The passband shows that the gain will vary by some amount plus or minus the desired gain. Similarly, the *stopband* shows that the gain may vary from the ideal (often the ideal is zero; in the figure, it is shown as 0.1 for the sake of example). Finally, the *transition band* links the two and indicates how "sharp" the filter cutoff actually is. This is an amount on either side of the cutoff frequency f_c and often is the most critical parameter in designing a filter. The passband ripple is often a very important parameter, depending on the application. Finally, note in the figure the fact that the *worst case* is what we often have to design

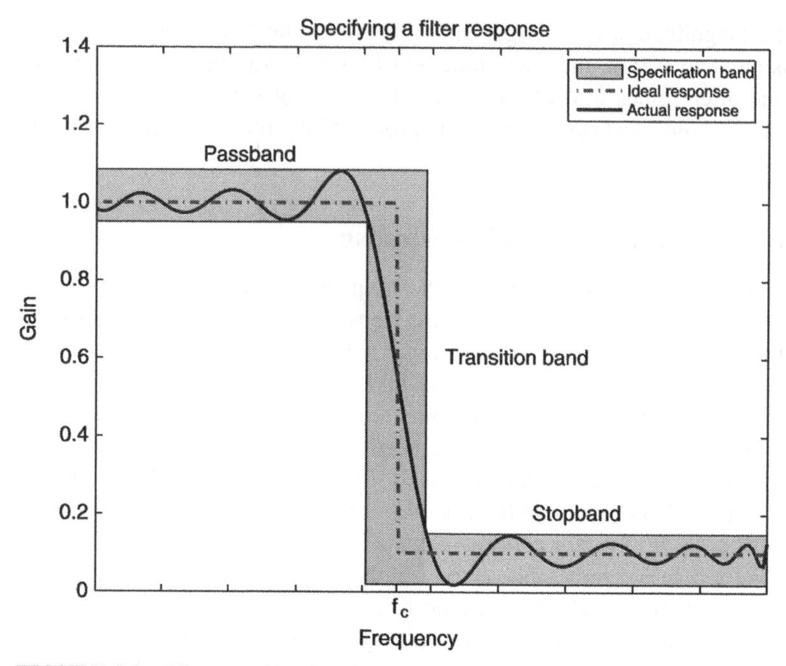

FIGURE 8.5 Filter specification for a low-pass filter. f_c is the cutoff frequency. Because the filter is not "ideal," we must specify the tolerance in the passband, in the stopband, and in the transition band.

toward—the highest ripple (or "lobe") in the stopband is the closest to the passband and thus is most likely to adversely affect the performance of the system in an overall sense.

8.4.3 Filter Design as Pole–Zero Placement: A Notch Filter

We now put these concepts together with a concrete example. Suppose we wish to design a digital filter to remove a certain frequency but pass all others. This often occurs when we want to remove power line "hum" in audio systems, for example, or in processing the ECG signal described earlier. The 50- or 60-Hz interfering signal is obviously unwanted, and the signal of interest (audio or biomedical potential) is present in nearby frequency bands, and of a very small magnitude.

To design such a filter, we need to have a sampling frequency in mind. This is chosen depending on the physical constraints of the system. As always, the Nyquist frequency of $f_s/2$ corresponds to π in our pole–zero and frequency diagrams. Let the normalized frequency that we wish to remove be ω_n rad/s (i.e., normalized to the sampling frequency of ω_s or 2π radians per sample).

As seen previously, the gain is the product of the zero distances divided by the product of the pole distances, for any point on the unit circle. The angle of the point defines the frequency. We place a zero at an angle ω_n on the unit circle, with

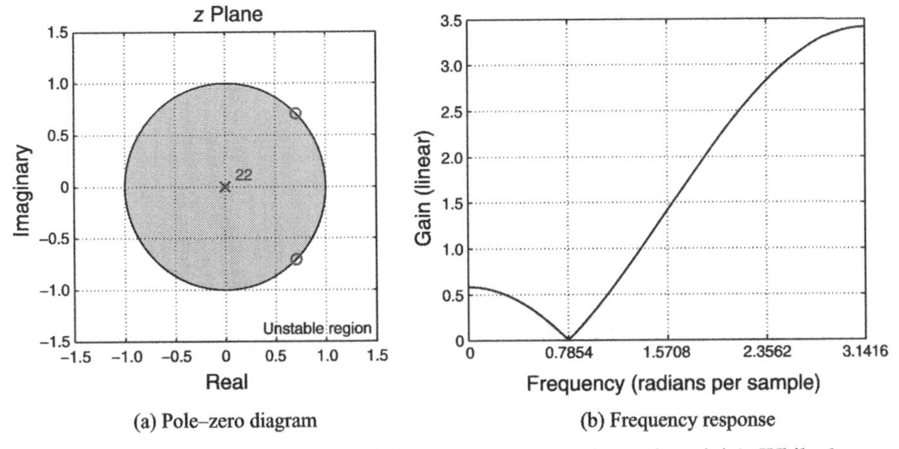

(a) Pole–zero diagram (b) Frequency response

FIGURE 8.6 A notch filter with zeros only (and simple poles at the origin). While there is a notch-like response around the desired frequency, the overall shape of the response is not what we would wish for.

the idea being to make the zero vector distance equal to zero at that frequency. We must also place a complex conjugate zero at angle $-\omega_n$ so that we have a realizable transfer function (i.e., no complex coefficients in the difference equation).

We then need two poles, so as to make the system causal; that is, we must remove the z^2 terms, which will result from employing only two zeros. We put the poles at the origin so as not to affect the frequency response. The pole–zero plot is then as shown in Figure 8.6, with the corresponding frequency response. The filter does indeed have zero gain at the required notch (reject) frequency, but it is clear that the passband response (at frequencies away from the notch) has a nonuniform gain, well away from the required unity gain. Additionally, it is also easily seen that the gain on either side of the notch is not controlled very well (the notch is not very "sharp").

To address these issues, we place poles just near the zeros. The rationale for this is that since the gain is the product of the zero distances divided by the product of the pole distances, at any point on the unit circle away from the zero–pole combination, these distances will be approximately equal, and hence cancel.

This new pole–zero configuration is shown in Figure 8.7a. Of course, we must take care to place the pole inside the unit circle so as to ensure stability. This is particularly important to keep in mind if sampling and coefficient quantization errors, or finite-precision arithmetic errors, mean that the actual value in practice may drift outside the unit circle. The resulting frequency response is evidently much improved, as shown in Figure 8.7b. The problem now is that the gain in the passband is not quite unity, as closer inspection of the figure will show. We need to apply some theory to see why this is so, and if the situation can be improved.

The transfer function of the filter so far is

$$G(z) = \frac{\left(z - 1e^{j\omega_n}\right)\left(z - 1e^{-j\omega_n}\right)}{\left(z - r_p e^{j\omega_n}\right)\left(z - r_p e^{-j\omega_n}\right)}, \tag{8.2}$$

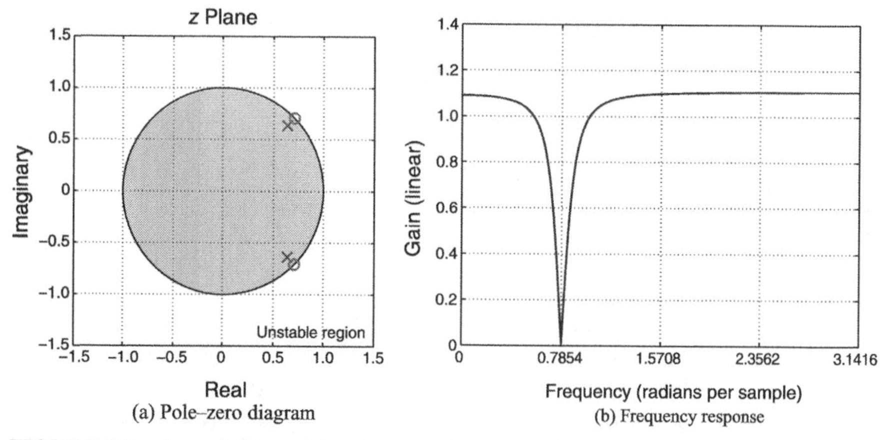

(a) Pole–zero diagram (b) Frequency response

FIGURE 8.7 A notch filter with poles near the zeros. The response is closer to the desired characteristic; however, the gain is not exactly equal to one over regions away from the notch as we desire.

where the angle of the pole–zero pair is ω_n and the radius of the pole is r_p. Expanding this and using $e^{j\theta} + e^{-j\theta} = 2\cos\theta$ gives

$$
\begin{aligned}
G(z) &= \frac{z^2 - ze^{-j\omega_n} - ze^{+j\omega_n} + 1}{z^2 - zr_p e^{-j\omega_n} - zr_p e^{+j\omega_n} + r_p^2} \\
&= \frac{z^2 - z\left(e^{j\omega_n} + e^{-j\omega_n}\right) + 1}{z^2 - zr_p\left(e^{j\omega_n} + e^{-j\omega_n}\right) + r_p^2} \\
&= \frac{z^2 - 2z\cos\omega_n + 1}{z^2 - 2zr_p\cos\omega_n + r_p^2}.
\end{aligned}
\tag{8.3}
$$

Now, inspection of the gain plot shows that we need a constant gain, K, so as to adjust the gain of the filter to one. We need this to occur for all points on the unit circle, except in the vicinity of the notch itself. Since all points on the unit circle are described by $|z| = 1$ (a radius of one, at any angle), we need to multiply by the gain factor K and evaluate

$$
KG(z)\big|_{z=1} = 1. \tag{8.4}
$$

So this becomes

$$
K\frac{z^2 - 2z\cos\omega_n + 1}{z^2 - 2zr_p\cos\omega_n + r_p^2} = 1. \tag{8.5}
$$

We can solve this for K at a given value of pole radius r_p and angle (relative notch frequency) ω_n. Figure 8.8 shows the resulting response, which exhibits the desired characteristics.

Finally, the parameter r_p sets the pole radius. It can be used to adjust the "sharpness" of the notch (in communications systems, this is also called the Q factor). If we make r_p larger—that is, move the poles closer to the zeros—the type

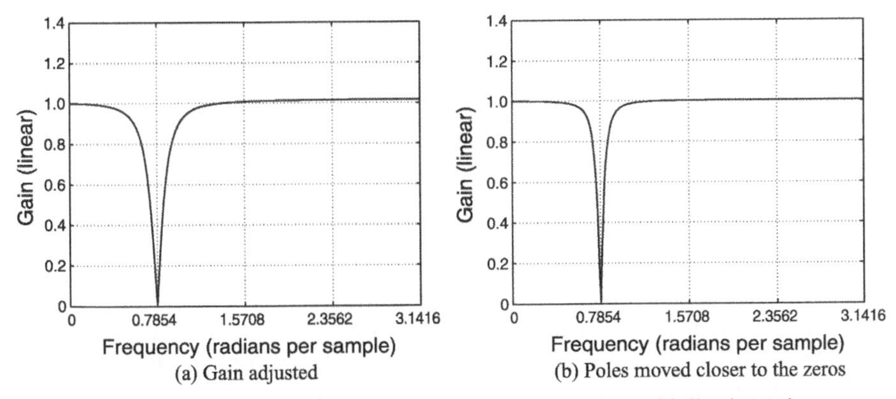

FIGURE 8.8 A notch filter with poles and zeros and with the multiplicative gain adjusted. The adjustment of the gain is shown in (a), and we see that the gain is close to unity away from the vicinity of the notch. Further adjustment of the pole position is shown in (b)—this yields a sharper notch if desired.

of response we obtain is shown in Figure 8.8b. Comparing the responses in Figure 8.8a,b, it is clear that the width of the notch is reduced as $r_p \to 1$.

So let us look at what we have achieved. We have determined the polynomials for the numerator, denominator, and the multiplicative gain. The poles and zeros may be derived by factorizing the polynomials. The difference equation is found by letting the transfer function equal the output/input ratio, and from the difference equation, we can determine the code required to implement the digital filter. We can be confident that the frequencies presented to the filter (singly, or in combination) will be passed through or reduced in amplitude according to our frequency response plot.

For the specific case of the notch filter described in this section, the following steps were undertaken:

Notch Filter Design Steps

1. From the signal bandwidth and notch frequency f_n, determine a suitable sampling frequency, f_s.
2. Calculate the notch angle on the unit circle ω_n. Remember that $f_s/2 \leftrightarrow \pi$.
3. Place the zeros on the unit circle and the poles at an angle, ω_n, with radius less than one.
4. Determine $G(z)$ from the above.
5. Plot the frequency response using $z = e^{j\omega}$ for $\omega = 0 \ldots \pi$ and iterate the above steps as necessary.
6. Let $Y(z)/X(z) = G(z)$. Solve for $Y(z)$.
7. Convert $Y(z)$ into a difference equation; find $y(n)$ in terms of $x(n)$, $x(n-1)$, $y(n-1)$, and so on.
8. Code a loop to sample the input $x(n)$, calculate $y(n)$, output $y(n)$, and buffer the samples.

In general, the determination of $G(z)$ is the most difficult step, and there are many ways to approach this. The notch filter serves as a stepping-stone to other design approaches considered in subsequent sections. First, though, we need to consider how to broadly categorize the various filter response types.

8.5 FILTER RESPONSES

The "notch" filter is just of many filter types we may need to use in practice. We now examine some other common filter response types.

The most fundamental and commonly encountered type of filter is the *low-pass* filter. An ideal low-pass filter is depicted in Figure 8.9. Frequencies up to a certain limiting frequency are passed through from input to output, with some gain factor; frequencies above this are eliminated altogether because the gain is zero. Most "real-world" signals tend to be composed of predominantly lower-frequency components, and so low-pass filters are commonly used to reduce the perception of noise, which often covers the entire bandwidth (or a large proportion of it). The converse, a *high-pass* filter, is shown in Figure 8.10. Again, this is an idealized representation. Only frequencies above the cutoff are passed through.

In some applications, such as telecommunications, we wish to transmit several frequencies on the one channel. Each is borne by a carrier frequency, and the receiver must "pick out" only one particular channel (frequency band) of interest. Obviously, this means that the frequency range of the filter must be varied as different channels are selected—something which is quite easy to realize using digital filters: It is simply a matter of changing the filter coefficients. This operation, in its simplest form, is termed *bandpass* filtering, as depicted in Figure 8.11.

The converse of the bandpass filter is the bandstop filter, as depicted in Figure 8.12. This type of filter finds application in rejecting narrow-band noise (e.g., such as power supply interference).

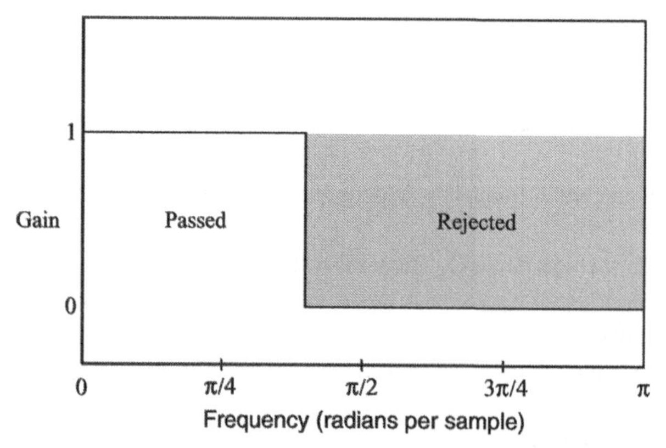

FIGURE 8.9 A low-pass filter with an idealized response. Any input component whose frequency is below the cutoff frequency is passed through with a gain of unity and hence appears at the output; any component whose frequency is above the cutoff is multiplied by a gain of zero and hence is removed entirely from the output.

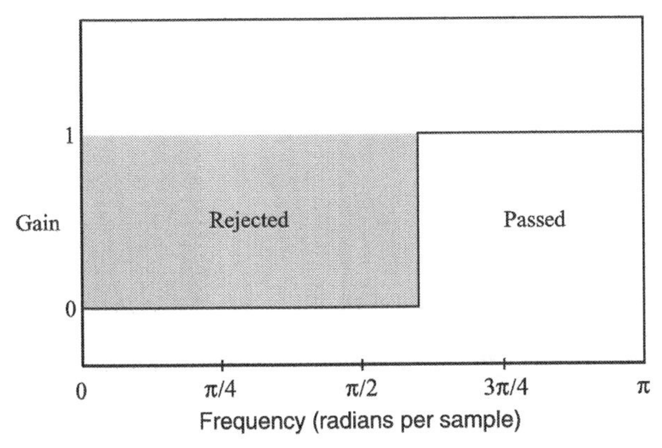

FIGURE 8.10 A high-pass filter with an idealized response. Frequency components above the cutoff are passed through with a gain of unity; frequency components below this are removed entirely.

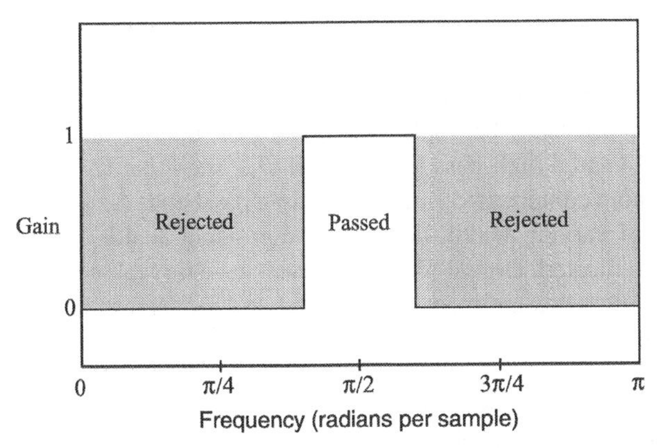

FIGURE 8.11 A bandpass filter with an idealized response. Only frequencies falling between the upper and lower frequency cutoff points are passed through from input to output; all other frequency components present in the input are eliminated from the output.

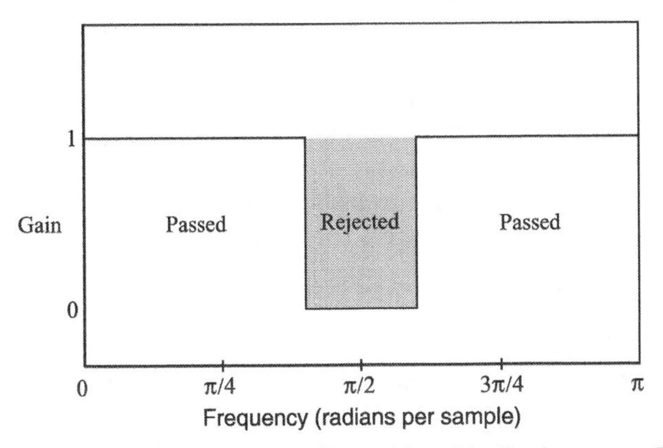

FIGURE 8.12 A bandstop filter with an idealized response. This is the complement of the bandpass filter—it removes all frequency components falling within a certain stopband.

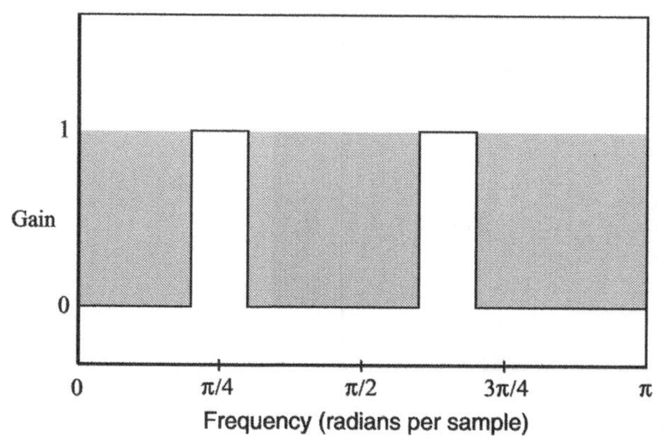

FIGURE 8.13 A filter with more than one passband. It is effectively a hybrid of the fundamental types encountered previously. Using digital signal processing, such a filter is a little more complicated than a low-pass filter, in either design or implementation.

With analog electronics, bandpass filters may be implemented using a cascade of a high-pass filter with cutoff Ω_h and a low-pass filter with cutoff Ω_l, such that $\Omega_l > \Omega_h$. Likewise, a bandstop filter may be implemented using the parallel addition of a low-pass filter (cutoff Ω_l) and a high-pass filter (cutoff Ω_h), such that $\Omega_l < \Omega_h$.

If we desire anything more complicated, such as multiple passbands (as shown in Fig. 8.13), or passbands of varying width, then the design using analog signal processing becomes very complicated. Digital filters, however, have no real increase in complexity when the specifications are more complicated and are in fact able to realize virtually any filter response characteristic that may be required. When we say that they have no increase in complexity, this is meant in terms of physical hardware. Of course, the software complexity (the difference equation implementation and the associated memory storage for samples and coefficients) may be greater. But this comes at essentially very little extra cost.

As might be expected, the idealized filters outlined above are not realizable in practice. In effect, they would require a transfer function of infinite order. However, another advantage of digital filters is that once the basic "A/D–processing–D/A" system is available, increasing the order of the filter (and hence making the characteristic closer to the ideal) is simply a matter of increasing the number of coefficients. Although this may require a faster processor and/or more memory, such a cost is usually not excessive.

A great many filter design algorithms are available to choose from. Often, these are in the form of computer-assisted packages. The initial decision steps are

1. to choose whether the filter has poles, zeros, or both; and
2. to choose the order of the filter.

The former may be viewed as a "pole–zero placement" problem: We must place the poles and/or zeros at specific locations in order to realize a certain frequency response. The relationship between poles, zeros, and the frequency response of a system was introduced in Section 5.8.

The question of whether the above steps lead to an acceptable performance of the filter usually requires an iterative solution: a lower-order filter may be tried, which may or may not satisfy the design requirements. Finally, the transfer function leads directly to the difference equation coefficients (a_k's and b_k's).

8.6 NONRECURSIVE FILTER DESIGN

We have now discussed the common filter types and examined in detail the specific design case of a notch filter. In order to design for the more general responses shown, we need a more generally applicable technique. One such method is the FIR approach. FIR filters may be contrasted with infinite impulse response (IIR) filters, which are considered in the next chapter.

The difference equation describing an FIR filter is

$$y(n) = \sum_{k=0}^{N-1} b_k x(n-k) \tag{8.6}$$

with a z transform of

$$H(z) = \sum_{k=0}^{N-1} b_k z^{-k}. \tag{8.7}$$

This is shown in block diagram form in Figure 8.14. We need a method to determine the b_k filter coefficients.

8.6.1 Direct Method

We now introduce the direct method for FIR filter design. It is based on the theory of the Fourier transform but requires some additional steps to make a useful filter in practice.

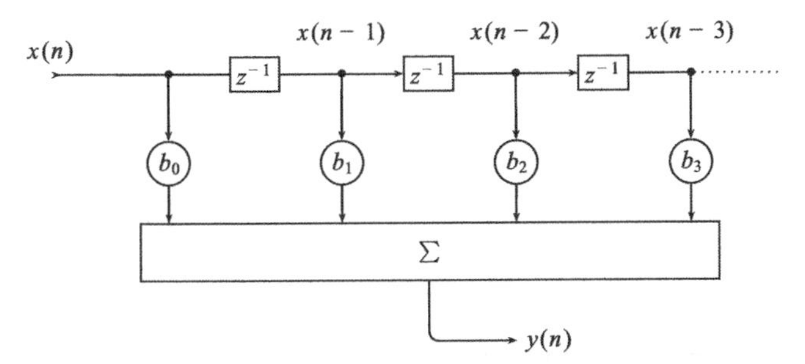

FIGURE 8.14 Block diagram representation of an FIR filter. The signal is delayed by each of the z^{-1} blocks and is multiplied by a coefficient, b_k, at each stage. Finally, the result of the delayed weighted products is summed to form the output $y(n)$.

The time domain impulse response of a filter corresponding to a given (desired) frequency response may be calculated from the inverse Fourier transform of the desired frequency response:

$$h_d(n) = \frac{1}{2\pi}\int_{-\pi}^{\pi} H_d(\omega)e^{jn\omega}\,d\omega. \tag{8.8}$$

The samples $h_d(n)$ from the above are time domain values, as indicated by the index n. These are the time domain samples that would have the frequency response $H_d(\omega)$. How does that help us design a filter? The conceptual leap is that we use these numbers as weighting coefficients in a difference equation to form the filter itself.

To design a filter using Equation 8.8, we must obviously know $H_d(\omega)$. However, note that the integration limits are $\pm\pi$. So, the desired passband in the 0 to π range must be mirrored into the negative frequency range of 0 to $-\pi$. This is illustrated in Figure 8.15 for a low-pass filter and in Figure 8.16 for a bandpass filter.

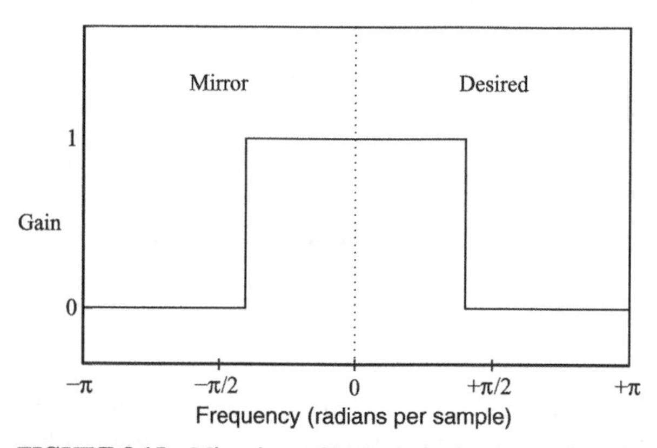

FIGURE 8.15 Mirroring a filter's desired response for a low-pass filter. The desired response from zero to the low-pass cutoff is mirrored in the negative frequency range.

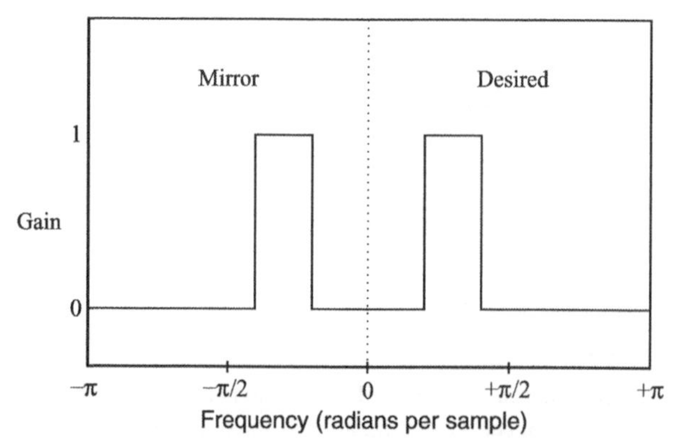

FIGURE 8.16 Mirroring a filter's desired response for a bandpass filter.

The reason for this may be understood by considering the desired time response coefficients $h_d(n)$. These will be used as filter coefficients and must be real numbers by definition; that is, we need terms of the form $b_k x(n-k)$, where $x(n-k)$ are the samples and b_k are the coefficients. Since for any α and its complex conjugate α^*, the product $(z-\alpha)(z-\alpha^*)$ will have only real coefficients, so any complex number products in $G(z)$ must occur in pairs in order to cancel out and give real coefficients.

The impulse response $h_d(n)$ thus computed will be infinite in extent. In a practical filter, the order must be limited. This is obtained by truncating the impulse response. Assuming N is odd, the restriction of the infinite-length impulse response to a finite length may be accomplished by calculating $h_d(n)$ over the range

$$-\frac{N-1}{2} \le n \le +\frac{N-1}{2}$$

The MATLAB script shown next implements the above design equation. First, the desired frequency response $H_d(\omega)$ in the band 0 to $f_s/2$ (corresponding to 0 to π radians) is mirrored over the range $-\pi$ to 0. The integration is then performed as a simple rectangular numerical integration.

We then take the real part of the resulting $h_d(n)$ values. This is because the results of the integration may have small imaginary values due to rounding. The computed impulse response $h_d(n)$ exists over positive and negative time, which cannot be realized in a true filter because it implies future sample references (or, equivalently, positive powers of z in the transfer function). The solution to this problem is simply to delay the impulse response by half the length of the filter. For odd N, this is $(N-1)/2$ samples. We then use these values as the coefficients of the FIR filter.

```
dw = pi/400;
w = -pi:dw:pi;
N = 21;         % order; must be odd
H = zeros(size(w));
wLow = pi/4;
wHigh = pi/2;
% 0 to pi
i = find((w >= wLow) & (w <= wHigh));
H(i) = ones(size(i));
% -pi to 0 is the mirror image
i = find((w >= -wHigh) & (w <= -wLow));
H(i) = ones(size(i));
nlow = - (N - 1)/2;
nhigh = (N - 1)/2;
K = 1/(2 * pi);
for n = nlow : nhigh
    h(n - nlow + 1) = K * sum(H. * exp(j * w * n) * dw);
end
h = real(h);  % compensate for rounding errors
```

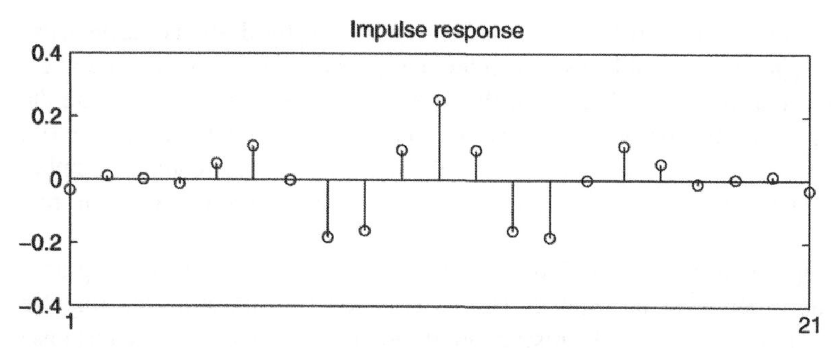

FIGURE 8.17 The impulse response of the initial FIR filter design. $N = 21$, $\omega_l = \pi/4$, $\omega_u = \pi/2$.

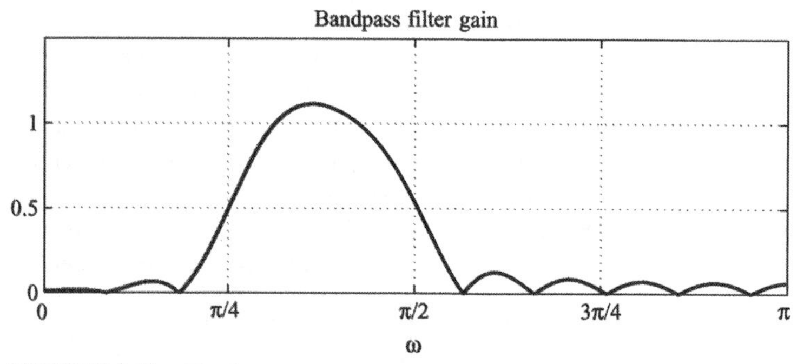

FIGURE 8.18 The frequency response of the initial FIR filter design, for $N = 21$, $\omega_l = \pi/4$, and $\omega_u = \pi/2$. Clearly, the response is adequate only as a first approximation to the desired passband and must be improved somewhat.

Using the above truncation for $N = 21$ and cutoff range $\omega_l = \pi/4$ to $\omega_u = \pi/2$, the impulse response shown in Figure 8.17 results. Clearly, the impulse response is symmetrical.

The frequency response, obtained as in Section 5.7 by setting $z = e^{j\omega}$, is shown in Figure 8.18. The frequency response exhibits a relatively poor approximation to that desired. There are two underlying issues: first, the sharpness of the transition bands and the attenuation of the stopbands; and second, the oscillatory nature of the passband and stopband frequency response characteristics. We will deal with each of these issues in turn.

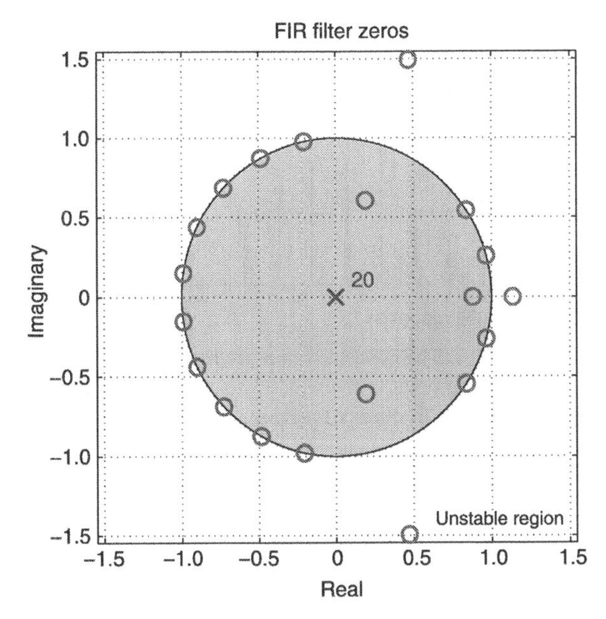

FIGURE 8.19 The zeros of the transfer function for an FIR filter with $N = 21$, $\omega_l = \pi/4$, and $\omega_u = \pi/2$.

Recall the earlier discussion on filter design as a pole–zero placement problem. Since this is an FIR filter, it has only zeros (and simple poles due to the causality requirement). The locations of the zeros are shown in Figure 8.19. Note that there are 20 zeros for a filter with 21 coefficients (terms) in the filter, and we need a further z^{-20} term to make the filter causal.

The problem with the response as calculated is that it is a poor approximation to the desired ideal response. One obvious step is to increase the order of the filter. The second nonobvious step is to "taper" the impulse response coefficients. Using the truncation method outlined above, we are effectively multiplying the filter coefficients by a rectangular window. Using a Hamming window of the form shown in Figure 8.20, we multiply each impulse response coefficient by the corresponding window coefficient. The Hamming window equation is[1]

$$\omega_k = 0.54 + 0.46\cos\left(\frac{2k\pi}{N}\right). \tag{8.9}$$

This may be implemented in MATLAB using the following code to augment that shown previously:

[1]If n is taken as $-(N-1)/2$ to $+(N-1)/2$, the Hamming window is as above. If n is 0 to $N-1$, the Hamming equation uses −0.46.

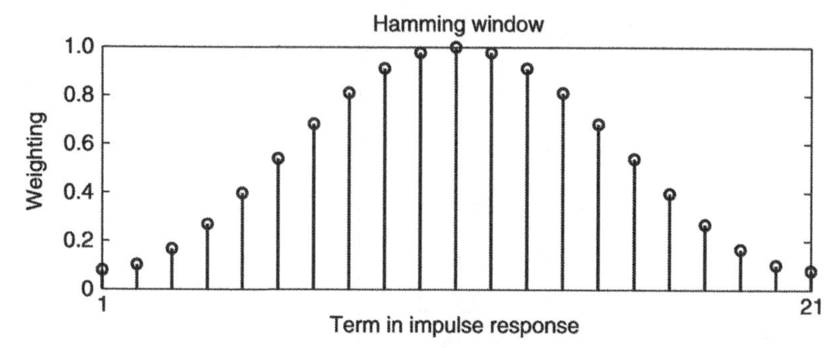

FIGURE 8.20 A 21-point Hamming window. The computed impulse response coefficients are multiplied by the corresponding point in the window to yield the filter coefficients. This has the effect of smoothing the frequency response.

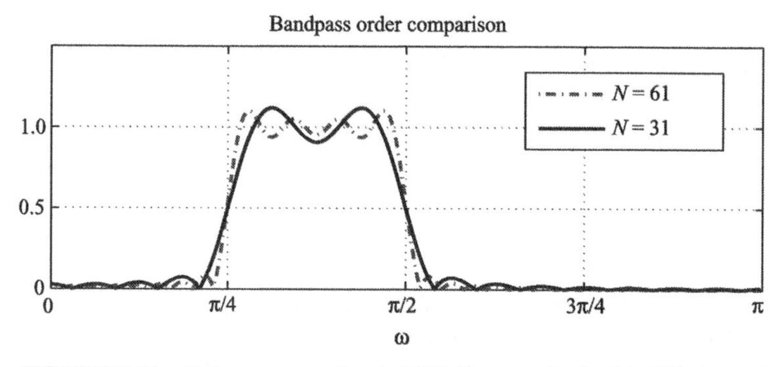

FIGURE 8.21 Gain responses for the FIR direct method with differing orders. $N = 31$ and $N = 61$, $\omega_l = \pi/4$, $\omega_u = \pi/2$.

```
k = nlow : nhigh;
w = 0.54 + 0.46 * cos(2 * pi * k/N);
hw = h. * w;
```

Figures 8.21 and 8.22 show the effect of increasing the order and of windowing the coefficients, respectively. The increased order does indeed sharpen the response edges, at the expense of additional ringing in the passband and the stopband. If a window is used, it will tend to reduce this ringing; however, it also serves to dull the sharpness of the response edges. Therefore, the two techniques are normally used in tandem. One of the window types as discussed in Section 7.11 may be used.

One characteristic of FIR filters is that they provide a nonflat response in the passband and stopband, and the window clearly helps to flatten the response.

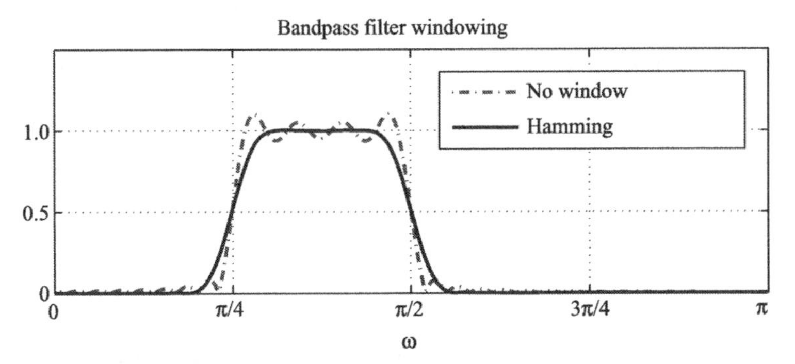

FIGURE 8.22 Gain responses for the FIR direct method with and without a Hamming window. $N = 201$, $\omega_l = \pi/4$, $\omega_u = \pi/2$.

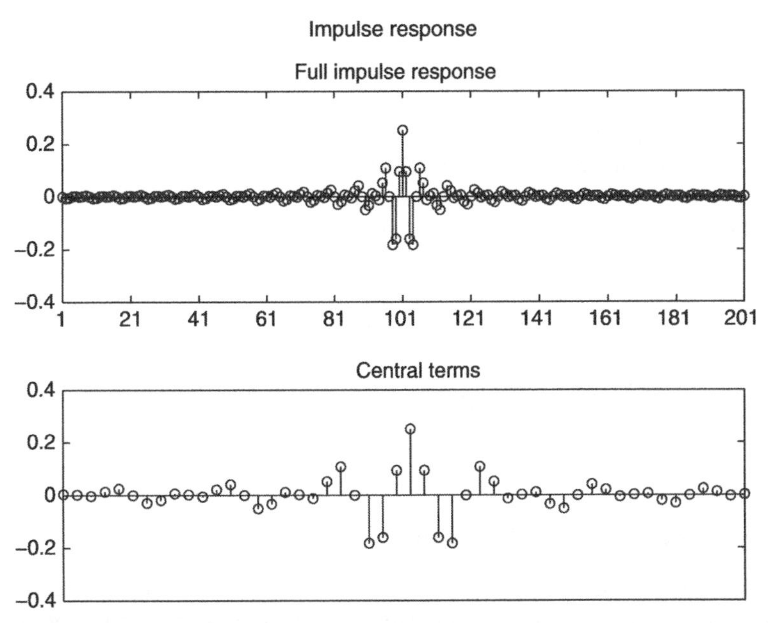

FIGURE 8.23 The FIR impulse response for $N = 201$, $\omega_l = \pi/4$, and $\omega_u = \pi/2$. The delay inherent in such a long filter is self-evident. The lower panel shows the center portion of the impulse response—clearly, the filter is symmetric about the middle of the filter.

However, the trade-off for obtaining a smoother passband response is that the transition bands between the stopband and the passband become wider.

Increasing the order certainly gives a sharper response; however, this comes at the expense of increased filter order. Figure 8.23 shows the impulse response (and, with a window as desired, the filter coefficients) for a filter of order 201. Not only are greater memory and computational resources required but the delay will also increase linearly with the filter order. It should be noted, however, that there are various "fast" algorithms available to speed up the FIR filter computation. These usually involve the fast Fourier transform (FFT) algorithm, and are very important in real-time implementations. These are explained in Section 8.9.

To summarize, for the FIR design method discussed, the following steps were undertaken:

FIR Filter Design Steps

1. From the characteristics of the signal to be filtered, determine a sampling frequency, f_s.
2. Determine the shape of filter characteristic required (low-pass, high pass, or custom).
3. Choose an initial filter order, N.
4. Select a window function such as the Hamming window (Equation 8.9) and determine coefficients w_k.
5. Sketch the filter response shape $H_d(\omega)$ from 0 to π and the mirrored response from $-\pi$ to 0.
6. Apply the FIR design equation (Equation 8.8) to determine $h_d(n)$.
7. Delay the $h_d(n)$ by half the filter order for causality.
8. Multiply the w_k by the delayed $h_d(n)$ to obtain filter coefficients b_k.
9. Plot the frequency response using $z = e^{j\omega}$ for $\omega = 0 \ldots \pi$.
10. Iterate the above steps, increasing N as necessary.
11. Let $Y(z)/X(z) = G(z)$. Solve for $Y(z)$.
12. Convert $Y(z)$ into a difference equation; find $y(n)$ in terms of $x(n)$, $x(n-1)$, $y(n-1)$, and so on.
13. Code a loop to sample the input $x(n)$, calculate $y(n)$, output $y(n)$, and buffer the samples.

8.6.2 Frequency Sampling Method

This method is quite similar in approach to the direct window method. Instead of the desired frequency response being completely specified, it is specified only as samples of the frequency response at regular intervals. Thus, the inverse discrete Fourier transform (DFT) yields the impulse response and, as before, these values are used as the coefficients of an FIR filter. For $n = -(N-1)/2$ to $n = +(N-1)/2$ and N assumed odd, the design equation is

$$h(n) = \frac{1}{N} \sum_{k=-\frac{N-1}{2}}^{+\frac{N-1}{2}} H(k) e^{j\frac{2n\pi k}{N}}. \tag{8.10}$$

Using this method, the desired frequency response is also sampled (the $H(k)$). The disadvantage is that we need to resample the frequency response, so care would have to be taken that no information is lost in choosing the number of samples to use.

8.7 IDEAL RECONSTRUCTION FILTER

It was noted in Chapter 3 that an ideal reconstruction filter—one which makes the reconstructed analog signal equal to the original—has an impulse response found by taking the inverse Fourier transform of the signal:

$$x(t) = \frac{1}{2\pi}\int_{-\infty}^{\infty} X(\Omega)e^{j\Omega t}\, d\Omega. \tag{8.11}$$

The scaling factor $1/2\pi$ comes about because of the use of radian frequency (remember that $\Omega = 2\pi f$ and $f = 1/\tau$). Since we want a band-limited reconstruction without aliasing, this becomes the integration as shown in Figure 8.24.

$$h(t) = \frac{1}{\Omega_s}\int_{-\frac{\Omega_s}{2}}^{+\frac{\Omega_s}{2}} X(\Omega)e^{j\Omega t}\, d\Omega. \tag{8.12}$$

This simplifies to

$$\begin{aligned} h(t) &= \frac{\sin\frac{\pi t}{T}}{\frac{\pi t}{T}} \\ &= \operatorname{sinc}\frac{\pi t}{T}. \end{aligned} \tag{8.13}$$

Thus, the impulse response at each sample instant $t = nT$ must be

$$h(t - nT) = \operatorname{sinc}\frac{\pi(t - nT)}{T}. \tag{8.14}$$

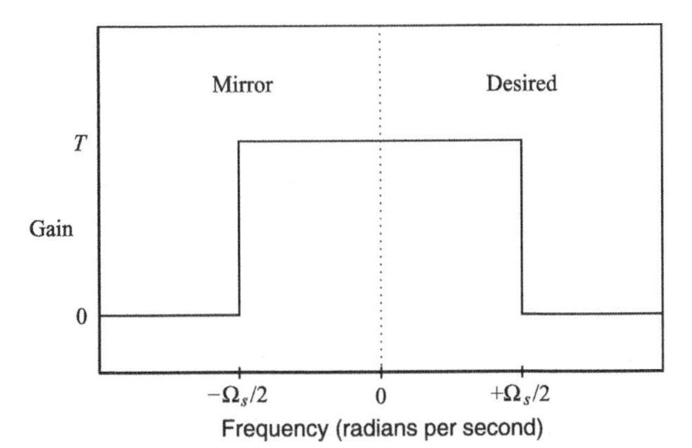

FIGURE 8.24 The frequency response of an ideal low-pass reconstruction filter. The gain is unity over the range up to half the sampling frequency. For the purpose of the impulse response derivation, we must mirror the response in the negative frequency range.

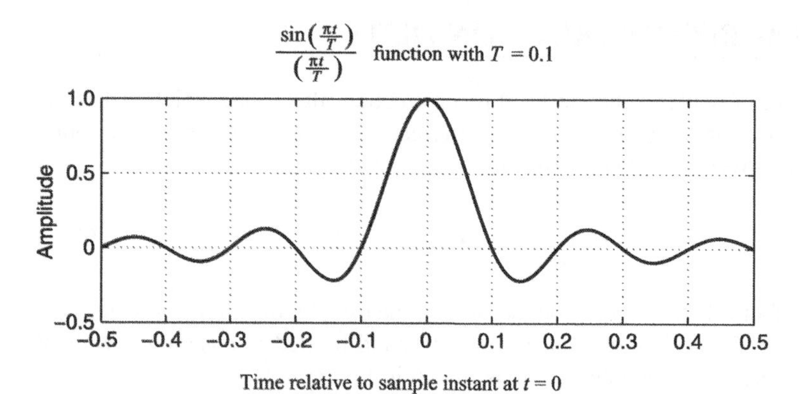

FIGURE 8.25 The "sinc" function sin $(\sin(\pi t/T)/(\pi t/T))$ for the ideal reconstruction of a sample at $t = 0$, with a sampling period of $T = 0.1$. Note how the function is exactly zero at the sampling instants and that the magnitude of the peak is one.

This "sinc" function for a sampling period of $T = 0.1$ is shown in Figure 8.25. Note that in the reconstruction function $h(t)$, the zero crossings are at nT. This may easily be seen from the above equation for $h(t - nT)$—the sine function is zero when $\pi t/T$ is 0, π, 2π, ... , which is whenever $t = nT$ for any integer n. What this means is that the reconstruction is done by a sample-weighted sinc function where, at the sampling instants, the sinc impulses from the other samples do not interfere with the current sample. In between the sampling instants, the sinc function provides the optimal interpolation.

8.8 FILTERS WITH LINEAR PHASE

In general, a signal is composed of several frequency components, and a filter has both *gain* and *phase* responses. Often, the gain response is what we are interested in, but sometimes the phase response is also crucial—either because we want a certain phase response for an application or because the incidental phase response of a filter which has the desired magnitude response may produce undesired side effects.

The types of FIR filters discussed have a linear phase—the phase angle increases linearly with frequency. IIR filters do not in general have this characteristic. This has significance in some applications such as filters for digital communications, as demonstrated below.

Figures 8.26 and 8.27 show the gain and phase responses for IIR and FIR filters, respectively. The design of IIR filters will be considered in detail in Chapter 9. For now, it is sufficient to understand that these are (arbitrarily) designed for a

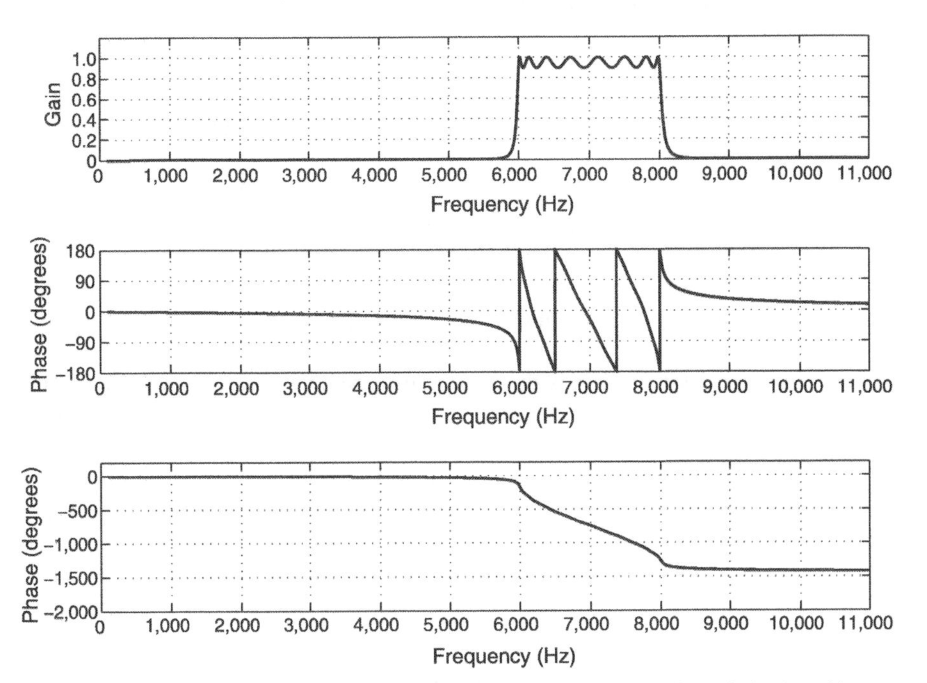

FIGURE 8.26 Gain and phase responses for an eighth-order recursive Chebyshev filter with a bandpass response. The gain is shown on a linear scale. Below that is the phase as calculated and the "unwrapped" phase. The design of recursive Chebyshev filters will be discussed in Chapter 9.

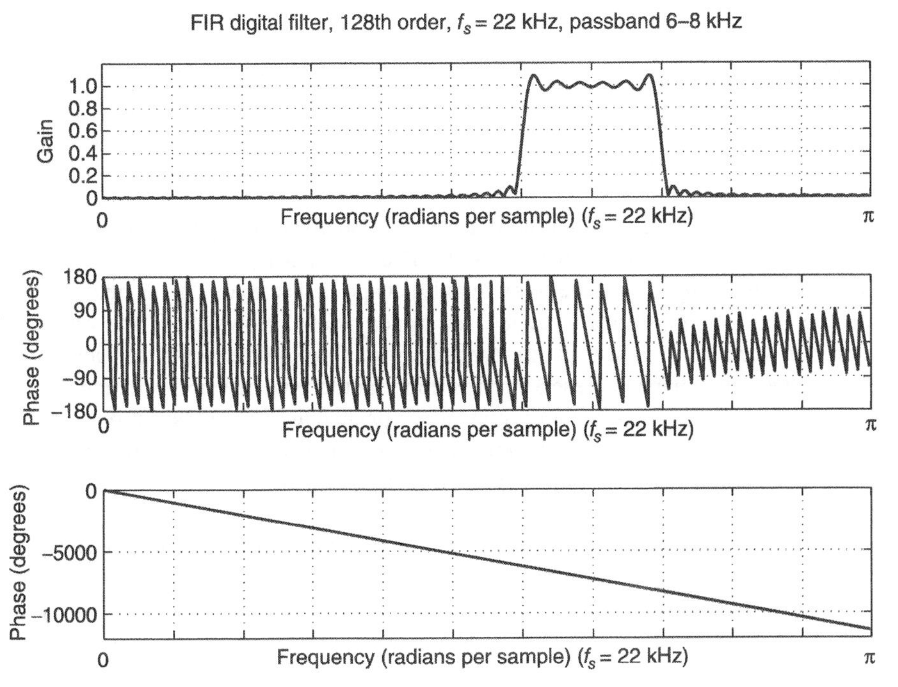

FIGURE 8.27 Gain and phase responses for a 128th order FIR nonrecursive filter with a bandpass response. The gain is shown on a linear scale. Below that is the phase as calculated and the unwrapped phase.

bandpass response for this example. The analog design clearly has a nonlinear phase, and this is translated into the discrete-time filter through transformation.

Note that these figures show the direct phase as calculated from the angle $\angle H$ in each case. The phase will normally "wrap" when crossing a 180° boundary. For example, a phase for successive frequencies of −178°, −179°, −180°, and −181° would actually be calculated as −178°, −179°, −180°, and +179°. The phase is a smooth function, and the discontinuity is an artifact of the fact that we are considering a circular function. The figures show the unwrapped phase for easier comparison, but this is not normally done when showing filter responses since the phase wrapping is understood.

The *phase delay* of filter is the amount of time delay of *each* frequency component. Filters with a nonlinear phase cause phase distortion because frequency components in the signal will each be delayed by an amount which is *not* proportional to frequency, thereby altering their time relationship. Mathematically, suppose a sinusoid of frequency Ω and phase φ_1 goes into a system at time t_1. Sometime later, at time t_2, it emerges with a differing phase, φ_2. For simplicity, assume the input and output magnitudes are the same. Furthermore, the frequency is assumed to be unchanged as it passes through the linear system. So the input is described by $\sin(\Omega t_1 + \varphi_1)$ and the output is $\sin(\Omega t_2 + \varphi_2)$. For these to be equal,

$$\begin{aligned}\Omega t_1 + \varphi_1 &= \Omega t_2 + \varphi_2 \\ \Omega(t_2 - t_1) &= \varphi_1 - \varphi_2 \\ \Omega \Delta t &= -\Delta\varphi \\ \Delta t &= \frac{-\Delta\varphi}{\Omega}.\end{aligned} \tag{8.15}$$

What this indicates is that the relative time change in the waveform from input to output is proportional to the phase difference. Moreover, for the same relative time change, if the phase difference $\Delta\varphi$ is (say) doubled, the frequency has to be doubled. The conclusion is that the phase must be proportional to the frequency.

To illustrate this, consider the Fourier series analysis in Figure 8.28, where we examine the effect of phase more closely. If a constant phase is added to all the Fourier series components, the relative pulse shape is distorted. This is because the relative time relationship is changed, as argued above. However, if the phase is altered in such a way that it is directly proportional to the frequency of each component, the pulse shape is preserved. It is important to note that in this example, the *magnitudes* of each sinusoidal component are unchanged, whatever the phase. The phase simply determines the position in time relative to the frequency. The term *group delay* is used to define the average time delay of the *composite* signal at each frequency.

The important point to note is that IIR filters invariably have a nonlinear phase response. In fact, analog filters can only approximate a linear phase response. On the other hand, the FIR filters discussed here always have a linear phase response, and thus composite waveforms do not suffer from this type of time domain distortion.

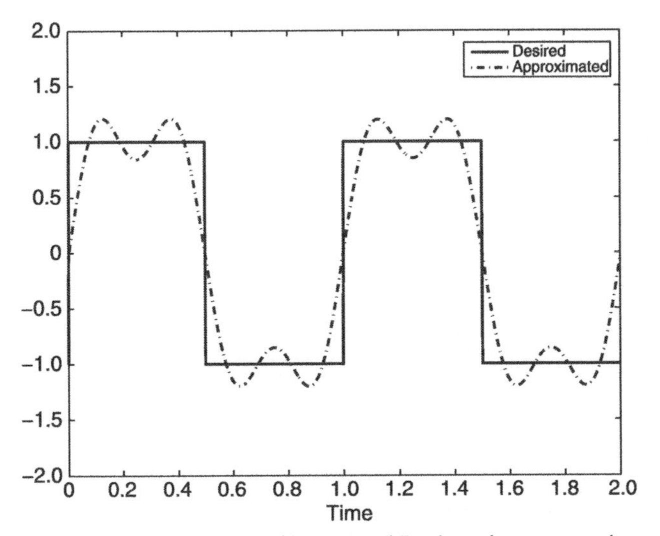

(a) The pulse waveform and its truncated Fourier series representation

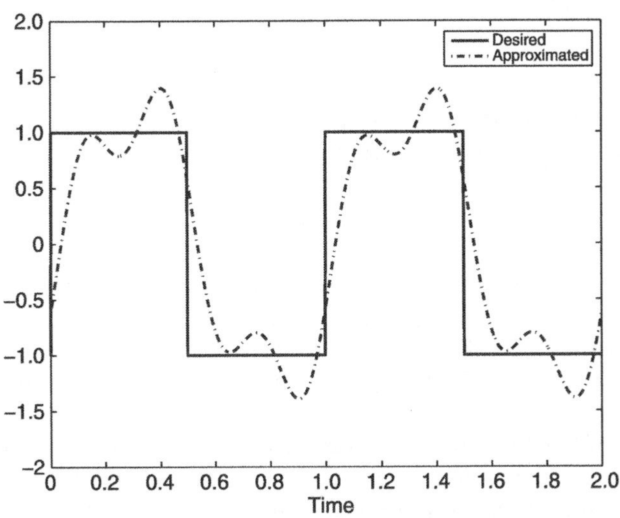

(b) The truncated Fourier series representation with a constant phase added to each component

FIGURE 8.28 The effect of phase on a pulse waveform. All three have the same Fourier series component magnitudes; hence, each magnitude frequency spectrum is identical. However, the phase relationship between components is altered—in sequence, they are correct phase, constant phase, and phase proportional to frequency. Clearly, the pulse shape is altered when the phase is *not* proportional to frequency.

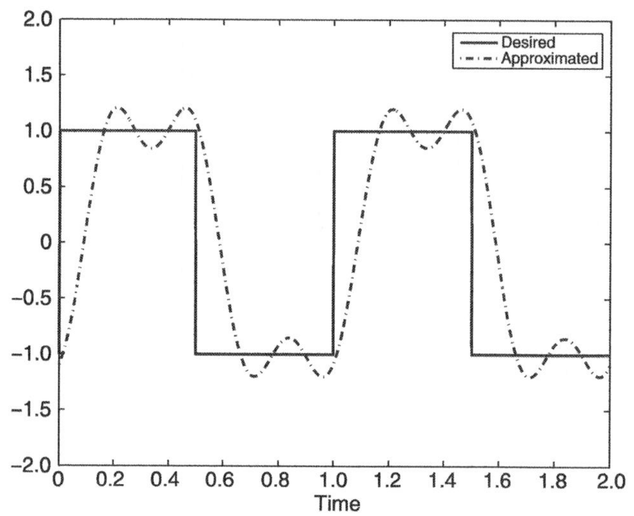

(c) The series with a phase proportional to frequency added to each component

FIGURE 8.28 (*continued*)

8.9 FAST ALGORITHMS FOR FILTERING, CONVOLUTION, AND CORRELATION

So-called fast algorithms are common in signal processing, particularly when real-time constraints are present and/or very large data set computations are required. Simply throwing faster hardware at the problem is not always necessary (nor wise). Fast algorithms exist, which can significantly speed up the computations required for a filter. The class of algorithms based on the FFT can be traced back over some time (see Stockham [1966] and Helms [1967]), but their relevance and importance is as great as ever.

The design of FIR filters has been examined in previous sections, and despite all their advantages, a significant problem is that the filter order tends to be fairly large to achieve reasonable performance (i.e., the difference equation requires a large number of terms). FIR filter operation consists of multiplication and addition operations, one per each term calculated, and if there are a large number of terms, then the computational cost scales up accordingly.

In this section, we examine some ways in which we can speed up the process of FIR filtering (and by extension, convolution, and correlation). This permits some filtering operations to be run in real time when it would otherwise not be possible. Alternatively, it permits less expensive hardware to be utilized for the same task. Even non-real-time operations may benefit—the end-user performance becomes more acceptable, and large data set operations can be completed in seconds rather than in minutes (or minutes rather than hours).

In the process of developing the fast algorithms, it will become apparent that they are also useful for correlation operations. Recall that correlation (Section 6.3.1)

has a great many practical applications such as pattern recognition and system identification, so developing fast algorithms will benefit these too.

8.9.1 Convolution Revisited

Convolution was discussed in Section 5.9, and Equation 8.16 shows the basic form of the convolution operation mathematically:

$$y(n) = \sum_{k=0}^{P-1} h_k x(n-k). \tag{8.16}$$

So let us visualize this. Figure 8.29 shows the operation of convolution for short sequences denoted by vectors x and **h**. The operation of Equation 8.16 may be

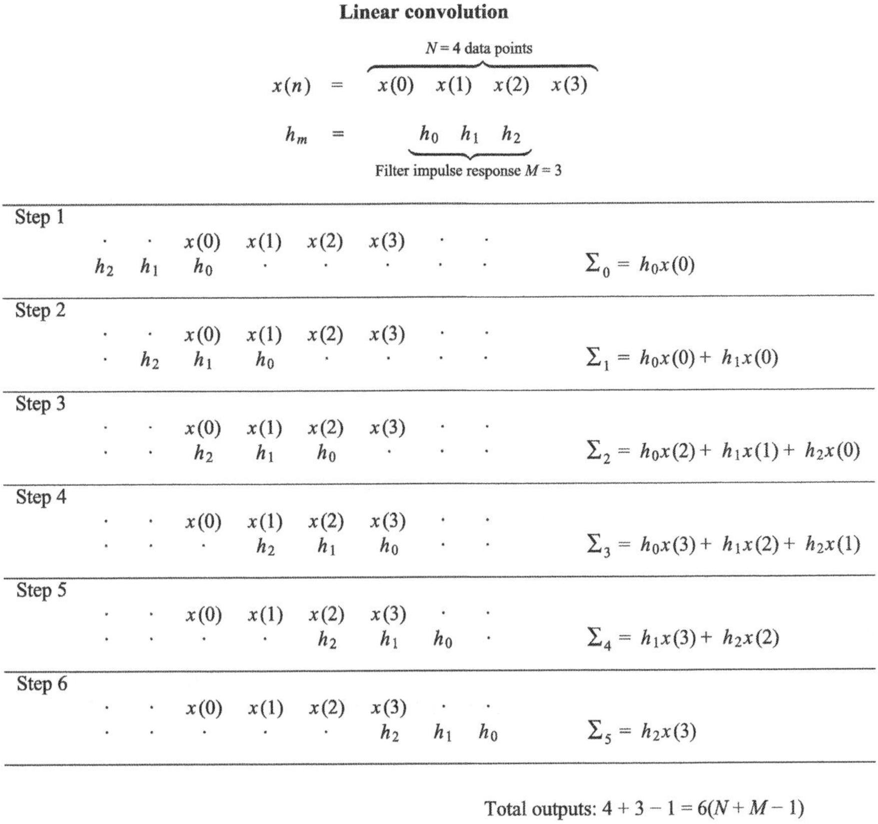

Linear convolution

$N = 4$ data points

$x(n) = x(0) \quad x(1) \quad x(2) \quad x(3)$

$h_m = h_0 \quad h_1 \quad h_2$

Filter impulse response $M = 3$

Step 1	·	·	$x(0)$	$x(1)$	$x(2)$	$x(3)$	·	·	
	h_2	h_1	h_0	·	·	·	·	·	$\Sigma_0 = h_0x(0)$
Step 2	·	·	$x(0)$	$x(1)$	$x(2)$	$x(3)$	·	·	
	·	h_2	h_1	h_0	·	·	·	·	$\Sigma_1 = h_0x(0) + h_1x(0)$
Step 3	·	·	$x(0)$	$x(1)$	$x(2)$	$x(3)$	·	·	
	·	·	h_2	h_1	h_0	·	·	·	$\Sigma_2 = h_0x(2) + h_1x(1) + h_2x(0)$
Step 4	·	·	$x(0)$	$x(1)$	$x(2)$	$x(3)$	·	·	
	·	·	·	h_2	h_1	h_0	·	·	$\Sigma_3 = h_0x(3) + h_1x(2) + h_2x(1)$
Step 5	·	·	$x(0)$	$x(1)$	$x(2)$	$x(3)$	·	·	
	·	·	·	·	h_2	h_1	h_0	·	$\Sigma_4 = h_1x(3) + h_2x(2)$
Step 6	·	·	$x(0)$	$x(1)$	$x(2)$	$x(3)$	·	·	
	·	·	·	·	·	h_2	h_1	h_0	$\Sigma_5 = h_2x(3)$

Total outputs: $4 + 3 - 1 = 6(N + M - 1)$

FIGURE 8.29 Illustrating the steps in involved in convolution. In this case, we have specifically labeled it "linear" convolution to distinguish from circular convolution (as explained in the text). One of the sequences—in this case, the vector **h**—is time reversed and can be imagined as sliding past the fixed vector x. As the sequence is moved, the sum of the product of the overlapping terms is calculated. The total number of output terms is seen to be the sum of the lengths of each input sequence, minus one.

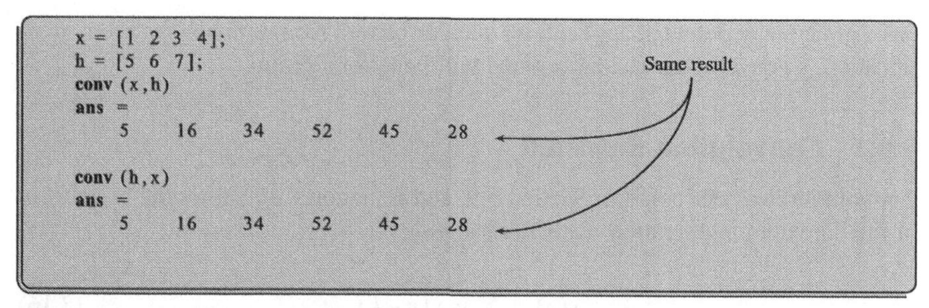

FIGURE 8.30 Illustrating linear convolution using MATLAB. The **h** sequence is time reversed and slides past the fixed sequence x, with a vector dot product calculated at each step. If we reverse the order of the inputs (i.e., hold the second sequence fixed and reverse/slide the first), the same result ought to be produced. This is shown in the second convolution operation above—the outputs are clearly identical.

visualized by time reversing the second sequence, sliding it all the way to the left, and then calculating the sum of product of terms which overlap. The sequence is then stepped one to the right, and the process is repeated. The time reversal may be understood from the equation. Since the output term is at index n, the index into the input vector x is $(n - k)$. Thus, as k in incremented in evaluating the summation, the index into x goes further back in time $x(n)$, $x(n-1)$, $x(n-2)$, ... $x(n-P+1)$. For example, in Step 2 of Figure 8.29, we multiply h_0 and h_1 by $x(1)$ and $x(0)$, respectively, and in Step 3, we have $[x(2)\ x(1)\ x(0)]$ multiplied by $[h_0\ h_1\ h_2]$. Imagining this as a time reversal helps in visualizing the process.

Figure 8.30 shows a concrete example of the process of convolution using simple numerical sequences. The result can be easily verified via manual calculation. Note that the inputs to the convolution process (the x and y vectors) may be interchanged without changing the end result. Finally, note that the number of output terms is the sum of the lengths of the input sequences, less one. This can be seen to be true since nonzero output values require an overlap of the input vectors of at least one (see Step 6 in Fig. 8.29).

8.9.2 FIR Filtering as Convolution

Now we turn to the filtering operation since that is the primary motivation for our investigation. The basic operation of an FIR filter is given in Equation 8.17:

$$\begin{aligned} y(n) &= b_0x(n)+b_1x(n-1)+\cdots+b_Nx(n-N) \\ &= \sum_{k=0}^{N} b_k x(n-k). \end{aligned} \tag{8.17}$$

For each output point $y(n)$, we multiply the delayed inputs $x(n-k)$ by the weighting coefficients, b_k, as determined by the design procedure. Note that the highest order of the filter delay is $(n - N)$, and thus we may say that the filter order is N (i.e., a z^{-N} term will result in the transfer function). However, there are $(N + 1)$ coefficients

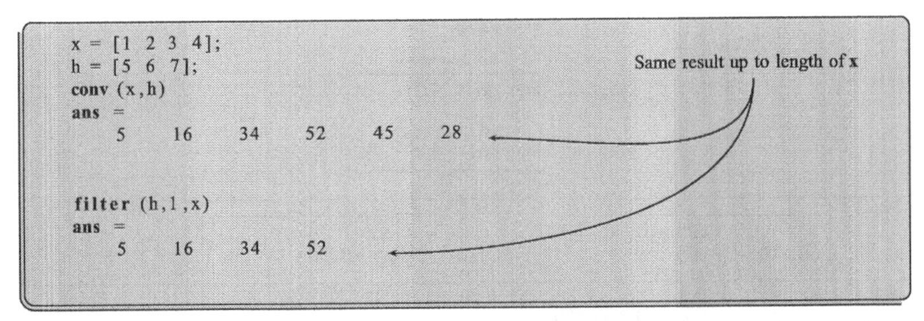

FIGURE 8.31 Convolution is equivalent to nonrecursive FIR filtering, as this example illustrates. Here we use the **conv()** operator, which produces nearly identical results to the **filter()** operation. The initial outputs are the same, but the filtering operation yields a shorter result—this is because the "tail" where there is no full overlap for the last terms is not calculated. In effect, this is the overlap produced by convolution. The number of terms not calculated is the length of the second sequence minus one.

since the coefficients start at zero. This is a minor but important point to keep in mind for when we subsequently examine the convolution operator.

Now if we compare Equations 8.17 and 8.16, we see that they are almost identical. We have an output $y(n)$ at time instant n, which is formed as the weighted sum of current and past inputs as denoted by the $x(n-k)$ term. The weighting coefficients in the filter are denoted as b_k's, and in the convolution as h_k's. However, the same multiply–add operations and indexing order appears in both. Thus, we conclude that FIR filtering is in fact identical to convolution of the data sequence **x** with the impulse response **h**, and furthermore that the coefficients actually form the impulse response itself. This is illustrated in Fig. 8.31.

8.9.3 Relationship between Convolution and Correlation

Correlation was discussed in Chapter 6. Recall that the operation of correlation may be described mathematically as

$$R_{xy}(k) = \frac{1}{N}\sum_{n=0}^{N-1} x(n)y(n-k) \quad k\text{:offset}, \tag{8.18}$$

where, again, k is the positive or negative offset (or lag).

Figure 8.32 illustrates these steps graphically, for short data sequences x and y. In the case of convolution, we imagined these as a data sequence and an impulse response, whereas now we consider them to be two data sequences (which may in fact be the same one, in the case of autocorrelation). Note that in the diagram, the second sequence is *not* time reversed. This may be understood by reference to the correlation equation, and in particular the $y(n-k)$ term. The offset or lag k is fixed for each output, and the summation is over index n. Thus, compared to Equation 8.16, we can see that the fundamental difference is that in the case of correlation, the second sequence is *not* reversed, whereas with convolution, it *is* reversed. Apart

Correlation

$N=4$ x data points

$x(n) = x(0)\ x(1)\ x(2)\ x(3)$

$N=4$ y data points

$y(n) = y(0)\ y(1)\ y(2)\ y(3)$

Step 1	· · · $x(0)\ x(1)\ x(2)\ x(3)$ · · · $y(0)\ y(1)\ y(2)\ y(3)$ · · · · · ·	$\Sigma_0 = x(0)y(3)$
Step 2	· · · $x(0)\ x(1)\ x(2)\ x(3)$ · · · · $y(0)\ y(1)\ y(2)\ y(3)$ · · · · ·	$\Sigma_1 = x(0)y(2) + x(1)y(3)$
Step 3	· · · $x(0)\ x(1)\ x(2)\ x(3)$ · · · · · $y(0)\ y(1)\ y(2)\ y(3)$ · · · ·	$\Sigma_2 = x(0)y(1) + x(1)y(2) + x(2)y(3)$
Step 4	· · · $x(0)\ x(1)\ x(2)\ x(3)$ · · · · · · $y(0)\ y(1)\ y(2)\ y(3)$ · · ·	$\Sigma_3 = x(0)y(0) + x(1)y(1) + x(2)y(2) + x(3)y(3)$
Step 5	· · · $x(0)\ x(1)\ x(2)\ x(3)$ · · · · · · · $y(0)\ y(1)\ y(2)\ y(3)$ · ·	$\Sigma_4 = x(1)y(0) + x(2)y(1) + x(3)y(2)$
Step 6	· · · $x(0)\ x(1)\ x(2)\ x(3)$ · · · · · · · · $y(0)\ y(1)\ y(2)\ y(3)$ ·	$\Sigma_5 = x(2)y(0) + x(3)y(1)$
Step 7	· · · $x(0)\ x(1)\ x(2)\ x(3)$ · · · · · · · · · $y(0)\ y(1)\ y(2)\ y(3)$	$\Sigma_6 = x(3)y(0)$

Total outputs: $2 \times 4 - 1 = 7$ $(2N - 1)$

FIGURE 8.32 The steps required to produce the correlation of sequences x and y. The second sequence slides past the first, with a vector dot product calculated at each step. In effect, this is the same as the convolution operation, except that the second sequence is *not* reversed.

from that, the multiply–add–shift steps are the same, as will be seen by comparing Figures 8.29 and 8.32.

The similarities and differences between convolution and correlation are illustrated in Figure 8.33. Correlation can be effected by first time reversing one of the sequences, followed by convolution (which itself reverses one of the sequences). Manual calculation of the results for Figure 8.33 will verify the correct results in each case (convolution and correlation).

8.9.4 Linear and Circular Convolution

Note that convolution produces a result whose length depends on the length of both the input vectors. Recall also that the DFT and its faster version, the FFT, produce a result which has the same length as the input block. In order to understand the fast convolution algorithms, it is necessary to investigate a variant of the convolution operation discussed so far.

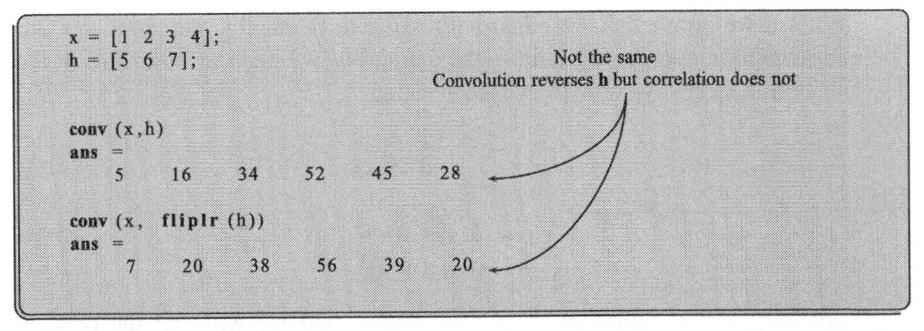

FIGURE 8.33 The differences between convolution and correlation are illustrated in this numerical example. Convolution flips one of the sequences, whereas correlation does not. In this example, we use the convolution operator to calculate correlation by flipping the second sequence before it is passed to the calculation, since the convolution operator will reverse it back.

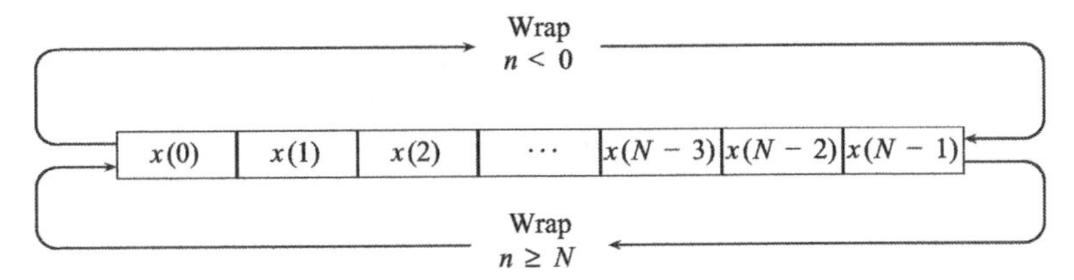

FIGURE 8.34 The operation of circular addressing for a buffer of N samples is shown. When incrementing the index pointer, the index travels to the right until it reaches the end of the array, at which time it reverts back to zero. Similarly, decrementing the index pointer to zero and below wraps the pointer back to sample $N - 1$. This is termed modulo addressing, where the modulus is N.

The convolution equation and shorthand notation is

$$y(n) = \sum_{k=0}^{P-1} h_k x(n-k) \tag{8.19}$$

$$\mathbf{y} = \mathbf{h} * \mathbf{x}. \tag{8.20}$$

We have used the symbol $*$ to denote convolution.[2] The indexing in this equation traverses linearly from 0 to $(P - 1)$. What if we kept the indexing to the length of the input blocks so as to produce an output block of the same length as the input block? Specifically, what if we used modulo addressing and "wrapped" the indexes as indicated in Figure 8.34? In this figure, incrementing the address beyond the limit of the array wraps the address index back to zero. We will use the notation $\tilde{x}(k)$ to indicate that the array is circularly addressed by index k.

[2]Not to be confused with complex conjugation, where the * is used as a superscript (r^*), or the optimal value of a parameter h, which is also denoted h^*.

What if we now employ circular addressing in calculating convolution? Our operating equation appears essentially the same, but we need to take care in the implementation to always use circular addressing:

$$y(n) = \sum_{k=0}^{P-1} h_k \tilde{x}(n-k) \tag{8.21}$$

$$y(n) = h(n) * \tilde{x}(n) \tag{8.22}$$

$$y(n) = h(n) \circledast x(n). \tag{8.23}$$

Unsurprisingly, this is called *circular convolution*, and we will call the original convolution *linear convolution* to distinguish it where necessary. The $*$ denotes regular or linear convolution; the $\tilde{x}(n)$ denotes circular addressing; and $\circledast$ denotes circular convolution.

Figure 8.35 shows the calculation that occurs with circular convolution, in a similar way to how we illustrated linear convolution previously. The important point to note is that the number of output points is equal to the length of the longest input sequence (in this case, x). The indexing of the x samples is performed modulo the length of the array. So, when linear convolution would call for indexing of sample $x(4)$ (shown in Step 1 of the figure), the index wraps back to $4 - N = 4 - 4 = 0$.

Note, however, that the results of linear convolution and circular convolution are not identical. This may be seen by comparing Figure 8.29 and Figure 8.35. Upon

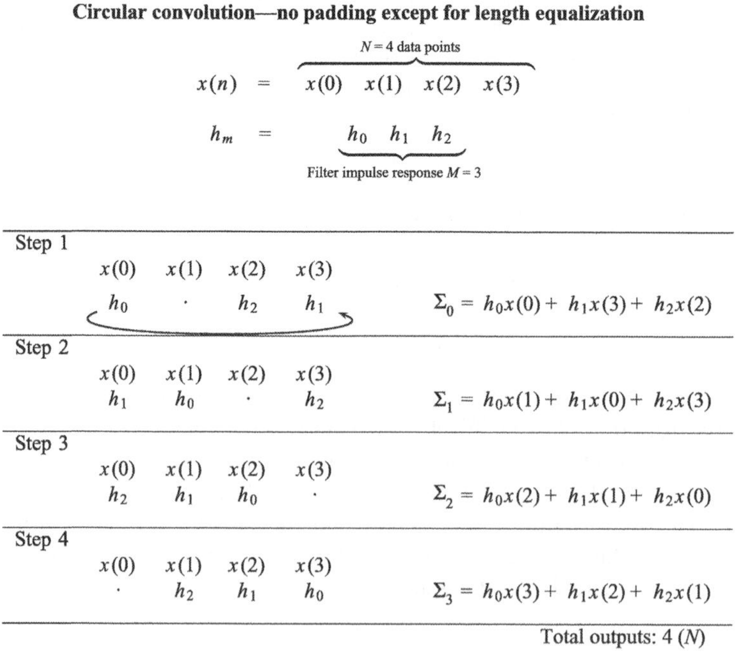

Circular convolution—no padding except for length equalization

$N = 4$ data points

$x(n) = x(0)\ \ x(1)\ \ x(2)\ \ x(3)$

$h_m = h_0\ \ h_1\ \ h_2$

Filter impulse response $M = 3$

Step 1	$x(0)$	$x(1)$	$x(2)$	$x(3)$	
	h_0	·	h_2	h_1	$\Sigma_0 = h_0x(0) + h_1x(3) + h_2x(2)$
Step 2	$x(0)$	$x(1)$	$x(2)$	$x(3)$	
	h_1	h_0	·	h_2	$\Sigma_1 = h_0x(1) + h_1x(0) + h_2x(3)$
Step 3	$x(0)$	$x(1)$	$x(2)$	$x(3)$	
	h_2	h_1	h_0	·	$\Sigma_2 = h_0x(2) + h_1x(1) + h_2x(0)$
Step 4	$x(0)$	$x(1)$	$x(2)$	$x(3)$	
	·	h_2	h_1	h_0	$\Sigma_3 = h_0x(3) + h_1x(2) + h_2x(1)$

Total outputs: 4 (N)

FIGURE 8.35 Circular convolution when no zeros are padded except to the shorter sequence (if need be), so as to make the two sequences equal in length.

Circular convolution—pad with one zero

$N = 4$ data points

$$x(n) = x(0) \quad x(1) \quad x(2) \quad x(3)$$

$$h_m = h_0 \quad h_1 \quad h_2$$

Filter impulse response $M = 3$

Step 1	$x(0)$	$x(1)$	$x(2)$	$x(3)$	0	
	h_0	·	·	h_2	h_1	$\Sigma_0 = h_0 x(0)$
Step 2	$x(0)$	$x(1)$	$x(2)$	$x(3)$	0	
	h_1	h_0	·	·	h_2	$\Sigma_1 = h_0 x(1) + h_1 x(0)$
Step 3	$x(0)$	$x(1)$	$x(2)$	$x(3)$	0	
	h_2	h_1	h_0	·	·	$\Sigma_2 = h_0 x(2) + h_1 x(1) + h_2 x(0)$
Step 4	$x(0)$	$x(1)$	$x(2)$	$x(3)$	0	
	·	h_2	h_1	h_0	·	$\Sigma_3 = h_0 x(3) + h_1 x(2) + h_2 x(1)$
Step 5	$x(0)$	$x(1)$	$x(2)$	$x(3)$	0	
	·	·	h_2	h_1	h_0	$\Sigma_4 = h_1 x(3) + h_2 x(2)$
						Total outputs: 5 $(N + 1)$

FIGURE 8.36 Circular convolution showing padding with one zero.

closer inspection, we see that some of the results at some of the steps are in fact the same—steps 3 and 4 produce the same result because the same x and h terms overlap. What is occurring is that the modulo addressing is wrapping the second sequence around (aliasing) so that incorrect output terms are calculated. How could we prevent this and still use circular convolution? The answer is to prevent the wrapping causing any "damage," and we can do this by the simple expedient of padding some zeros at the end of the array. Figure 8.36 shows the calculation steps for circular convolution, with one zero appended. There are still some terms calculated in the output which are not present in the linear convolution. Inspection shows that this occurs because the h sequence overlaps more than one of the x terms. So if we pad with two zeros, as shown in Figure 8.37, the maximum overlap is only one (with the rest of the h terms being multiplied by zero), and the correct linear convolution results.

The conclusion we reach is that we must pad the longer sequence with $(M - 1)$ zeros, where M is the length of the shorter sequence. It goes without saying that the shorter sequence is also padded with zeros, but this is to equalize its length with the (padded) longer sequence. These zeros are shown as dots $(\cdot)$ in the figures.

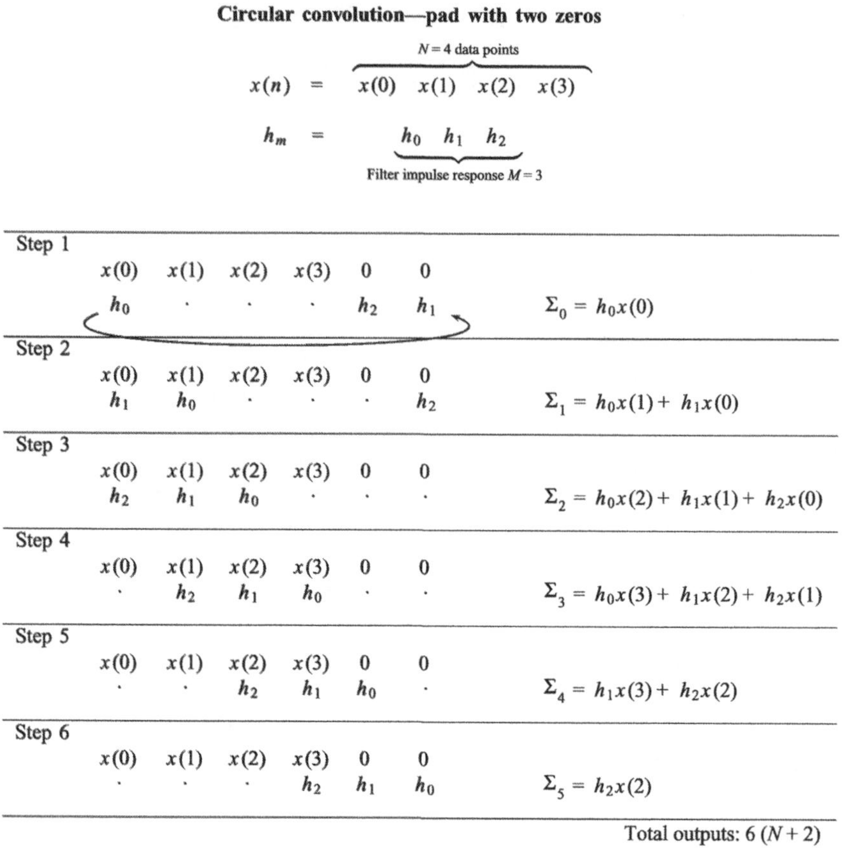

Circular convolution—pad with two zeros

$N = 4$ data points

$x(n) = x(0) \quad x(1) \quad x(2) \quad x(3)$

$h_m = h_0 \quad h_1 \quad h_2$

Filter impulse response $M = 3$

Step 1	$x(0)$	$x(1)$	$x(2)$	$x(3)$	0	0	
	h_0	·	·	·	h_2	h_1	$\Sigma_0 = h_0x(0)$
Step 2	$x(0)$	$x(1)$	$x(2)$	$x(3)$	0	0	
	h_1	h_0	·	·	·	h_2	$\Sigma_1 = h_0x(1) + h_1x(0)$
Step 3	$x(0)$	$x(1)$	$x(2)$	$x(3)$	0	0	
	h_2	h_1	h_0	·	·	·	$\Sigma_2 = h_0x(2) + h_1x(1) + h_2x(0)$
Step 4	$x(0)$	$x(1)$	$x(2)$	$x(3)$	0	0	
	·	h_2	h_1	h_0	·	·	$\Sigma_3 = h_0x(3) + h_1x(2) + h_2x(1)$
Step 5	$x(0)$	$x(1)$	$x(2)$	$x(3)$	0	0	
	·	·	h_2	h_1	h_0	·	$\Sigma_4 = h_1x(3) + h_2x(2)$
Step 6	$x(0)$	$x(1)$	$x(2)$	$x(3)$	0	0	
	·	·	·	h_2	h_1	h_0	$\Sigma_5 = h_2x(2)$

Total outputs: 6 $(N + 2)$

FIGURE 8.37 Circular convolution showing padding with two zeros. In this case, the number of zeros is one less than the length of the shorter sequence. This is the important case where the end of the **h** sequence does not wrap around to the start of the x sequence in the last step(s). As a result, the circular convolution result is the same as the linear convolution.

8.9.5 Using the FFT to Perform Convolution

So now we have an understanding of linear convolution, circular convolution, and how the two are related. The final piece in the puzzle is to determine how this helps in developing a fast algorithm.

Figure 8.38 shows a numerical example of convolution and introduces the fast approach, which we will develop further. The first line shows the standard convolution. The second line calculates the output

$$\mathbf{y} = \text{IDFT}(\text{DFT}(\mathbf{x}) \cdot \text{DFT}(\mathbf{h})), \tag{8.24}$$

where DFT( ) calculates the DFT of the input vector, IDFT( ) calculates its inverse, and · is the vector dot product. Before we look into why the DFT has been employed, let us examine the outputs as calculated in Figure 8.38. With no zeros appended to

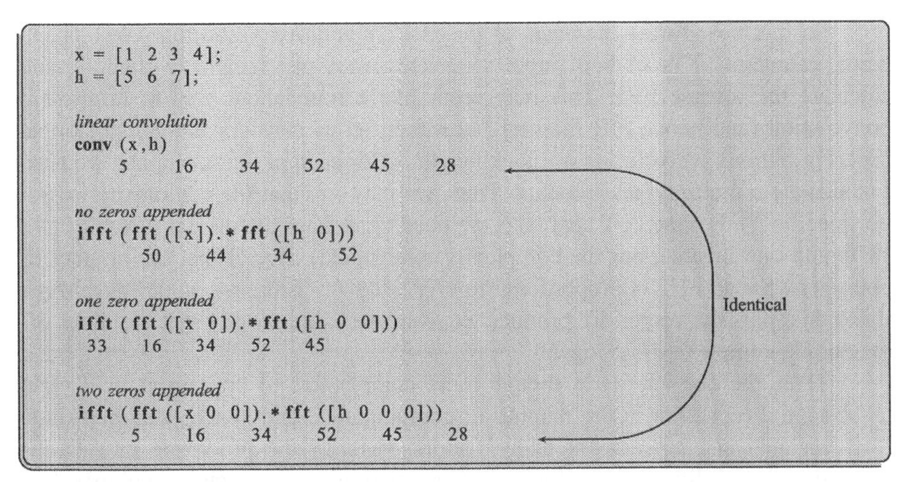

FIGURE 8.38 A numerical example of fast convolution using the FFT. Note that we need to pad the second sequence with zeros so as to be equal in length to the first. To produce a "correct" linear convolution, the first sequence must be padded with P zeros, where P is one less than the length of the second sequence. Failure to correctly pad with zeros produces erroneous results.

x, the four output points calculated appear to have only the last two output values (34 and 52) in common with the "correct" convolution. With a zero appended, three values correspond (34, 52, and 45). Finally, with two zeros appended, the full convolution is correctly calculated. Note that in the figure, we have employed the `fft()` library function in MATLAB (an FFT rather than a DFT), and this will form the basis of the fast algorithm. Furthermore, the **h** sequence has zeros appended so that it has a length equal to the x vector (which itself may be augmented with zeros). This is so that the vector dot product can be calculated (recall that for N input samples, the FFT calculates N output values).

Let us look further into Figure 8.38 and the first FFT calculation. Although the last two values output (34 and 52) correspond to the true convolution, the first two (50 and 44) do not correspond. However, looking more carefully, it may be observed that $50 = 45 + 5$ and $44 = 28 + 16$. In other words, the last two correct convolution values are added to the first two to produce the (erroneous) convolution. In fact, this operation is not linear convolution but corresponds to circular convolution as introduced in the previous section. The same can be said of the calculation with one zero appended—the "incorrect" value of 33 is in fact $28 + 5$, which again is circularly wrapped.

So, from this numerical example, it appears that Equation 8.24 is actually calculating the circular convolution. From what we have examined, circular convolution can in fact be made into linear convolution by appending the appropriate number of zero values to the longer sequence—the appropriate number is in fact the length of the second sequence minus one, as shown previously.

So this now forms the basis of the fast convolution algorithm—we append zeros, calculate FFTs of both input sequences, multiply sample by sample, and calculate the inverse FFT. This may seem like a roundabout way to implement convolution (and hence FIR filtering); however, let us consider the computational cost. For a direct convolution of block length N, for each of the N outputs, we need to calculate a multiply–add product. Thus, we may say that the complexity is proportional to N^2. For the FFT approach, we need to perform three FFTs (actually, two FFTs and one inverse, but the complexity is identical). Recall that the number of operations for an FFT is proportional to $(N/2)\log_2 N$. Thus, we would have three times this, plus the vector dot product. So, it is reasonable to say that the order-of-magnitude complexity is $N\log_2 N$.

Thus, the fast approach, though more complicated to implement, will yield a substantial reduction in the number of computations, and hence execution time. Such an analysis depends on many factors, such as the processor instructions (e.g., the availability of a multiply–add instruction), how complex numbers are handled, the memory speed, whether the blocks can be held in faster cache memory, and so forth.

The actual improvement achieved depends on the system environment, but there is no doubt that even for relatively small blocks, the speedup is significant (and becomes even more pronounced for larger and larger blocks).

Now that the technique has been introduced, it is appropriate to place the theory on a firmer footing. Recall that convolution is

$$y(n)=\sum_{m=0}^{M-1} h_m x(n-m) \tag{8.25}$$

and that the DFT is

$$X(k)=\sum_{n=0}^{N-1} x(n)e^{-j2nk\pi/N}. \tag{8.26}$$

So, the DFT of $y(n)$ is

$$Y(k)=\sum_{n=0}^{N-1} y(n)e^{-j2nk\pi/N}. \tag{8.27}$$

Substituting the convolution equation for $y(n)$,

$$\begin{aligned} Y(k)&=\sum_{n=0}^{N-1}\left(\sum_{m=0}^{M-1} h_m x(n-m)\right)e^{-j2nk\pi/N} \\ &=\sum_{m=0}^{M-1} h_m \sum_{n=0}^{N-1} x(n-m)e^{-j2nk\pi/N}. \end{aligned} \tag{8.28}$$

Multiplying the above by $e^{-j2mk\pi/N}e^{j2mk\pi/N}$ (which does not change the value, since we are effectively multiplying by one) and regrouping the summations, we have

$$\begin{aligned} Y(k) &= \sum_{m=0}^{M-1} h_m \sum_{n=0}^{N-1} x(n-m) e^{-j2nk\pi/N} \cdot e^{-j2mk\pi/N} e^{j2mk\pi/N} \\ &= \sum_{m=0}^{M-1} h_m \underbrace{\sum_{n=0}^{N-1} x(n-m) e^{-j2(n-m)k\pi/N}}_{\text{Shifted DFT}} e^{-j2mk\pi/N} \\ &= \sum_{m=0}^{M-1} h_m \tilde{X}(k) e^{-j2mk\pi/N}. \end{aligned} \tag{8.29}$$

For the "shifted DFT," it does not matter where we start the indexing if the samples in x are circularly addressed. The index $(n - m)$ used to address samples in x must correspond to the argument for the complex exponential—$j(n-m)k\pi/N$. This just means that the indexing is delayed by m samples. Since we are circularly addressing the array, the end summation is unchanged. As before, we denote circular addressing of any array $x(n)$ by $\tilde{x}(n)$.

The last line in the above derivation is just the product of the DFT of **h** with the DFT of x using circular indexing, so we write it as

$$\begin{aligned} Y(k) &= \sum_{m=0}^{M-1} h_m \tilde{X}(k) e^{-j2mk\pi/N} \\ &= H(k) \cdot \tilde{X}(k). \end{aligned} \tag{8.30}$$

What this shows is that the DFT of the output can be calculated using the product of the DFTs of the two inputs. Furthermore, if we use circular addressing, the output is not the linear convolution of the two inputs but rather the circular convolution of the two inputs. Mathematically, we have calculated

$$y(n) = h(n) * \tilde{x}(n) \tag{8.31}$$

$$y(n) = h(n) \circledast x(n) \tag{8.32}$$

where $\circledast$ means circular convolution. This is where Equation 8.23 comes from. Moreover, it justifies our use of the FFT in Equation 8.24.

Using the FFT, which, as we have already seen, provides substantial computational savings, is the justification for using the FFT in the fast convolution algorithm.

8.9.6 Fast Convolution and Filtering for Continuous Sampling

One problem which may arise in many situations is the need to process samples in real time. In the above, the fast algorithm for convolution is based on block-oriented processing; that is, we must buffer a block of samples and then process them all at once. The first result is not available until the entire block has been processed. This does not occur if we use "direct" implementation of the filtering/convolution equation, since each output is calculated independently. But the direct calculation is slower to execute.

There are two common approaches to this problem. Both require buffering of a block, but we can choose the block size to be appropriate to the system design. Specifically, the time delay will depend on the block size—the longer the block size, the longer the delay. A longer block size will yield faster average computation per sample but requires more memory and introduces a longer delay. What we do in practice is to employ an appropriate sized block for buffering.

Unfortunately, this has the potential to introduce discontinuities at the block boundary. Specifically, if we refer back to Figure 8.31, the last two samples in the calculated output ($M - 1$ in the general case) need to be factored into the first two outputs for the subsequent block. This is because the tail of the second vector will overlap both blocks at the boundary.

The two approaches to handle this problem are termed "overlap–add" and "overlap–save." These are both based on the FFT fast convolution approach but handle the block discontinuity slightly differently.

Figure 8.39 shows the *overlap–add* method. We have used an input size of 6, and because the second sequence is three samples long, we have an overlap of two samples. This means that the size for per-block processing, and hence the block length passed to the FFTs, is $6 + 2 = 8$ (recall that the FFT requires the number of samples to be a power of two). Figure 8.39 shows the convolution of the concate-

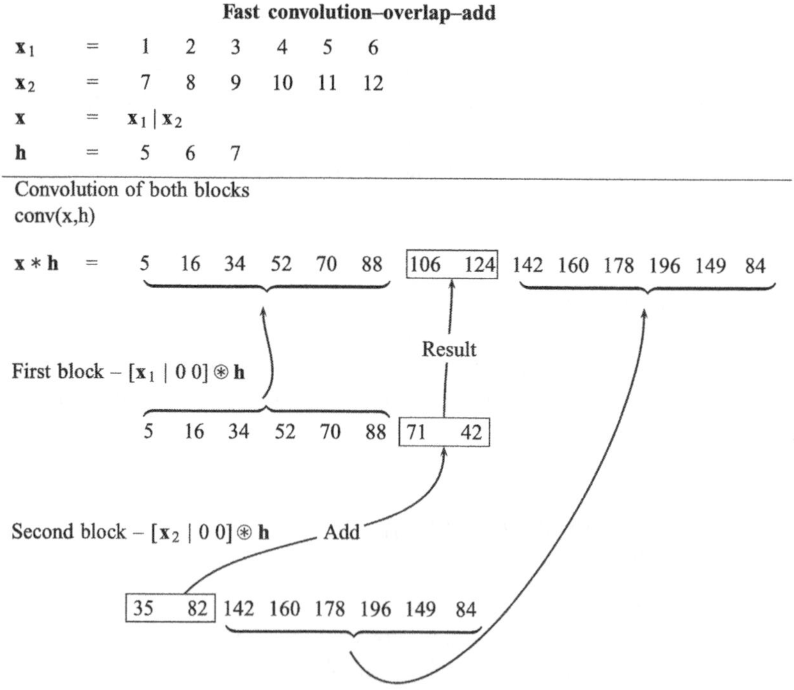

Fast convolution–overlap–add

$\mathbf{x}_1$ = 1 2 3 4 5 6

$\mathbf{x}_2$ = 7 8 9 10 11 12

$\mathbf{x}$ = $\mathbf{x}_1 | \mathbf{x}_2$

$\mathbf{h}$ = 5 6 7

Convolution of both blocks
conv(x,h)

$\mathbf{x} * \mathbf{h}$ = 5 16 34 52 70 88 [106 124] 142 160 178 196 149 84

Result

First block – $[\mathbf{x}_1 \mid 0\ 0] \circledast \mathbf{h}$

5 16 34 52 70 88 [71 42]

Second block – $[\mathbf{x}_2 \mid 0\ 0] \circledast \mathbf{h}$ Add

[35 82] 142 160 178 196 149 84

FIGURE 8.39 Convolution of sequential blocks using the overlap–add procedure. If the length of **h** is M, then $(M - 1)$ zeros are appended, followed by circular convolution. The $(M - 1)$ terms at the end are saved and, when added to the first $(M - 1)$ terms of the next block, produce the correct convolution for the continuous sequence.

nated blocks x_1 and x_2. This is the benchmark for the correct result. However, we want to produce outputs in a shorter time frame—remember that in practice, the sequence may, in fact, be essentially infinite in length. So for the first block, we append two zeros and calculate the (fast) convolution. As shown earlier, the last two outputs [71 42] are the residual which must be added to the first two samples of the subsequent block, so we must save them.

In the next step, the second block x_2 again has two zeros appended, and with these six new samples, the circular convolution is performed. The first two samples are not used directly in the output but must be added to the last two values from the previous calculation—in this case, [35 82] is added to [71 42] to produce the two output samples, [106 124]. The following six output samples are taken from the last six values of the current convolution block (i.e., 142, 160, … , 84). The process as shown for the second block is then repeated—take new samples, append zeros, calculate block, add two samples from the beginning of this block to the end of the previous block. Note that although it seems the very first block is treated differently, in fact, it is identical if we imagine the overlap from the previous block to be [0 0]—that is, a zero vector upon initialization.

A second method for performing fast convolution and filtering on continuous blocks is the *overlap–save* approach, as illustrated in Figure 8.40. This method does not require the saving and adding of values in the region of overlap, as does the overlap–add method. It takes advantage of a problem noted earlier—the fact that the higher samples are wrapped around and "aliased" into the lower samples. Referring to the figure, we begin with two zeros prepended to the first six samples. A block length of 8 is then produced using the fast FFT-based convolution. The first two (aliased) samples are then discarded. For the second block, we prepend not two zeros, but the last two samples from the previous input block (in this case, the values [5 6]). Together with the six new samples, we have another calculation block of 8. The first two samples are again discarded. The process repeats as per the second block: Overlap the two samples from the previous input block, calculate, and discard the first two outputs.

For this reason, this approach has sometimes been referred to as "overlap–discard" or "overlap–scrap," both of which arguably better describe the algorithm. To generalize the procedure, we must remember that the two overlapped and discarded samples are in fact $(M - 1)$ samples for an impulse response vector $\mathbf{h}$ of length M.

8.10 CHAPTER SUMMARY

The following are the key elements covered in this chapter:

- a review of *filter specifications* in the frequency domain
- the role of *FIR* filters
- how filters may be *implemented*, including *fast algorithms*
- the link between fast filtering algorithms and algorithms for *fast convolution* and *fast correlation*

Fast convolution–overlap–save

$\mathbf{x}_1$ = 1 2 3 4 5 6

$\mathbf{x}_2$ = 7 8 9 10 11 12

$\mathbf{x}$ = $\mathbf{x}_1 \mid \mathbf{x}_2$

$\mathbf{h}$ = 5 6 7

Convolution of both blocks
conv(x,h)

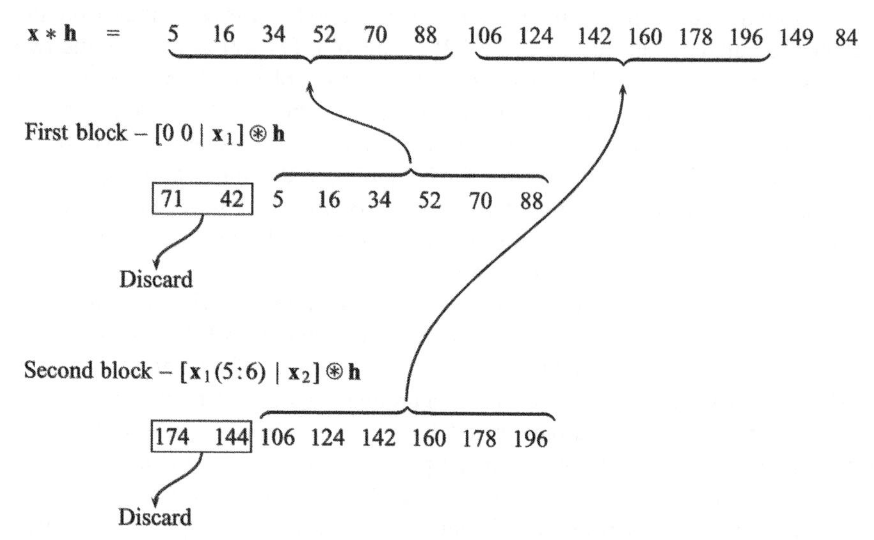

FIGURE 8.40 Convolution of sequential blocks using the overlap–save procedure. If the length of **h** is M, then $M - 1$ zeros are prepended, followed by circular convolution. The $M - 1$ terms at the start are discarded. For subsequent blocks, the last $M - 1$ terms from the previous block are prepended, and the convolution/discard process is repeated. This produces the correct convolution for the continuous sequence.

PROBLEMS

8.1. By sketching the individual frequency responses, explain how bandstop and bandpass filters may be realized using cascades of low- and high-pass filters (refer to Section 8.5).

8.2. A bandpass filter is required to pass frequencies from 8 to 16 kHz using a sample rate of 96 kHz.

- **(a)** For this sample rate, draw a frequency response plot showing how the required characteristic would map to normalized sampling frequencies in radians per sample.
- **(b)** Derive an expression for the filter coefficients.
- **(c)** Explain how you would test the frequency response of the filter using the coefficients derived.
- **(d)** Write a MATLAB script to evaluate the frequency response of the filter.

(e) What would you expect to happen as the filter order is increased? If the coefficients were windowed?

(f) Under what circumstances would a sample rate of 48 kHz be acceptable?

8.3. An ultrasonic signal of maximum bandwidth 40 kHz is to be processed for the purpose of measurement of distance. Would a sampling rate of 48 kHz be suitable? Would 96 kHz be suitable? Explain your answer in each case. If an interfering signal of 10 ± 1 kHz is known to interfere with the measurement sensor, determine the normalized band edge frequencies for a bandstop filter, in radians per sample.

8.4. It was stated that for every complex-valued pole (or zero), we must also place a complex conjugate pole (or zero) on the z plane so that we have a realizable transfer function (i.e., no complex coefficients in the difference equation).

(a) Create a transfer function with one pole at angle ω_p and radius $r < 1$. Derive the difference equation and show that it requires complex coefficients.

(b) To the above, add a pole at the complex conjugate (i.e., angle $-\omega_p$ at the same radius). Derive the difference equation and show that it requires only real coefficients.

8.5. Design a notch filter with two notches to cater for half-wave rectified sine wave interference from a power line. Plot the frequency response to verify that the response is correct.

8.6. Design an FIR filter which has two passbands: one from $\pi/10$ to $2\pi/10$ and another from $7\pi/10$ to $8\pi/10$. By plotting the frequency response magnitude, verify that the filter meets the passband specifications (choose the filter order as you consider to be appropriate). Try different filter orders and add a Hamming window to improve the response.

8.7. A moving-average filter simply computes the arithmetic average of the last N samples.

(a) Write the difference equation for a three-term averaging filter.

(b) How many poles does the system have? How many zeros?

(c) Plot the pole–zero diagram and from that, estimate the shape of the frequency response, from just above zero to the Nyquist frequency (π radians per sample). What shape would you classify the frequency response as? Is that what you would expect from intuition?

8.8. Repeat the above moving-average filter exercise for a generic N-term filter and show that the transfer function is

$$G(z) = \frac{1}{N}\left(\frac{1-z^{-N}}{1-z^{-1}}\right).$$

8.9. A Hilbert transform filter is a filter that passes all frequencies within a specified band with a phase shift of –90°.

(a) If the passband is 0 to π, verify that the desired frequency response is

$$H_d(\omega) = \begin{cases} -j & : \quad 0 < \omega < \pi \\ j & : \quad -\pi < \omega < 0 \end{cases}. \tag{8.33}$$

(b) Show that the corresponding FIR filter coefficients may be found from

$$h_n = \begin{cases} \dfrac{1}{n\pi}(1 - \cos n\pi) & : \; n \neq 0 \\ 0 & : \; n = 0 \end{cases}. \tag{8.34}$$

Note: Treat $n = 0$ as a special case before performing the integration.

8.10. As in the previous question, a Hilbert transform filter is a filter that passes all frequencies within a specified band with a phase shift of –90°. Implement the numerical integration in order to calculate the FIR filter coefficients for $N = 201$. Plot the gain and phase response. Compare the phase response to that of a bandpass filter over the same range. Verify that the phase shift of $\pi/2$ is present, as expected.

8.11. A differentiator is a filter whose output approximates the time derivative of the input signal. It has an ideal magnitude response that is proportional to frequency.

(a) If the passband is 0 to π, verify that the desired frequency response is

$$H_d(\omega) = \jmath\omega \quad : \quad -\pi < \omega < \pi. \tag{8.35}$$

(b) Show that the corresponding FIR filter coefficients may be found from

$$h_n = \begin{cases} \dfrac{\cos n\pi}{n} & : \; n \neq 0 \\ 0 & : \; n = 0 \end{cases}. \tag{8.36}$$

Note: Treat $n = 0$ as a special case before performing the integration.

8.12. As in the previous question, a differentiator is a filter whose output approximates the time derivative of the input signal. Implement the numerical integration in order to calculate the FIR filter coefficients for $N = 201$. Plot the gain and phase response. Verify that the gain response is as expected and explain the shape of the response.

8.13. Verify the result shown in Figure 8.39 for two blocks, and continue for a third block. That is, the FFT-based overlap–add should yield the same result as the directly calculated convolution.

8.14. Verify the result shown in Figure 8.40 for two blocks and continue for a third block; that is, the FFT-based overlap–save should yield the same result as the directly calculated convolution.

8.15. Section 8.9.5 noted that the order of magnitude for computations required for direct convolution and filtering was N^2, versus $N \log_2 N$ for the fast FFT-based methods. Consider N over the range 4, 8, 16, … , 1,024 and compare these quantities. Where, approximately, is the point at which the fast approach becomes viable in terms of reduced complexity?

CHAPTER 9

RECURSIVE FILTERS

9.1 CHAPTER OBJECTIVES

On completion of this chapter, the reader should be able to

1. design a *Butterworth* or *Chebyshev* prototype *analog low-pass filter* and plot its response;
2. *scale* an analog low-pass filter to a given cutoff and transform it to a *high-pass* or *bandpass characteristic*;
3. take an analog filter design and create a *digital filter approximation*; and
4. understand the possible *shortcomings of the conversion* from an analog to digital filter design and compensate for those where possible.

9.2 INTRODUCTION

This chapter covers recursive filters, that is, filters with "memory." This is another class of filter; in some ways, recursive filters are complementary to the nonrecursive filters discussed previously, and they may be an appropriate alternative in many situations. Recursive filters are generally more difficult to design but often result in lower-order filters as compared to nonrecursive filters. This means that they require less memory and fewer computational resources to implement, as well as having an output after a shorter time period. However, the trade-off is a more involved design process. This chapter considers the design of recursive filters using the classical method of analog filter prototyping. Although a brief introduction to continuous systems is given, some prior exposure to continuous systems theory is desirable.

9.2.1 Defining a Recursive Filter

With the exception of the notch filters introduced in Chapter 8, all of the filters discussed so far have a finite-duration impulse response. This means that once the input is removed, the output will decay to zero. One might say that this would be true of any digital filter, and in a practical sense, that is the case (assuming the transfer function is stable; sometimes, we want a "marginally stable" system in order to build a digital oscillator). If we have pole at anywhere other than the origin, the output, theoretically, never reaches zero.

Digital Signal Processing Using MATLAB for Students and Researchers, First Edition. John W. Leis.

For example, consider

$$\frac{Y(z)}{X(z)} = \frac{z}{z-0.9},$$

which has a corresponding difference equation,

$$y(n) = x(n) + 0.9y(n-1).$$

For any $x(n)$ applied then removed (i.e., subsequent $x(n) = 0$), the output $y(n)$ will follow a decaying pattern and never reach zero (theoretically, at least). For example, if $x(0) = 1$ and subsequent $x(n) = 0$, then $y(n) = 1, 0.9, 0.81, \ldots$. For this reason, this class of system is termed infinite impulse response (IIR).

The converse, where the impulse response does tend to a final settling value—in theory and in practice, has zeros and no poles (or possibly only simple poles at $z = 0$, which represent a time delay). For example,

$$\frac{Y(z)}{X(z)} = \frac{z+0.9}{z}$$

has a corresponding difference equation,

$$y(n) = x(n) + 0.9x(n-1).$$

Strictly speaking, this transfer function *does* have a pole, but it is only a single pole at $z = 0$. If the input $x(n)$ has some (nonzero) value, and is then removed in subsequent samples, the output will equate to zero eventually. In the present case, if we have $x(n) = 1, 0, 0, \ldots$, then the corresponding output is $1, 0.9, 0, 0, \ldots$ and zero ever after. This is a finite impulse response (FIR) system.

FIR difference equations are nonrecursive, and hence there is no problem with instability. IIR filters, on the other hand, have recursive coefficients—a type of "memory" of past outputs—and, as a consequence, are shorter than FIR filters for the same or similar frequency responses. The recursive coefficient can be seen in the preceding examples, where we have a term corresponding to previous outputs ($y(n-1)$ in the above) in the IIR system but not in the FIR system.

In general, IIR filters have the following characteristics:

1. They are usually based on discrete (digital) approximations of analog filters.
2. They have a nonlinear phase, a property which may or may not be important in any given application.
3. A perfectly flat frequency response is possible, in either the passband or the stopband.
4. IIR filters are, in general, not easy to construct if an arbitrary frequency response is desired.

Because IIR filters are based on continuous-time filter designs, we will provide an overview of the necessary background on continuous-time systems in general. We do not attempt to give an exhaustive coverage, nor be mathematically rigorous, since such topics are covered in existing texts. Rather, we a give a concise summary of the salient points of analog transfer functions before investigating the digital signal processing aspects, so as to serve as a reference point for the reader.

This is followed by a consideration of continuous-to-discrete conversion methods. There is no one "right" method for converting continuous ("analog") IIR filters into discrete-time transfer functions, and we consider the two most widely used methods, which are known as the bilinear transformation (BLT) and the impulse invariant method (IIM).

Since the usual approach is to design a low-pass filter and convert the design by changing the structure into a matching high-pass or bandpass response as required, subsequent sections will develop this concept and illustrate how MATLAB code may be developed to aid in both understanding the concepts involved and addressing the specific design problem.

9.2.2 Recursive Discrete-Time Filter Structure

An IIR filter is shown in block diagram form in Figure 9.2. Because such filters are based on analog designs, they must be converted into discrete form. This can be rather difficult, and in practice, only an approximation can be achieved. Realistically, any approximation will have shortcomings, and in order to produce good designs, it is important to understand what these shortcomings are.

Figure 9.1 shows the standard FIR filter structure, which has already been studied in the previous chapter. The output $y(n)$ at sample n is a weighted sum of the input $x(n)$ and past inputs $[x(n-1), x(n-2), \ldots]$. The weighting coefficients $[b_0, b_1, \ldots]$ are fixed at the design stage, as is the number of coefficients (thereby

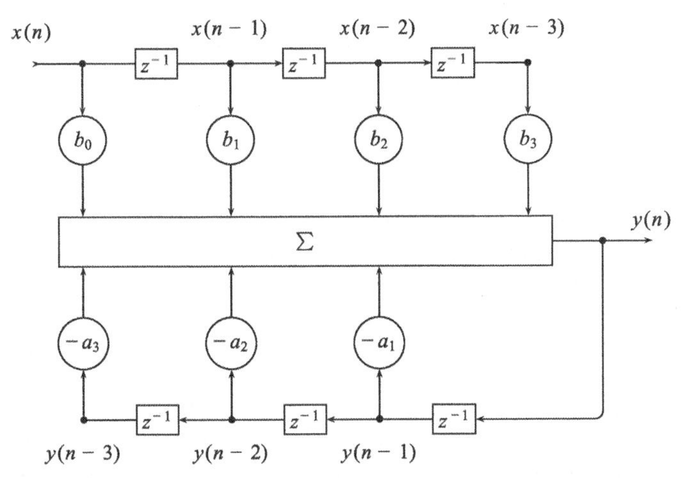

FIGURE 9.1 Block diagram representation of a finite impulse response (FIR) filter. The output is a linearly weighted summation of the input and delayed inputs. Four "taps" and weights are shown, and of course this can be extended into any number of taps and weights. Depending on the application at hand, this structure may extend to hundreds of weights (or more).

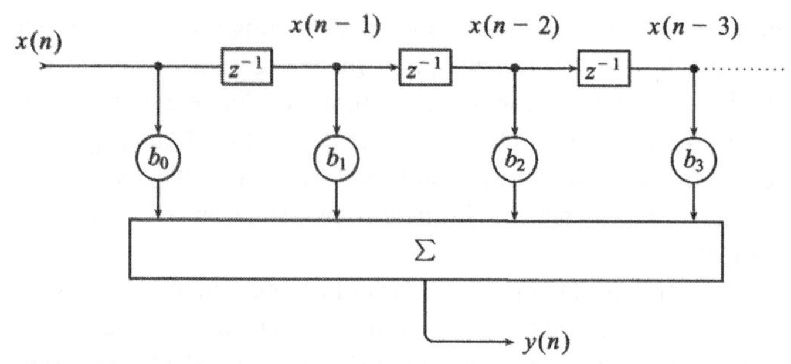

FIGURE 9.2 Block diagram representation of an infinite impulse response (IIR) filter. This is an extension of the FIR filter, in one important way: the feedback or recursion of the output. This constitutes a "memory" effect and usually allows the filter order to be lower for a similar frequency response specification. Note that this representation implies $a_0 = 1$ in Equation 9.2. This simply means that all other coefficients need to be normalized by this value if $a_0 \neq 1$.

determining the order of the filter). The filter operation simply consists of calculating a new output, $y(n)$, at each sample instant nT, with the the multiply–add operations required to compute the output defined by the difference equation. The block diagram can be converted into an appropriate weighted-sum transversal structure (sometimes called "tapped delay line") defined by

$$y(n) = b_0 x(n) + b_1 x(n-1) + \cdots + b_N x(n-N). \tag{9.1}$$

The IIR filter may be considered as an extension of the FIR structure, wherein the output is fed back to the summation. This is shown in Figure 9.2. The output $y(n)$ at time nT depends not only on the present and past inputs but also on past outputs. The IIR difference equation in general is

$$\begin{aligned} y(n) = &(b_0 x(n) + b_1 x(x-1) + \cdots + b_N x(n-N)) \\ &- (a_1 y(n-1) + a_2 y(n-2) + \cdots + a_M y(n-M)). \end{aligned} \tag{9.2}$$

It may be imagined that the IIR filter has a memory of past outputs. This is why, in general, fewer coefficients are required for an IIR filter to achieve a similar response to an FIR filter. Moreover, the design process for IIR filters is almost always more involved, as we will see shortly.

The usual approach for designing FIR filters is a direct one: We can go from the desired frequency response to the impulse response, and then to the difference equation coefficients. However, because of the memory effect of IIR filters, such a design approach is not possible for recursive filters. This is because the filter coefficients themselves become the impulse response in the FIR case, but the same is not true for the IIR case. Instead, it is usual to start with an analog filter equation (Laplace or s domain) and to convert that to a discrete-time structure. The following sections deal with each of these issues in turn. First, however, we briefly review the Laplace transform in both the time domain and the frequency domain.

9.3 ESSENTIAL ANALOG SYSTEM THEORY

An analog or continuous-time system is one which has a smoothly varying response over any time range. The vast majority of the "real" world is analog in nature. Almost any quantity which one cares to mention (sound, vibration, light, temperature) varies continuously over time, and we hope that our sampling of the quantity is sufficiently precise in time so as to be able to capture any event of importance. Discrete-time or digital systems, on the other hand, rely on snapshots of the analog world at the sampling instants.

It is not surprising, then, that there is a vast quantity of literature developed over the last hundred years or more analyzing continuous-time systems and, in particular, electrical systems. Electrical systems are represented as circuit elements (resistors, capacitors, inductors) with an observed quantity (current, voltage, charge) varying smoothly over time. The theory and mathematical tools for this are well developed, and some approaches to discrete-time systems rely on extensions or modifications of the analog system theory. Perhaps the best-known mathematical tool for analyzing electrical networks is the Laplace transform.

9.3.1 The Laplace Transform

The solution of analog systems using differential equations to model their observed behavior—and hence to predict their future states—is the staple of much of the electrical and mechanical system theory. Differential equations are difficult to solve algebraically, and the Laplace transform has a rich history as a tool in aiding their solution. Mathematically, the Laplace transform is defined as an operation on an observed system characteristic $f(t)$ using the notation $F(s) = \mathcal{L}\{f(t)\}$, where the transformation operation is defined as

$$F(s) = \int_0^{\infty} e^{-st} f(t)\,dt. \tag{9.3}$$

Note that the standard notation is to use lowercase for the time function, and uppercase for the transformed function. Thus, we have the Laplace transform pair $f(t) \leftrightarrow F(s)$ for a given $f(t)$.

A key result which can be derived from the definition is the differentiation property. If we are prepared to assume that a system is initially at rest (i.e., zero initial conditions or $f(0) = 0$), then the operation of differentiation of $f(t)$ can be replaced with a multiplication of the corresponding Laplace transform by the operator s. Mathematically,

$$\begin{aligned} f(t) &\rightarrow F(s) \\ \frac{df(t)}{dt} &\rightarrow sF(s) \end{aligned} \tag{9.4}$$

for zero initial conditions ($f(t) = 0$ for $t < 0$). Thus, the operation of differentiation is replaced by a simple multiplication.

Intuitively, the operation of integration is the "opposite" of differentiation in calculus, and so it is satisfying to note that integration can be represented in a similar way by division by s. Assuming zero initial conditions, we have

$$\begin{aligned} f(t) &\to F(s) \\ \int f(t)\,dt &\to \frac{F(s)}{s}. \end{aligned} \tag{9.5}$$

Of course, the above results have a sound mathematical basis, but we do not present the proofs here.

Another key result in linear systems and signal processing is the convolution property. Recall that the output of a linear system is the convolution integral performed on (or, simply, the convolution of) the input signal and the system's impulse response. This complex integral operation can be replaced by a simple multiplication of the respective Laplace transforms. If we define the input, output, and system impulse response as

$$\begin{aligned} x(t) &\to X(s) \\ y(t) &\to Y(s) \\ h(t) &\to H(s), \end{aligned}$$

then, for zero initial conditions, the system output may be calculated via convolution and Laplace transforms:

$$y(t) = \int_0^t h(\tau)x(t-\tau)\,d\tau \tag{9.6}$$

$$Y(s) = X(s)H(s). \tag{9.7}$$

Here we have used a common convention: $x(t)$ for the input to a system, $y(t)$ for the output, and $h(t)$ for the impulse response of a system. We can mathematically determine the Laplace transform of known signals (those signals which have an equation defining their amplitude at a given time). Tables of standard transforms are widely available; a few common transforms are listed in Table 9.1.

TABLE 9.1 Some Laplace Transforms

Description	Signal $f(t)$	Transform $F(s)$
Step at $t = 0$	$u(t)$	$\frac{1}{s}$
Step at $t = \tau$	$u(t-\tau)$	$\frac{e^{-s\tau}}{s}$
Decaying exponential for $t \geq 0$	e^{-at}	$\frac{1}{s+a}$
Sine wave with frequency Ω for $t \geq 0$	$\sin\Omega t$	$\frac{\Omega}{s^2+\Omega^2}$
Cosine wave with frequency Ω for $t \geq 0$	$\cos\Omega t$	$\frac{s}{s^2+\Omega^2}$

In order to calculate a system's output in response to an input, we need to know the impulse response of the system. Then, the Laplace multiplication operation can be applied to determine the output using $Y(s) = X(s)H(s)$. Then, we need to map from $Y(s)$ back to $y(t)$ using algebraic expansion and the standard tables mentioned above. As an alternative to mathematically deriving transforms of systems and applying the convolution property, we can iteratively approximate the integrations implicit in the $1/s$ operation, as will be explained in the following section.

9.3.2 Time Response of Continuous Systems

The response of a continuous-time function consists of the values of $y(t)$ from $t = 0$ until some specified time, given the input $x(t)$ and the coefficients of the transfer function. We now wish to develop a method for modeling the time response of an arbitrary system. Specifically, we want to determine the response of a system to a given input, using only the impulse response.

If the system is linear and time invariant (i.e., the output can be calculated via superposition and the coefficients do not change over time), the impulse response completely characterizes the system. To develop this idea, suppose the transfer function is

$$\frac{Y(s)}{X(s)} = \frac{2s+1}{s^2+2s+4}.$$

Then, the coefficients $b = [2\ 1]$ and $a = [1\ 2\ 4]$ completely specify the system. We can rewrite this so as to have only negative powers of s by multiplying the numerator and the denominator as follows:

$$\begin{aligned}\frac{Y(s)}{X(s)} &= \frac{2s+1}{s^2+2s+4} \\ &= \frac{s^{-2}}{s^{-2}} \cdot \frac{0s^2+2s^1+1s^0}{s^2+2s^1+4s^0} \\ &= \frac{0s^0+2s^{-1}+1s^{-2}}{1+2s^{-1}+4s^{-2}}.\end{aligned}$$

This is easily rearranged to yield the output as a function of the input:

$$\begin{aligned}Y(s)\left(1+2s^{-1}+4s^{-2}\right) &= X(s)\left(0s^0+2s^{-1}+1s^{-2}\right) \\ Y(s) &= X(s)\left(0s^0+2s^{-1}+1s^{-2}\right) - Y(s)\left(2s^{-1}+4s^{-2}\right).\end{aligned}$$

Since we only have negative powers of s, we can apply the idea that division by the s operator corresponds to integration. So, for any signal $x(t)$,

$$\begin{aligned}X(s) &= s^0 X(s) \rightarrow x(t) \\ \frac{X(s)}{s} &= s^{-1}X(s) \rightarrow \int x(t)\,dt \\ \frac{X(s)}{s^2} &= s^{-2}X(s) \rightarrow \iint x(t)\,dt\,dt.\end{aligned}$$

The integrals may be computed using a numerical approximation to the integral. The continuous integrations are converted into sums using

$$\int_{t=0}^{\tau} x(t)\,dt \approx \sum_{k=0}^{K-1} x(t_k)\,\delta t, \tag{9.8}$$

where the index K is chosen to correspond to the limit τ, and $x(t_k)$ is the value of $x(t)$ for a specific time $t = t_k$. The integration approach can be generalized as follows:

$$\begin{aligned}\frac{Y(s)}{X(s)} &= \frac{b_0 s^{M-1} + b_1 s^{M-2} + \cdots + b_{M-1} s^0}{a_0 s^{N-1} + a_1 s^{N-2} + \cdots + a_{N-1} s^0} \\ &= \frac{b_0 s^{M-1} + b_1 s^{M-2} + \cdots + b_{M-1} s^0}{s^{N-1}\left(a_0 + a_1 s^{-1} + \cdots + a_{N-1} s^{-N+1}\right)} \\ &= \frac{b_0 s^{M-N} + b_1 s^{M-N-1} + \cdots + b_{M-1} s^{-N+1}}{a_0 + a_1 s^{-1} + \cdots + a_{N-1} s^{-N+1}}.\end{aligned} \tag{9.9}$$

We need $y(t)$ on the left-hand side, and this is easily accomplished by dividing all b_m and a_n by a_0. So, without loss of generality, for $a_0 = 1$, we have

$$Y(s) = X(s)\left(b_0 s^{M-N} + b_1 s^{M-N-1} + \cdots + b_{M-1} s^{-N+1}\right) - Y(s)\left(a_1 s^{-1} + \cdots + a_{N-1} s^{-N+1}\right). \tag{9.10}$$

We are then in a position to implement our solution. Using the coefficients defined previously, we set up the simulation as follows. First, we need to set the maximum simulation time, a small time increment δt for the integrations, and the number of samples in the output. For an impulse at $t = 0$, we need the input initialized as $x(0) = 1/\delta t$. The highest power of s is N (assuming more poles than zeros), and thus we need N integral terms `xi` and `yi`.

```
b = [2 1];
a = [1 2 4];
tmax = 10;
dt = tmax/1000;
L = round(tmax/dt);
x = zeros(L, 1);
x(1) = 1/dt;
y = zeros(L, 1);
t = zeros(L, 1);
b = b(:)/a(1);
a = a(:)/a(1);
M = length(b);
N = length(a);
xi = zeros(N, 1);
yi = zeros(N, 1);
```

In the above generalization, we have a term, s^{M-N}. This is implemented by shifting the b coefficients to the right in an array—remember that the leftmost term is s^0, then the next is s^{-1}, and so forth, each corresponding to a shift-right by one position in the array.

```
if(N > M)
    % denominator order greater, pad with zeros on LHS to
    % make same length
    bp = [zeros(N - M, 1); b];
else
    bp = b;
end
ap = [0; a(2:N)];  % cancel a0 multiplication by setting
                   % to zero
```

We are then in a position to implement the iteration of the differential equations using the following loop. The key steps are the weighting of each input and output term by their respective coefficients, which is accomplished using `sum(bp.*xi) - sum(ap.*yi)`, followed by the update of the integrals of $y(t)$ and $x(t)$ using a simple rectangular approximation to the area of the form `xi(i) + xi(i - 1)*dt`.

```
tcurr = 0;
for n = 1:L
    t(n) = tcurr;
    ynew = sum(bp.*xi) - sum(ap.*yi);
    y(n) = ynew;
    xi(1) = x(n);
    yi(1) = y(n);
    for i = 2:N
        xi(i) = xi(i) + xi(i - 1)*dt;
        yi(i) = yi(i) + yi(i - 1)*dt;
    end
    tcurr = tcurr + dt;
end
```

We now have a general method of approximating a continuous-time system in order to obtain the impulse response. An example of this is shown in Figure 9.3. This figure also shows the locations of the poles of the transfer function. As with z domain functions, the poles for an s-domain function are defined as the values of s that make the denominator equal to zero. The next step is to develop a general method for obtaining the frequency response for a given continuous-time system.

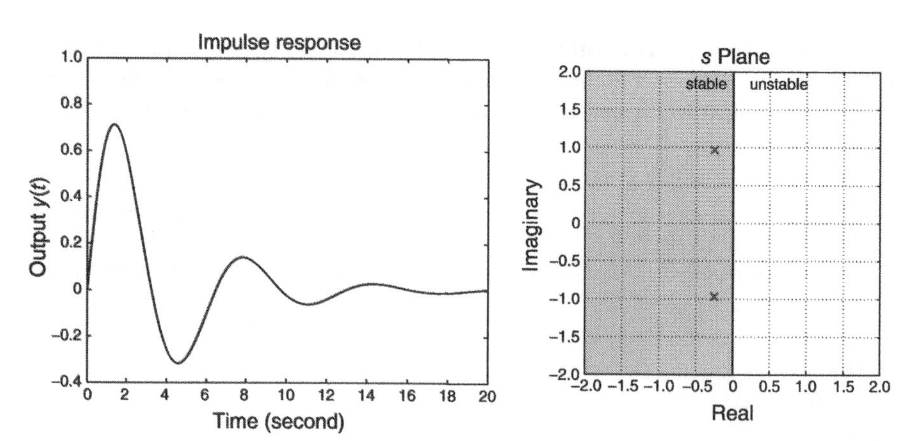

FIGURE 9.3 Impulse response of the system $\frac{1}{s^2+\frac{1}{2}s+1}$ and the corresponding poles.

9.3.3 Frequency Response of Continuous Systems

To determine the frequency response of a continuous-time linear system, we need to substitute $s \rightarrow \jmath\Omega$. The reason for this may be seen by comparing the Laplace transform $F(s)$ with the Fourier transform $F(\jmath\Omega)$:

$$F(s)=\int_0^\infty e^{-st}f(t)\,dt \tag{9.11}$$

$$F(\jmath\Omega)=\int_0^\infty e^{-\jmath\Omega t}f(t)\,dt. \tag{9.12}$$

When we come to implement the frequency response, we have each of the coefficients b_m and a_n multiplied by terms of the forms s^{M-1-m} and s^{N-1-n}. We do not need to convert to negative powers of s as in the time response calculation since we do not need to keep track of what the signal values were at any given time. We simply need to perform the substitution of $\jmath\Omega$ for s. In the numerator, this gives terms of the form $b_m(\jmath\Omega)^{M-1-m}$; the denominator has terms of the form $a_n(\jmath\Omega)^{N-1-n}$. We initialize using

```
b = b(:);
a = a(:);
M = length(b);
N = length(a);
bpow = [M - 1:-1:0];
apow = [N - 1:-1:0];
bpow = bpow(:);
apow = apow(:);
L = round(MaxOmega/dOmega);
H = zeros(L, 1);
Omega = zeros(L, 1);
```

The loop to calculate a frequency response magnitude $|H(j\Omega)|$ and phase $\angle H(j\Omega)$ at each frequency point Ω is then as follows. This gives us the complex value $H(j\Omega)$, but often we only want the magnitude $|H(j\Omega)|$.

```
% starting frequency
OmegaCurr = 0.01;
for n = 1:L
    Omega(n) = OmegaCurr;
    % j Omega terms
    NumFreq = ones(M, 1) * j * OmegaCurr;
    DenFreq = ones(N, 1) * j * OmegaCurr;
    % coefficient x (j Omega)^n terms
    Num = b.*(NumFreq.^bpow);
    Den = a.*(DenFreq.^apow);
    Hcurr = sum(Num)/sum(Den);
    H(n) = Hcurr;
    OmegaCurr = OmegaCurr + dOmega;
end
```

So we now have a general method for determining the frequency response of a continuous-time system. An example frequency response is shown in Figure 9.4, which again also shows the locations of the poles of the transfer function.

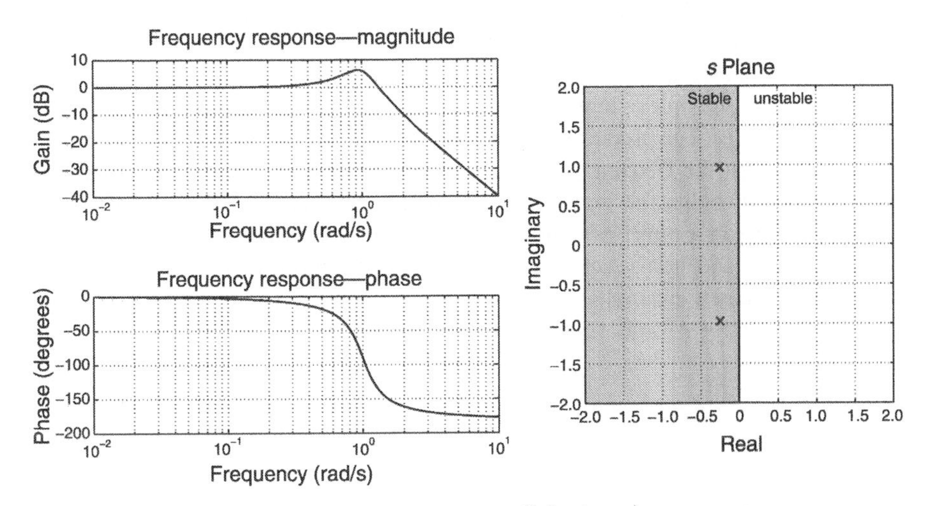

FIGURE 9.4 Frequency response of the system $1/\left(s^2 + \frac{1}{2}s + 1\right)$ and its poles.

Since we now know how to compute the frequency response from a general transfer function, we now return to the original problem: how to determine a general transfer function which gives us a desired frequency response.

9.4 CONTINUOUS-TIME RECURSIVE FILTERS

This section covers the "classical" design of recursive filters using two common approaches: the Butterworth and Chebyshev polynomials. It builds upon the foundation of continuous-time systems as outlined in the previous section and develops algorithms and codes to design filters using these approaches. The usual method employed is to design a low-pass filter with a unity cutoff frequency and then to extend to a low-pass filter of any desired frequency, or a different characteristic (high-pass, bandpass, or other). Methods for converting the analog filter design into a discrete-time implementation follow in subsequent sections.

9.4.1 Butterworth Filter Structure

Butterworth filters provide a maximally flat response in either the passband or stopband. They are described in general by the transfer function $G(s)$, defined as

$$|G(s)|^2 = \frac{1}{1+\left(\frac{s}{\Omega_c}\right)^{2N}}. \tag{9.13}$$

This is for a filter with a cutoff of Ω_c radians per second. We will see later that it is usual practice to design for $\Omega_c = 1$ and then to simply replace s with s/Ω_c to scale the filter to any desired cutoff as required. Indeed, it may be that we do not need a low-pass filter but another characteristic such as a high-pass or bandpass filter.

Note that this equation describes the magnitude *squared*, not the actual value $G(s)$ which we would like; that is, $G(s)$ is a complex number for any s and, as such, has magnitude and phase. So, we need to undertake a little more derivation to find the filter transfer function. Equation 9.13 defines not $G(s)$, as we would like, but rather $|G(s)|^2$. Now, for any complex number $a = re^{j\theta} = x + jy$, the complex conjugate is $a^* = re^{-j\theta} = x - jy$. The product of a complex number and its conjugate is $a\,a^* = |a|^2$. Recognizing that the frequency response is found when we put $s \rightarrow j\Omega$, what we need in order to solve the problem and to determine $G(s)$ is to find solutions for $G(s)G(-s)$, and to use only "half" of the solution. This is done by replacing all s^2 terms with $s\cdot(-s) = -s^2$ and then by selecting only half of the resulting factors (poles).

For a concrete example, consider the case of $N = 1$. The poles of $|G(s)|^2$ are the solutions of $1 + s^{2N} = 0$, so with $s^2 \rightarrow -s^2$, we have

$$\begin{aligned} 1+\left(-s^2\right)^N &= 0 \\ \therefore \quad \left(-s^2\right)^1 &= -1 \\ s^2 &= 1 \\ s &= \pm 1. \end{aligned}$$

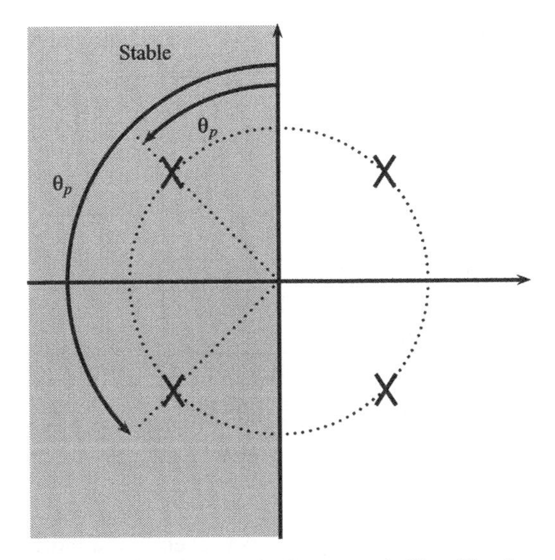

FIGURE 9.5 Roots of a Butterworth filter. For the general solution, we start at $\pi/2$ and rotate at an angle of $\theta_p = (2k+1)(\pi/2N)$ to obtain each solution. The solution repeats after every $2N$ roots found.

We choose the stable pole only ($s = -1$), which yields a transfer function of

$$G(s) = \frac{1}{s+1}.$$

Note that the setting of the cutoff frequency Ω_c to unity above does not compromise the solution: The cutoff frequency is simply a scaling factor which can be incorporated as desired (this will be examined in later sections).

Consider now a second-order example ($N = 2$) and assume $\Omega_c = 1$ again. The poles of $|G(s)|^2$ are at

$$\begin{aligned} 1+\left(-s^2\right)^2 &= 0 \\ \therefore \quad s^4 &= -1 \\ s^4 &= e^{\jmath(\text{odd number})\pi} \\ s &= e^{\jmath(\text{odd number})\frac{\pi}{4}}. \end{aligned}$$

In the above, we need to remember that -1 is $e^{\jmath\pi}$, and that an angle of π could be 3π, 5π, -1π, or in fact any odd multiple of π. This is where the (odd number) π term comes from—it is usually written as $(2k + 1)\pi$, where k is an integer.

The easiest way to visualize this result is to plot the factors (poles of the transfer function), as illustrated in Figure 9.5. Starting at a radius of one, we move anticlockwise at an angle of $1 \times 45°$ $\left(e^{\jmath\pi/4}\right)$, then $3 \times 45°$, $5 \times 45°$, and $7 \times 45°$.

When we come to $9 \times 45°$, we effectively end up back at $1 \times 45°$. The stable poles (the two with a negative real part) are at

$$s = \begin{cases} 1e^{j3\times\frac{\pi}{4}} \\ 1e^{j5\times\frac{\pi}{4}} \end{cases}.$$

So, the values of s which satisfy the design equation are

$$s = -\frac{1}{\sqrt{2}} \pm j\frac{1}{\sqrt{2}}.$$

Taking product of the (s – pole) terms, we have the transfer function

$$G(s) = \frac{1}{s^2 + \sqrt{2}s + 1}.$$

In the above, the $(-s^2)^2$ canceled the negative sign because N was even. For a more general solution (N odd or even), we again find poles of $|G(s)|^2$ without making any assumptions about N:

$$\begin{aligned} 1+\left(-s^2\right)^N &= 0 \\ \left(-s^2\right)^N &= e^{j(\text{odd number})\pi} \\ -s^2 &= e^{j(\text{odd number})\frac{\pi}{N}} \\ s^2 &= -e^{j(\text{odd number})\frac{\pi}{N}} \\ s &= je^{j(\text{odd number})\frac{\pi}{2N}} \\ s &= e^{j\frac{\pi}{2}}e^{j(\text{odd number})\frac{\pi}{2N}} \\ s &= e^{j\frac{\pi}{2}}e^{j(2k+1)\frac{\pi}{2N}} \quad k: \text{integer} \end{aligned} \tag{9.14}$$

Using Figure 9.5 again, we start at an angle of $\pi/2$ (90°) and travel anticlockwise by an odd multiple of $\pi/2N$.

We have examined the procedure for first- and second-order systems, and it becomes a little tedious for higher orders. More importantly, when we come to Chebyshev filters in the next section, an analytical solution as found in Equation 9.14 is much harder to derive. What we would like is a general solution for any order so that we may specify a signal filter of any order. To develop such an algorithmic solution, we again use a normalized cutoff, $\Omega_c = 1$, and write the expression for the poles as

$$\begin{aligned} 1+\left(-s^2\right)^N &= 0 \\ 1+(-1)^N s^{2N} &= 0. \end{aligned}$$

We can solve this for a given order N by first realizing that the above is actually

$$(-1)^N s^{2N} + 0s^{2N-1} + 0s^{2N-2} + \cdots + 1s^0 = 0. \tag{9.15}$$

The position in the coefficient array is determined by the power of s, which starts at $2N$. The value of this coefficient is $(-1)^N$, so we set up the coefficients and find the roots of the polynomial as follows:

```
% remember to set the order N
d = zeros(2 * N + 1, 1);
d(2 * N + 1 ) = 1;
d(1) = (-1)^(N);
dr = roots(d);
i = find(real(dr) < 0);
drs = dr(i);
p = poly(drs);
```

After setting up the coefficient array with the correct coefficients of s (with powers from $2N$ to 0), we find the roots of the Butterworth polynomial (denominator of $|G|^2$) using the **roots()** command. Recall that we have the product $G(s)G(-s)$, of which we only effectively want half of the terms. Obviously, we want a stable system, so we need the terms which define poles in the negative half of the s plane. That is easily done by selecting those whose real part is negative. With those stable coefficients, all that remains is to expand the roots into a polynomial using **poly()**. If we calculate all coefficient values for the denominator in this manner for $N = 1$–10, we obtain

```
N=  1:  1  1
N=  2:  1  1.41  1
N=  3:  1  2.00  2.00   1
N=  4:  1  2.61  3.41   2.61   1
N=  5:  1  3.24  5.24   5.24   3.24   1
N=  6:  1  3.86  7.46   9.14   7.46   3.86   1
N=  7:  1  4.49  10.10  14.59  14.59  10.10  4.49   1
N=  8:  1  5.13  13.14  21.85  25.69  21.85  13.14  5.13   1
N=  9:  1  5.76  16.58  31.16  41.99  41.99  31.16  16.58  5.76   1
N= 10:  1  6.39  20.43  42.80  64.88  74.23  64.88  42.80  20.43  6.39  1
```

For example, for $N = 2$, the Butterworth polynomial is $s^2 + 1.41s + 1$. Mathematically, the Butterworth polynomials in general are written as

$$\begin{aligned} B_1(s) &= s+1 \\ B_2(s) &= s^2+s\sqrt{2}+1 \\ B_3(s) &= s^3+2s^2+2s+1 \\ \vdots & \qquad \vdots \end{aligned} \tag{9.16}$$

The previous table of results can be checked using any standard table of Butterworth polynomials. Note, however, that such tables usually show the coefficients in factorized form, so it may be necessary to use the **conv** ( ) function to expand the factors to get the polynomial coefficients themselves. For example, the tabulated polynomial for $N = 3$ is usually $(s + 1)(s^2 + s + 1)$. To expand this, we take the polynomial coefficients of each factor and apply polynomial multiplication:

```
conv([1 1],[1 1 1])
ans =
      1 2 2 1
```

This specifies the polynomial coefficients, which agrees with the case above for $N = 3$. The coefficients are read from left to right, so that the rightmost term is the coefficient of s^0, then s^1 and so forth up to the highest order on the left (s^3 in this case).

Now we have a set of stable polynomials in terms of s. For example, if we take the polynomial coefficients for $N = 4$ and find the poles, we have

```
d = [1.00  2.61  3.41  2.61  1.00]
roots(d)
ans =
     -0.3818 + 0.9242i
     -0.3818 - 0.9242i
     -0.9232 + 0.3844i
     -0.9232 - 0.3844i
```

Clearly, these all have negative real parts, and as such would be stable in an analog (s-domain) transfer function. Note also that when we have complex poles, they always occur in conjugate pairs. This process is illustrated in Figure 9.6, which shows the location of the calculated poles in the s plane. Note that the poles lie in a circle, as has been derived algebraically (refer back to Equation 9.14, which gave the pole locations on the complex plane). Real poles may occur by themselves, but

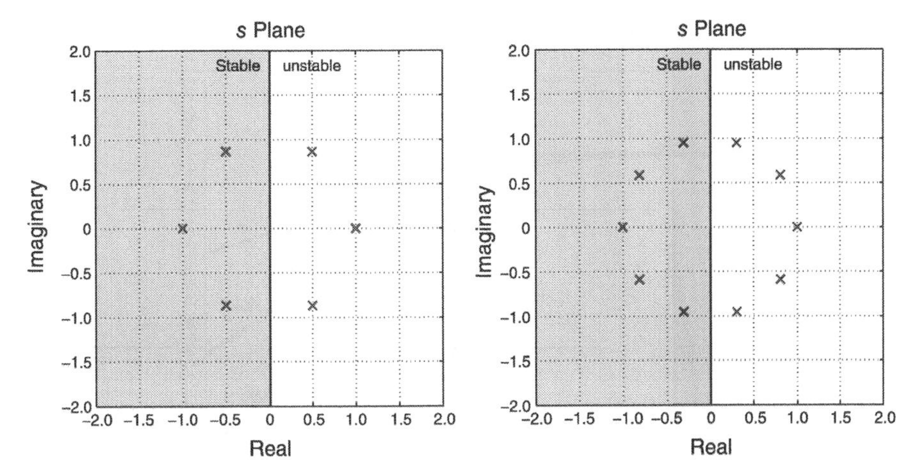

FIGURE 9.6 Poles of a low-pass Butterworth filter, as the order is increased. Note that all the poles are shown here, and these are actually the poles of $B(s)\ B(-s)$. For a stable filter, we select the poles of $B(s)$ to correspond to the left-half stable poles.

complex poles always occur in conjugate pairs. Finally, the poles are symmetric about the imaginary axis. Thus, the process of deriving $G(s)$ corresponds to selecting those poles in the left half of the s plane in the diagram.

Now that we have the filter polynomials, we can investigate the frequency response using the techniques developed in Section 9.3.3. As with all analog systems, this is found by substituting $j\Omega$ for s, where Ω is the desired frequency in radians per second. To convert to "real" frequency in cycles per second (Hz), recall that $\Omega = 2\pi f$, where f is in hertz. Figure 9.7 shows the responses of Butterworth low-pass filters using this approach as the order is increased. Note that the −3 dB point is also shown. This is the frequency where the gain is $1/\sqrt{2}$, since the gain in decibels at this point is $20\log_{10}\left(1/\sqrt{2}\right) \approx -3$ dB.

9.4.2 Chebyshev Filter Structure

Chebychev[1] filters also utilize a polynomial for approximating the filter transfer function. They provide a sharper roll-off than Butterworth filters, but at the expense of a ripple in the passband.[2] Chebyshev filters are described by the transfer function $G(s)$ defined as

$$|G(s)|^2 = \frac{K}{1+\varepsilon^2 T_N^2\left(\dfrac{s}{\Omega_c}\right)}. \tag{9.17}$$

[1]Chebyshev also made many other contributions to mathematics, including the Chebyshev or L_∞ distance. There are various other spellings arising from the translation, such as Chebycheff.

[2]We consider here so-called Chebyshev Type I filters. Type II filters have ripples in the stopband.

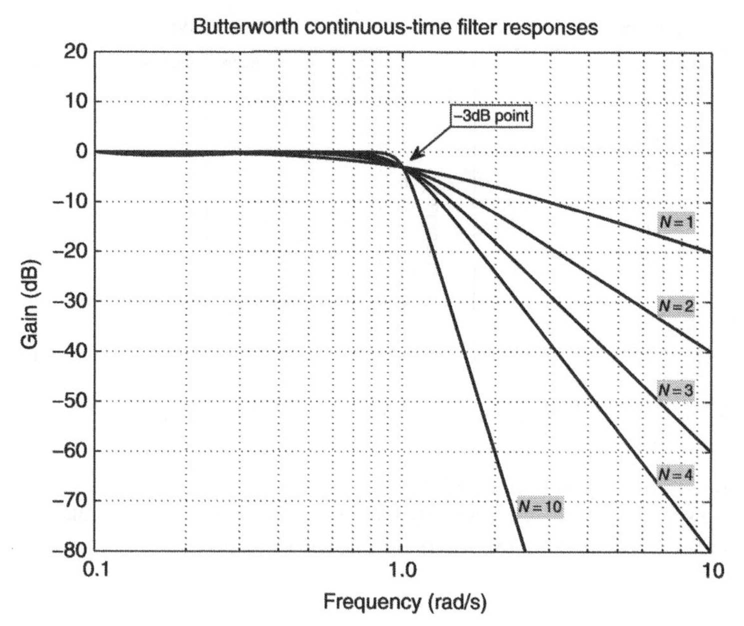

FIGURE 9.7 Butterworth continuous-time low-pass filter responses. The filter is normalized to a cutoff frequency of $\Omega_c = 1$. As we increase the order of the filter, the cutoff grows sharper. A characteristic of Butterworth filters is that the passband is flat (no ripples) over the passband of zero to Ω_c.

The function $T_N(\cdot)$ is the Nth order Chebyshev polynomial. Before describing the Chebyshev polynomial, we note that the poles of $|G(s)|^2$ can be described in a similar fashion to those of the Butterworth structure. The poles of the function $|G(s)|^2$ are the solutions of

$$1+\varepsilon^2 T_N^2\left(\frac{s}{\Omega_c}\right)=0. \tag{9.18}$$

The parameter ε controls the shape of the passband—this will be explained shortly. Letting for now $\varepsilon = 1$ and once again setting the cutoff frequency $\Omega_c = 1$, we have poles at

$$1+T_N^2(s)=0. \tag{9.19}$$

It is reasonable to ask at this stage what happens when we have Ω_c not equal to one, and when ε is not equal to one. These questions will be addressed in due course. For now, it is sufficient to know that Ω_c scales the frequency response (i.e., instead of a cutoff frequency of 1 rad/s, we have Ω_c rad/s), and second, that ε has an effect on the ripple in the passband. These issues are best addressed after the Chebyshev polynomials are explained. Furthermore, the additional questions of how to convert the filter if we do not want a simple low-pass filter, and how to convert from an analog filter into a discrete-time digital filter, will be expanded upon afterward.

So for now, we need to define the Chebyshev polynomial. There are many ways to approach this, and perhaps the simplest is via the recursive definition

$$\begin{aligned} T_0(x) &= 1 \\ T_1(x) &= x \\ T_N(x) &= 2xT_{N-1}(x) - T_{N-2}(x). \end{aligned} \tag{9.20}$$

The first two lines initialize the recursion, and the following line is applicable for all subsequent values of N. We can thus apply this recursion to obtain the polynomials as follows:

$$\begin{aligned} N = 2 \rightarrow T_2(x) &= 2x^2 - 1 \\ N = 3 \rightarrow T_3(x) &= 2x(2x^2 - 1) - x \\ &= 4x^3 - 3x \\ N = 4 \rightarrow T_4(x) &= 2x(4x^3 - 3x) - (2x^2 - 1) \\ &= 8x^4 - 8x^2 + 1 \\ N = 5 \rightarrow T_5(x) &= 16x^5 - 20x^3 + 5x. \end{aligned} \tag{9.21}$$

Figure 9.8 shows the shape of some low-order Chebyshev polynomials. There are a couple of points to note regarding the nature of the polynomials thus generated. First, the order of the polynomial equals N. For example, for $N = 4$, we have a polynomial starting with x^4. For odd values of N, the polynomial has no constant term, and for even values of N, we have alternating +1 and −1 constant terms in each polynomial. These observations will be useful shortly, when we need to find the poles of the transfer function (and thus the shape of the filter's frequency response).

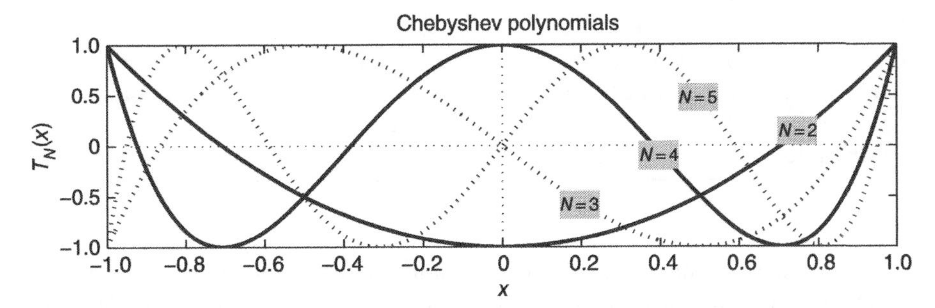

FIGURE 9.8 Some Chebyshev polynomials $T_N(x)$ as defined by the recursion in Equation 9.20. Because of the use of normalized frequencies s/Ω_c in the filter design equation, and the fact that $\Omega \geq 0$, the portion of interest corresponds to the range $x = 0$ to 1 above.

The derivation of the polynomial coefficients is tedious, and it would be helpful to have an automated solution. We can do this by implementing the recursion defined by Equation 9.20. The multiplication $2xT_{N-1}(x)$ at each stage is just a polynomial multiplication, since we can write it as $2(1x^1 + 0x^0)T_{N-1}(x)$. Since polynomial multiplication can be implemented as the convolution operation, we can use the **conv**() operator in MATLAB. If we have the recursion polynomial coefficients of the current stage in vector `tc`, then this multiplication for the subsequent stage can be calculated using `2*conv([1 0], tc)`, from which the $T_{N-2}(x)$ coefficients are subtracted (refer to Equation 9.20).

All the polynomial coefficients up to order M may be calculated in a matrix of coefficients T as shown next, with each row being a vector of coefficients. The recursion as defined in Equation 9.20 then yields all polynomial coefficients.

```
M = 10;
T = zeros(M + 1, M + 1);
T(1, M + 1) = 1;
T(2, M) = 1;
for n = 2:M
    t = 2*conv([1 0], T(n, :));
    t = t(2:M + 1 + 1);
    T(n + 1, :) = t - T(n - 1, :);
end
```

The output of the above (edited for clarity) is the T matrix:

```
N = 2 :  2   0 -1
N = 3 :  4   0 -3    0
N = 4 :  8   0 -8    0 1
N = 5 :  16  0 -20   0 5    0
N = 6 :  32  0 -48   0 18   0 -1
N = 7 :  64  0 -112  0 56   0 -7   0
N = 8 :  128 0 -256  0 160  0 -32  0 1
N = 9 :  256 0 -576  0 432  0 -120 0 9  0
N = 1 0: 512 0 -1280 0 1120 0 -400 0 50 0 -1
```

The first column is the value of N, and the values in each row are the coefficients of x^n in order of decreasing powers of x. For example, the second-last line with the zero-valued coefficients removed is

$$T_9(x) = 256x^9 - 576x^7 + 432x^5 - 120x^3 + 9x. \tag{9.22}$$

We can select the coefficients for order N from T_N using `tc = T(N + 1, M + 1 - N:M + 1);`.

Once we have the Chebyshev polynomials, we need incorporate them into a frequency domain specification. The discussion above created polynomials of the form $T_N(x)$, but we actually need a filter $G(s)$ defined by

$$|G(s)|^2 = \frac{K}{1+\varepsilon^2 T_N^2\left(\frac{s}{\Omega_c}\right)}, \tag{9.23}$$

where K is a gain factor. So, the remaining problems are the expansion of $T_N^2(\cdot)$, the square root operation, and the selection of the poles to use. These are all best considered together. The selection of the stable poles is not unlike the procedure used for Butterworth polynomials. To motivate the approach, consider the case of order $N = 2$ and $\varepsilon = 1$. The poles of the above equation will be defined by the denominator of Equation 9.17 to become

$$1+\varepsilon^2\left(2s^2-1\right)^2 = 0. \tag{9.24}$$

For the purposes of explanation, we will use $\varepsilon = 1$ for now, yielding

$$4s^4 - 4s^2 + 2 = 0. \tag{9.25}$$

This gives us the denominator polynomial in terms of s. Since we have the magnitude-squared function, we can use the same approach as with the Butterworth filter case. We need to replace $s^2 \rightarrow (s)(-s) = -s^2$.
The polynomial then becomes

$$4s^4 + 4s^2 + 2 = 0. \tag{9.26}$$

So now we take the polynomial coefficients, find the stable roots, and expand to obtain

```
d2 = [4 0 4 0 2];
d  = roots(d2)
d  =
    -0.3218 + 0.7769i
    -0.3218 - 0.7769i
    0.3218 + 0.7769i
    0.3218 - 0.7769i
i = find(real(d) < 0);
as = poly(d (i))
as =
   1.0000    0.6436    0.7071
```

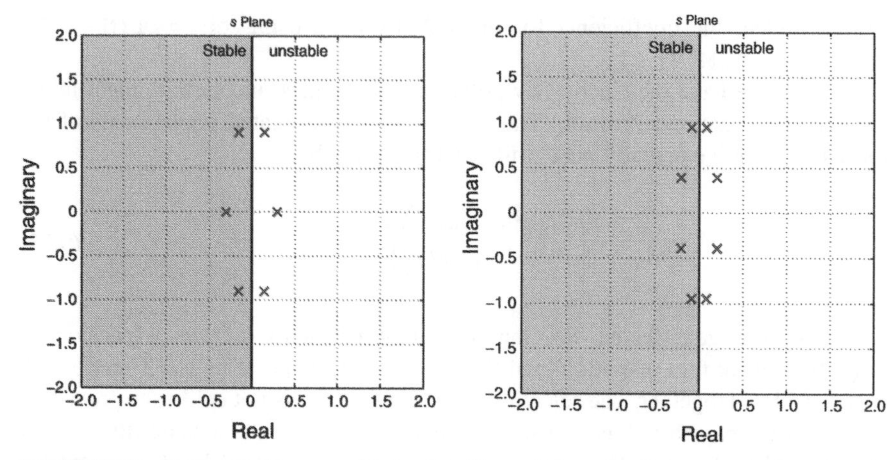

FIGURE 9.9 Poles of a low-pass Chebyshev filter, as the order is increased. Note that all the poles are shown here, and these are actually the poles of $G(s)G(-s)$. For a stable filter, we select the poles of $G(s)$ to correspond to the left-half stable poles.

Figure 9.9 shows the poles of Chebyshev filters of various orders. Note that, like the Butterworth case, we are calculating the magnitude squared, and hence have twice as many poles plotted as we have in the transfer function. To define the transfer function, we select the stable poles (those in the left half of the s plane) and expand out as a polynomial. Whereas the Butterworth poles formed a circle, the Chebyshev poles formed an ellipse with the major axis being the imaginary (vertical) axis.

To investigate the general solution, consider again the case of

$$\begin{aligned} T_2(s) &= 2s^2 - 1 \\ \therefore \quad T_2^2(s) &= (2s^2-1)(2s^2-1) \\ &= 4s^4 - 4s^2 + 1. \end{aligned}$$

So, replacing $s^2 \to -s^2$,

$$\overbrace{4((s)(-s))^2}^{n=2} - \overbrace{4((s)(-s))}^{n=1} + 2 = 0. \tag{9.27}$$

Inspecting the above, it is evident that every $s^{2n} = (s^2)^n$ term is replaced by $s^{2n}(-1)^n$. If we have the Chebyshev coefficients calculated as explained earlier, then we can find the poles using the following:

```
tc = tc(:) % polynomial coefficients for order N
% square, multiply by epsilon squared, add one (will be
% 2N + 1 coefficients)
d = conv(tc, tc);
d = d*epsilon^2;
d(2*N + 1) = d(2*N + 1) + 1;
% powers for s
spow = [2*N: - 1:0]';
% (s)(-s) terms
% s^2 -> -s.s = -s^2
% s^4 -> (-s.s)(-s.s) = s^4
% s^6 -> (-s.s)(-s.s)(-s.s) = -s^6
% spow/2 will always be an integer, since
% we have T()^2 where T is the chebyshev poly
snew = (-1)*ones(2*N + 1, 1);
svals = snew.^(spow/2);
% finally, the denominator coefficients
% this is the entire polynomial for |G|^2,
% so we take the stable (left - half s plane) roots
d2 = svals.*d;
% find roots of chebyshev |G|^2,
% then stable ones (real part negative)
d = roots(d2);
i = find(real(d)) < 0);
% expand to polynomial using only stable roots
as = poly(d(i));
```

The gain K of the Chebyshev filter is clearly dependent upon the roots of the denominator. Since the denominator will be factored into terms of the form $(s - p_i)$ for poles p_i, then the product when $s = 0$ (i.e., when $s = \jmath\Omega$ with $\Omega = 0$) will be the product of all the p_i factors (note that if p_i is complex, then there will be a corresponding term $(s - p_i^*)$ with complex conjugate pole, and the product will still be real). As will be seen when we plot the response curves, for an odd filter order, the gain is unity at zero frequency; for an even order, the gain is $1/\sqrt{1+\varepsilon^2}$. Putting this together, the gain is

$$c = \begin{cases} -\prod p_i & : \ N \quad \text{odd} \\ \dfrac{1}{\sqrt{1+\varepsilon^2}} \prod p_i & : \ N \quad \text{even} \end{cases}. \tag{9.28}$$

So how does the filter perform in practice—what is the actual frequency response like? Figure 9.10 shows the responses, again for low-pass filters of various orders. Compared with to the Butterworth responses derived previously, the Chebyshev

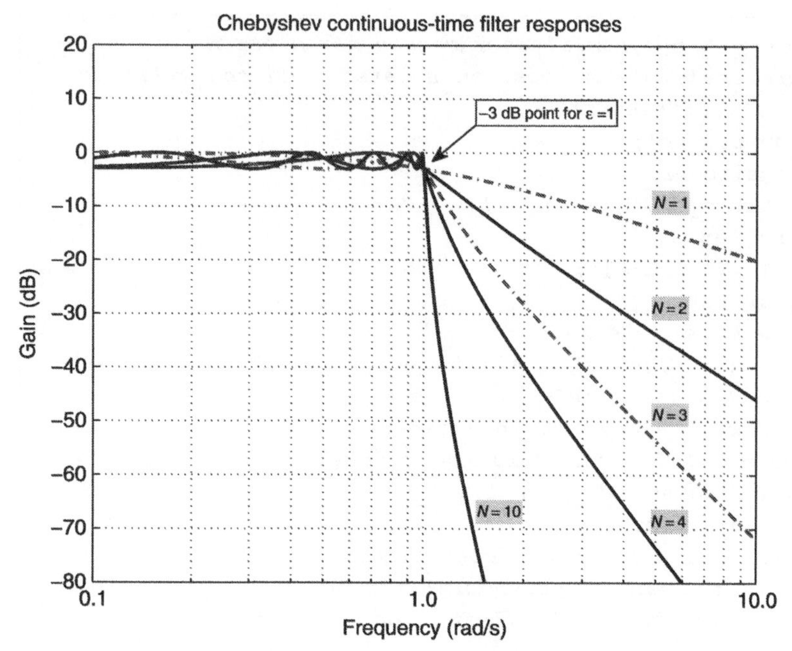

FIGURE 9.10 Chebyshev continuous-time low-pass filter responses. Clearly, the higher-order filter results in a sharper response, but the disadvantage is that more terms are required in the filter polynomial, hence greater complexity in order to realize the filter in practice.

responses are sharper (as will be seen by comparing the responses for the same order).

The faster roll-off rate of the Chebyshev filter comes at a price: the ripple in the passband.[3] This is an appropriate time to revisit the parameter ε, which we previously had set to unity, since it controls the amount of ripple in the passband. This may be seen by looking at the Chebyshev filter equation, where ε is multiplied by the polynomial coefficients, which in effect means multiplication by the shape illustrated in Figure 9.8. A larger ε in effect means a larger contribution from the polynomial evaluation, and this may be seen by looking at the filter responses in Figure 9.10.

We can quantify the amount of ripple in the passband, and the gain magnitude at the cutoff frequency, as follows. First, note that in the Chebyshev polynomials, $T_N(1) = 1$ for all N. Thus, the magnitude at the cutoff where $s = j\Omega_c = 1$ will be

$$|G|^2 = \frac{1}{1+\varepsilon^2}. \tag{9.29}$$

[3]For Type II filters, the ripple is in the stopband.

If we define the passband ripple to be δ, then the gain at the cutoff is $1 - \delta$. Thus, we must have

$$\begin{aligned} 1-\delta &= \frac{1}{\sqrt{1+\varepsilon^2}} \\ \therefore \quad \delta &= 1-\frac{1}{\sqrt{1+\varepsilon^2}}. \end{aligned} \tag{9.30}$$

Thus, for $\varepsilon = 1$, the gain at the cutoff is $1/\sqrt{2} \approx 0.707$, or -3 dB. For $\varepsilon = 0.5$, the gain is $1 - \delta \approx 0.9$ and the ripple is $\delta \approx 0.1$.

9.5 COMPARING CONTINUOUS-TIME FILTERS

The preceding sections introduced some analog filter types and their characteristics and their design. There are a great many analog filter design approaches; we have only introduced the most widely used methods. It is, however, worth comparing what we have studied: How are the responses different and why would one choose one design over another?

It would be expected that a higher order for a filter would result in a sharper transition, and this is shown for the Chebyshev filter in Figure 9.11. For Chebyshev filters, a higher order has the side effect of increasing the number of ripples in the passband—in any given application, this may be tolerable, or it may be totally unacceptable. If such ripples in the passband are undesirable, we may need to turn to a Butterworth design. In Chebyshev filters, the passband ripple is controlled by the

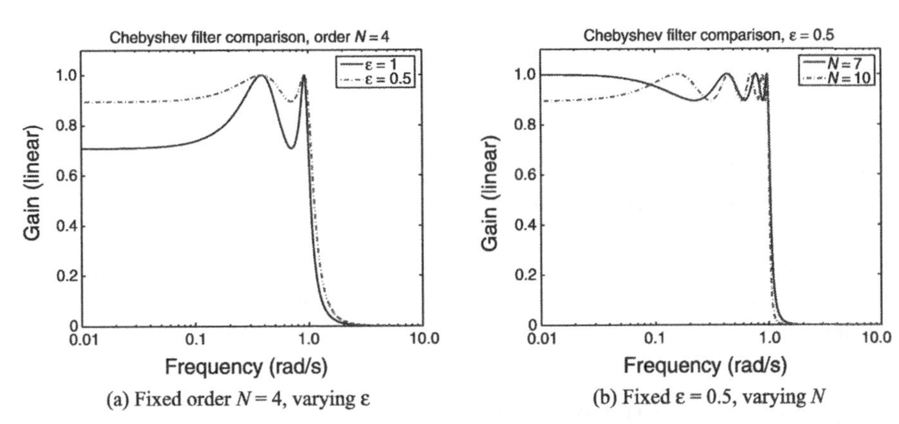

(a) Fixed order $N = 4$, varying ε (b) Fixed $\varepsilon = 0.5$, varying N

FIGURE 9.11 Comparison of the effect of the ripple parameter ε on the response of a Chebyshev filter (a), and (b) the effect of the filter order N. For $\varepsilon = 1$, the gain varies between $1/\sqrt{2} \approx 0.707$ and unity, for a ripple of 3 dB. For $\varepsilon = 0.5$, the gain varies between approximately 0.9 and unity, for a ripple of around 1 dB. Note that the vertical gain scale is linear rather than logarithmic (dB), in order to highlight the shape of the response.

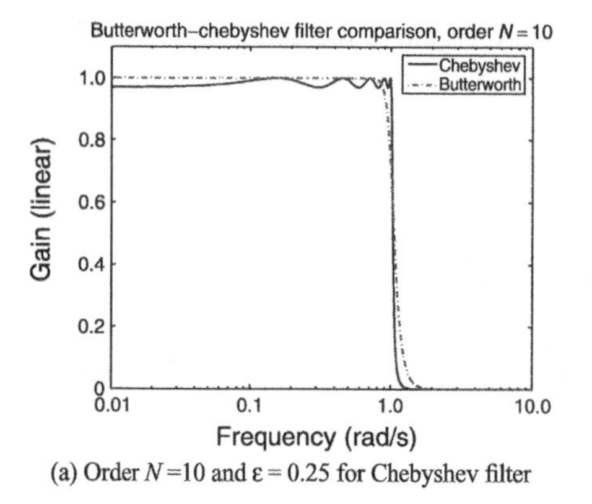

(a) Order $N = 10$ and $\varepsilon = 0.25$ for Chebyshev filter

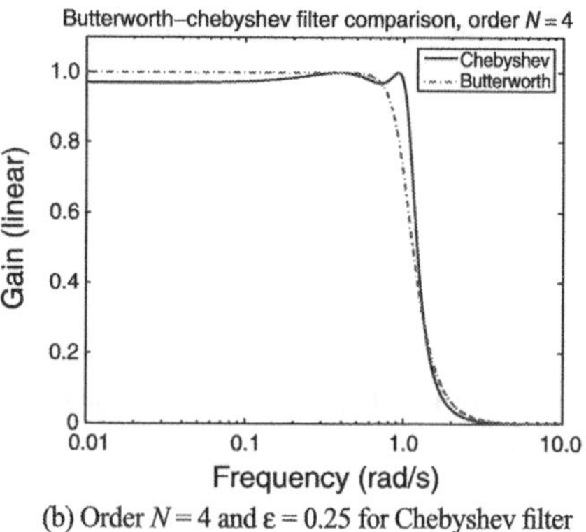

(b) Order $N = 4$ and $\varepsilon = 0.25$ for Chebyshev filter

FIGURE 9.12 Comparison of filter designs, using the Butterworth and Chebyshev polynomials. Clearly the Chebyshev filter has a sharper transition band, but has ripples in the passband. Note that the vertical gain scale is linear rather than logarithmic (dB), in order to highlight the shape of the response.

parameter ε. Figure 9.11 shows the effect of changing this parameter. The tradeoff is self-evident: A lower ripple may be obtained at the expense of a less sharp transition at the cutoff frequency.

Figure 9.12 compares the responses of Butterworth and Chebyshev design approaches. It is seen that for the same order, the Chebyshev has a sharper roll-off, which is usually a desirable characteristic. However, that comes at the expense of the passband ripple, which is inherent in the Chebyshev designs studied.

There are many other types of continuous-time filters. The following sections describe methods for converting from a continuous filter (function of s) into a discrete filter (function of z). These conversions are applicable to all types of s-domain filters.

9.6 CONVERTING CONTINUOUS-TIME FILTERS TO DISCRETE FILTERS

Once we have an analog filter prototype, we can convert it into a digital (discrete) filter and thus employ a difference equation to implement it. The following sections examine two commonly used approaches for the s to z conversion.

9.6.1 Impulse Invariant Method

The *impulse invariant method* is one of two commonly used approaches for converting a continuous-time function of s into a discrete-time function of z. The basic idea is, as the name suggests, to produce a discrete transfer function that has the same impulse response as the analog transfer function prototype.

The method works as follows. First, we decompose the continuous-time s-domain function into first-order sections using partial fractions. For each, we then write the z transform equivalent. It is necessary to scale by the sample period T to keep the zero-frequency gain the same. Mathematically, this means that if we set the discrete impulse response $h_d(n)$ equal to the scaled continuous response $h(nT)$,

$$h_d(n) = T\,h(nT)$$

then the sampled version $y_d(n)$ will approximate the true system $y(t)$ at the sampling instants (when $t = nT$); that is,

$$y_d(n) = y(nT).$$

Each first-order section is converted from the discrete domain to the digital domain using the mapping

$$\frac{1}{s+p} \rightarrow T\left(\frac{z}{z-e^{-pT}}\right); \tag{9.31}$$

that is, a general first-order section with a single pole at $s = -p$ is replaced with the corresponding discrete-time approximation. As well as the parameter p, we need to specify the sample period T (recall that this is the reciprocal of the sampling frequency f_s). In principle, *any* transfer function can be decomposed into the sum of such first-order sections. The mapping is applied in turn to each term, when the s-domain function is expressed as a sum of such first-order terms. Finally, the individual first-order terms in z can be algebraically recombined to form a single transfer function (numerator and denominator polynomials). To give an illustrative example, consider an integrator:

$$H(s) = \frac{1}{s}. \tag{9.32}$$

In this particular case, $p = 0$ in Equation 9.31. The discrete version becomes

$$H_d(z) = T\left(\frac{z}{z-1}\right). \tag{9.33}$$

So, the difference equation is

$$y(n) = x(n)T + y(n-1). \tag{9.34}$$

This is a rectangular approximation to the integrator—in words, we could express an integrator as

$$\begin{array}{ccccc} y(n) & = & y(n-1) & + & x(n)T \\ \text{New area} & = & \text{area so far} & + & \text{increment of area.} \end{array}$$

The frequency response is

$$H_d(e^{j\omega}) = \left(\frac{e^{j\omega}}{e^{j\omega}-1}\right)T. \tag{9.35}$$

We can generalize this approach to higher orders. Suppose we have a second-order system which is decomposed into a cascade of two first-order systems given by

$$\begin{aligned} H(s) &= \frac{1}{(s+1)(s+2)} \\ &= \frac{1}{s+1} + \frac{-1}{s+2}. \end{aligned} \tag{9.36}$$

The discrete approximation is

$$H_d(z) = \left(\frac{z}{z-e^{-T}} + \frac{-z}{z-e^{-2T}} \right) T. \tag{9.37}$$

The major problem with impulse invariant approximation is aliasing. Figure 9.13a shows the effect of the mirror-image response centered on f_s. The response in the center region is the summation of the two (the lower "real" response and the "folded" version). It is clear that in order to minimize this problem, a high sample rate has to be chosen.

The specific case shown in Figure 9.13b is a comparison of filters with the same analog transfer function with poles at $s = -1/\sqrt{2} \pm 1/\sqrt{2}$, but differing sampling frequencies for the digital versions. For the case of a 2-Hz sampling rate, the mirrored frequency response about f_s produces a tail which interferes with the response from 0 to $f_s/2$. Clearly, this introduces a certain amount of aliasing in the frequency response. The higher sampling rate (10 Hz) as shown in Figure 9.13b moves the mirrored response further out, and thus the overlap is negligible.

9.6.2 Corrected Impulse Invariant Method

The usual method as described above does not take into account the discontinuity in the impulse response at $t = 0$. In 2000, it was independently shown (Jackson 2000; Mecklenbraüker 2000) that the correct form is slightly different to what is "traditionally" given. The first-order section correspondence is normally defined as

$$\frac{K}{s+p} \to KT\left(\frac{z}{z-e^{-pT}} \right). \tag{9.38}$$

However, the first-order sections ought to be converted using

$$\frac{K}{s+p} \to K\frac{T}{2}\left(\frac{z+e^{-pT}}{z-e^{-pT}} \right). \tag{9.39}$$

What is the difference between these approaches? In practice, it is relatively small but worth examining. First, at DC (or zero frequency), we have $\omega = 0$ rad per sample. As usual, $z = e^{j\omega}$; hence, $z = 1$ at a frequency of zero. Further, we see many terms of the form e^{-pT} in both of the above expressions, where p is the pole in the

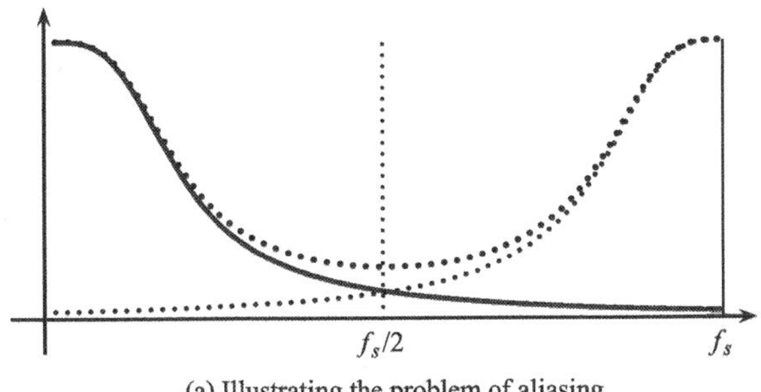

(a) Illustrating the problem of aliasing

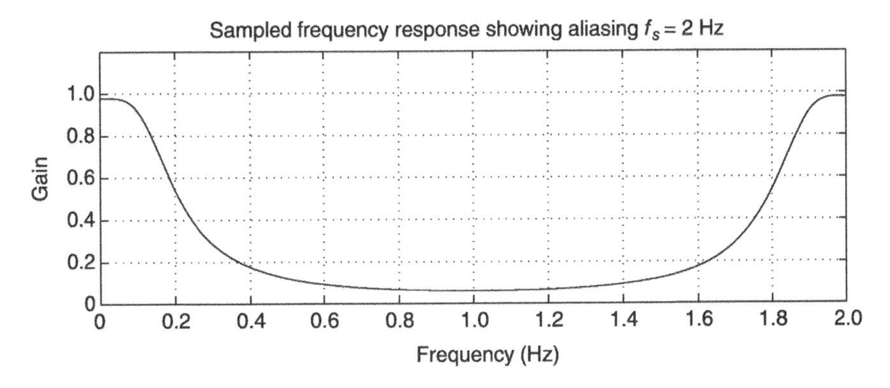

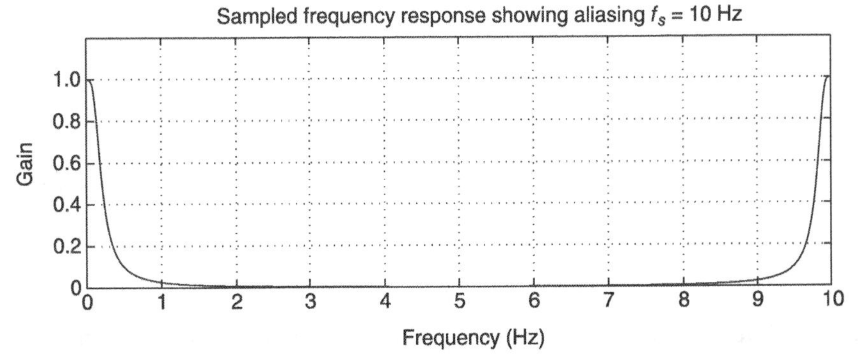

(b) Aliasing as it occurs for the same discrete transfer function, as the sample rate is increased

FIGURE 9.13 Aliasing in impulse invariant filters. The diagram at the top shows that the response from zero to $f_s/2$ is mirrored at f_s going back to zero. The observed response is the sum of these and hence is not entirely correct. The worst-case error occurs in the middle, at $f_s/2$. One way to minimize this unwanted side effect is to increase the sampling frequency f_s so that the second "aliased" response is shifted higher in frequency (further to the right).

continuous-time transfer function $G(s)$. Now observe that (as defined previously) p will be a positive number (since, for stability, the s poles must be negative). Using the expansion for

$$e^x = 1 + x + \frac{x^2}{2!} + \frac{x^3}{3!} + \cdots,$$

we have

$$\begin{aligned} e^{-pT} &= 1 - pT + \frac{(-pT)^2}{2} + \cdots \\ &\approx 1 - pT, \end{aligned} \tag{9.40}$$

where the last approximation is valid since f_s is generally large, and hence T is small, compared to the pole p. Thus, the original impulse invariant approximation (omitting the scaling constant K for clarity) is

$$\begin{aligned} T\left(\frac{z}{z - e^{-pT}}\right) &\approx \frac{T}{1-(1-pT)} \\ &= \frac{T}{pT} \\ &= \frac{1}{p}, \end{aligned} \tag{9.41}$$

whereas the revised approximation is

$$\begin{aligned} \frac{T}{2}\left(\frac{z + e^{-pT}}{z - e^{-pT}}\right) &\approx \frac{T}{2}\frac{(1+(1-pT))}{1-(1-pT))} \\ &= \frac{T}{2}\frac{(2-pT)}{pT} \\ &= \frac{1}{p} - \frac{T}{2}. \end{aligned} \tag{9.42}$$

Thus, the difference is proportional to T, and hence is generally small in practice. Furthermore, the difference only applies where the system response is discontinuous at $t = 0$. Thus, the theoretical difference does *not* apply to the second-order case shown in Figure 9.14, which has a second-order prototype of $1/\left(s^2 + \sqrt{2}s + 1\right)$. Figure 9.15 shows a case where the methods do in fact give differing results.

9.6.3 Automating the Impulse Invariant Method

In manipulating the algebra required for the impulse invariant transform, we first need to convert the given transfer function into a sum of fractions, for which we need partial fractions. As an example, suppose we have

$$\frac{1}{(s+1)(s+2)}. \tag{9.43}$$

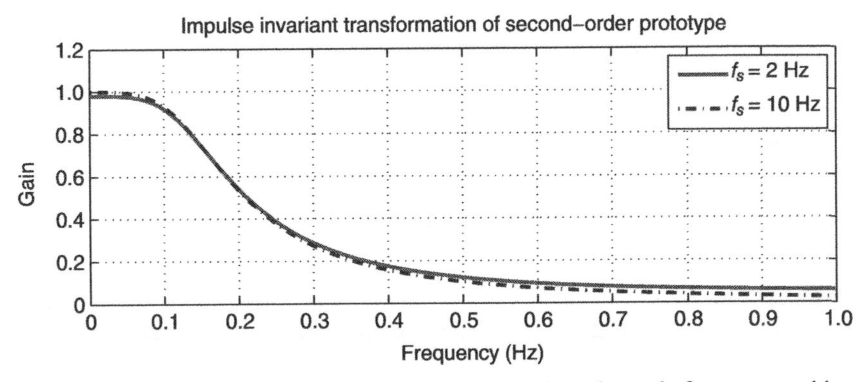

FIGURE 9.14 A practical demonstration of the selection of sample frequency and its effect on aliasing in impulse invariant filters.

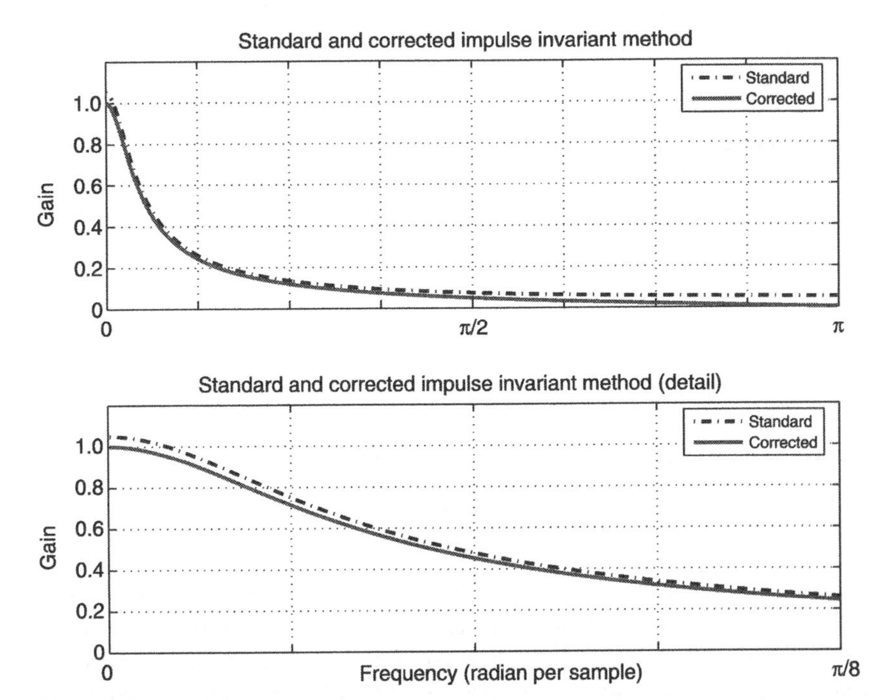

FIGURE 9.15 Original and corrected impulse invariant filter responses. The responses are shown in the sampled-frequency domain, with the lower panel showing a magnified view over the lower frequency range.

Then the corresponding sum of fractions is

$$\frac{1}{(s+1)(s+2)} = \frac{A}{s+1} + \frac{B}{s+2}. \tag{9.44}$$

We can expand the denominator as either a set of coefficients (using **conv()**) or roots (using **poly()**) to obtain the denominator polynomial.

```
as = conv([1 1], [1 2])
as =
    1     3     2
as = poly([-1 -2])
as =
    1     3     2
```

This is now in a form suitable for converting the polynomial ratio to partial fractions using the **residue()** function, noting that the numerator is unity (bs = 1).

```
[r p k] = residue(bs, as);
N = length(p);
```

Calling **residue()** gives us the poles and constant gain factor of each partial fraction term in p and r, respectively. In the general case, we wish to recombine all the terms into one transfer function, as a ratio of polynomials. To understand the approach, consider the following example. Suppose we have

$$\frac{1}{(s+1)(s+2)(s+3)} = \frac{A}{s+1} + \frac{B}{s+2} + \frac{C}{s+3}. \tag{9.45}$$

This can be solved using partial fractions to yield $A = \frac{1}{2}$, $B = -1$, and $C = \frac{1}{2}$. So, the recombination problem becomes

$$\frac{A}{s+1} + \frac{B}{s+2} + \frac{C}{s+3} = \frac{A(s+2)(s+3) + B(s+1)(s+3) + C(s+1)(s+2)}{(s+1)(s+2)(s+3)}. \tag{9.46}$$

The denominator becomes the multiplication of each term, while the numerator becomes the multiplication of the numerator of each term in turn by the *other* denominator terms. Noting that multiplication of the simple polynomial terms can be accomplished by the convolution operator **conv()**, we can generalize the operation to the specific case of the impulse invariant problem as defined in Equation 9.31.

The denominator of Equation 9.31 can be calculated as the expansion of all the individual terms:

```
az = [1];
for n = 1:N
    az = conv(az, [1 -exp(p(n)*T)]);
end
```

The numerator of Equation 9.31 may be calculated in a similar way. However, one must remember that all terms are multiplied out and added together, *except* the one corresponding to the current factor, as shown in the nested loops below:

```
bz = zeros(N + 1, 1);
for n = 1:N
    t = [r( n ) 0];
    % loop across other terms
    for k = 1:N
        if(k ~ = n)
            t = conv(t, [1 -exp(p(k)*T)]);
        end
    end
    bz = bz + t(:);
end
bz = bz(:)';
zgain = T;
```

For the corrected impulse invariant method, the appropriate modifications can be easily made to the above. Let us apply some test cases. The manual algebraic approach should of course agree with the MATLAB code as given above. For the first case, consider a very simple first-order filter:

$$G(s) = \frac{1}{s+1}. \tag{9.47}$$

Using the standard method, this yields directly

$$G(z) = \frac{Tz}{z - e^{-T}}. \tag{9.48}$$

Since by inspection $p = 1$, we know from linear systems theory that the time constant is 1 second. Thus, for simplicity, a much faster sample rate, say, $T = 0.1$ for the purpose of the example, is chosen to yield

$$G(z) = \frac{0.1z}{z - 0.9048}. \tag{9.49}$$

Using the corrected approach, the lookup of the first-order section becomes

$$G(z) = \frac{T}{2}\left(\frac{z + e^{-T}}{z - e^{-T}}\right).$$

With $T = 0.1$, we have

$$G(z) = 0.05\left(\frac{z + 0.9048}{z - 0.9048}\right).$$

As a second example, consider a cascade of two first-order sections given by the partial fraction expansion

$$\frac{1}{(s+1)(s+2)}=\frac{1}{(s+1)}+\frac{-1}{(s+2)}. \tag{9.50}$$

Using the value of p for each first-order section, the direct lookup gives

$$G(z)=\frac{Tz\left(e^{-T}-e^{-2T}\right)}{z^2-z\left(e^{-2T}+e^{-T}\right)+e^{-3T}}.$$

With $T=0.1$, we have

$$G(z)=0.1\frac{0.0861z}{z^2-1.7236z+0.7408}.$$

The corrected method yields the same result in this case.

9.6.4 Bilinear Transform Method

An alternative to the impulse invariant method (IIM) as discussed above is the bilinear transform (usually called BLT). This approach aims to obtain a direct discrete-time approximation using reasoning similar to that outlined earlier for the approximation of continuous systems.

Ultimately, what we want is some function of z which we can use as a substitute for s in the continuous transforms. The resulting transfer function in z will always be an approximation, but certain characteristics are desirable.

To motivate the development of such a substitution, consider a differentiator,

$$G(s)=s.$$

For an input $X(s)$ and an output $Y(s)$, this represents the differential equation

$$y(t)=\frac{dx(t)}{dt}.$$

An approximation to this is

$$y(nT)\approx\frac{x(nT)-x(nT-T)}{T};$$

so, the discrete-time approximation is

$$Y(z)=\frac{1-z^{-1}}{T}X(z).$$

The transformation from continuous to discrete is given by

$$s\rightarrow\frac{1-z^{-1}}{T}.$$

Using this method of "backward differences," the imaginary (frequency) axis in the s plane is *not* mapped into the unit circle in the z plane. So, although it may be a reasonable approximation in the time domain, the frequency mapping is not desirable. Furthermore, we would like to ensure that a stable system in s will be a stable system in z when converted.

Another mapping, which is almost as simple, does in fact map the imaginary (frequency) axis in the s plane onto the unit circle in the z plane. It is called the bilinear transform (BLT) and is defined as the substitution

$$s \rightarrow \frac{2}{T}\left(\frac{z-1}{z+1}\right). \tag{9.51}$$

Why does the mapping have this form? To answer that question, consider an integrator,

$$G(s) = \frac{1}{s}.$$

Using the BLT, this becomes, after some algebraic manipulation,

$$\begin{aligned} G(z) &= G(s)\big|_{s=\frac{2}{T}\left(\frac{z-1}{z+1}\right)} \\ &= \frac{T}{2}\left(\frac{1+z^{-1}}{1-z^{-1}}\right) \\ \therefore Y(z) - z^{-1}Y(z) &= \frac{T}{2}X(z)\left(1+z^{-1}\right) \\ \therefore y(n) &= \frac{T}{2}(x(n)+x(n-1)) + y(n-1). \end{aligned} \tag{9.52}$$

This is a trapezoidal approximation to the integrator—in words,

$$\begin{array}{ccccc} y(n) & = & y(n-1) & + & \left(\dfrac{x(n)+x(n-1)}{2}\right)T \\ \text{New area} & = & \text{area so far} & + & \text{increment of area.} \end{array}$$

The increment of area is the trapezoidal approximation—half sum of parallel sides times perpendicular distance. This is illustrated in Figure 9.16. The integrator above results in a transfer function:

$$G(z) = \frac{T}{2}\left(\frac{z+1}{z-1}\right);$$

that is, a pole at $z = 1$ and a zero at $z = -1$. The zero means that the response at $\omega = \pi$ is exactly zero.

In effect, the range of continuous frequencies Ω from 0 to ∞ is mapped into discrete frequencies ω from 0 to π.

Furthermore, poles in the stable section of the s plane (the left half of the plane) are mapped into the inside of the unit circle in the z plane (which, as always, is the stable region). This is a considerable advantage because a stable continuous system will map to a stable discrete system.

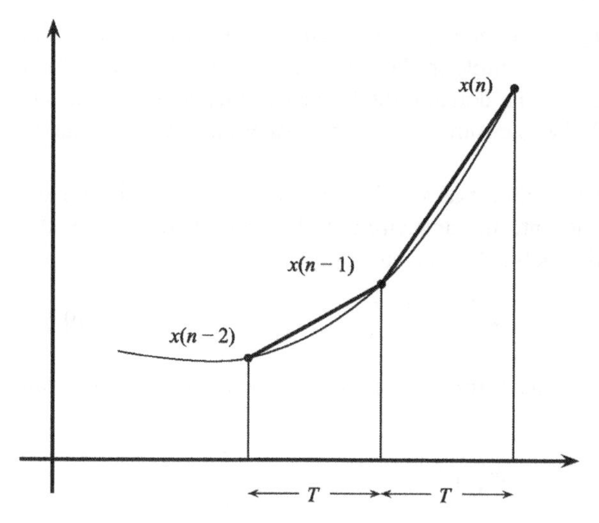

FIGURE 9.16 The bilinear transform of an integrator. In effect, the discrete operation implements the trapezoidal rule for numerical integration.

To demonstrate the use of the BLT, consider a first-order lag:

$$G(s) = \frac{1}{s+1}.$$

Using the BLT, this becomes

$$G(z) = \frac{T}{T+2}\left(\frac{z+1}{z+\frac{T-2}{T+2}}\right).$$

The frequency mapping may be analyzed as follows. The s domain has $s = \jmath\Omega$, and the z domain has $z = e^{\jmath\omega}$. Using the BLT

$$s \rightarrow \frac{2}{T}\left(\frac{z-1}{z+1}\right), \tag{9.53}$$

the system in question becomes

$$\jmath\Omega = \frac{2}{T}\left(\frac{e^{\jmath\omega}-1}{e^{\jmath\omega}+1}\right).$$

This may be simplified to yield

$$\Omega = \frac{2}{T}\tan\frac{\omega}{2}. \tag{9.54}$$

This equation relates the discrete and analog frequencies. For a small ω, this reduces to the familiar $\omega = \Omega T$, as shown in Figure 9.17; that is, a linear mapping from continuous frequencies (radian per second) to discrete frequencies (radians per sample).

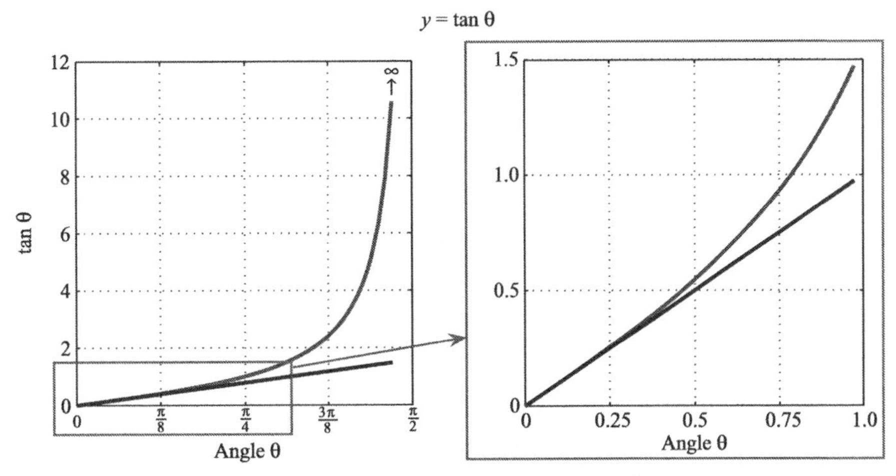

FIGURE 9.17 The function $y = \tan\theta$, showing linearity for $\theta \lessapprox 1$.

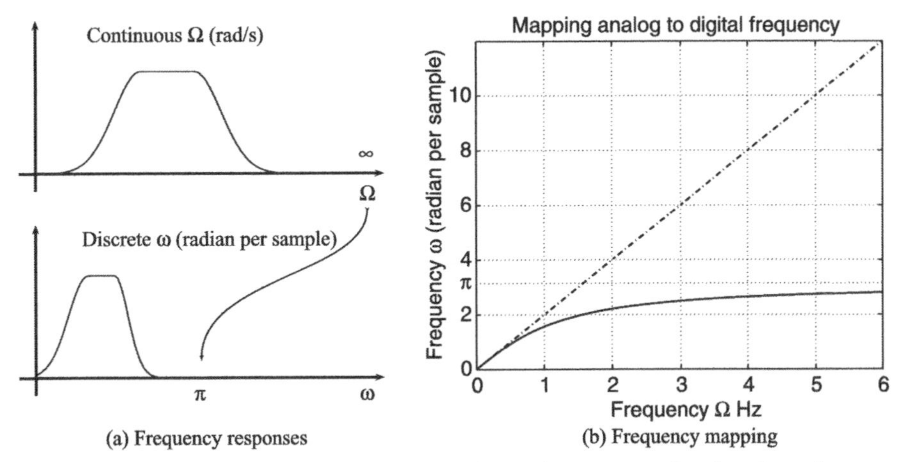

(a) Frequency responses (b) Frequency mapping

FIGURE 9.18 Illustrating the mapping from the continuous or analog domain to the discrete-time domain.

The frequency mapping from Ω to ω is illustrated in Figure 9.18, resulting in a "warping" of the frequency axis. The upper plot in Figure 9.18a shows the frequency response in terms of "real" frequency (radians per second), with the corresponding lower plot showing the response in the digital domain (radians per sample). Clearly, this mapping results in a "squashing" of the desired frequency response into a much smaller range. The mapping process is shown in Figure 9.18b. For low frequencies (relative to the sample rate), the mapping from continuous to discrete is approximately linear, whereas as the analog frequency becomes higher, it is mapped into π radians per sample in the discrete domain.

Another way to look at this is to consider a system described by a simple first-order lag,

$$G(s) = \frac{1}{s+a}.$$

This system has gain $1/a$ and cutoff frequency $\Omega_c = a\,\text{rad/s}$. Using the BLT, the transformed system becomes

$$\begin{aligned}
G(z) &= \frac{1}{\dfrac{2}{T}\left(\dfrac{z-1}{z+1}\right)+a} \\
&= \frac{(z+1)}{\dfrac{2}{T}(z-1)+a(z-1)} \\
&= \frac{z+1}{z\left(\dfrac{2}{T}+a\right)+\left(a-\dfrac{2}{T}\right)} \\
&= \frac{z+1}{\left(a+\dfrac{2}{T}\right)\left(z+\dfrac{a-\dfrac{2}{T}}{a+\dfrac{2}{T}}\right)}.
\end{aligned} \tag{9.55}$$

Now going back and putting $z = e^{j\omega}$,

$$\begin{aligned}
G(e^{j\omega}) &= \frac{e^{j\omega}+1}{e^{j\omega}\left(a+\dfrac{2}{T}\right)\left(a-\dfrac{2}{T}\right)} \\
&= \frac{e^{j\omega}+1}{e^{j\omega}\left(a+\dfrac{2}{T}\right)\left(a-\dfrac{2}{T}\right)} \\
&= \frac{e^{j\frac{\omega}{2}}\left(e^{+j\frac{\omega}{2}}+e^{-j\frac{\omega}{2}}\right)}{e^{j\frac{\omega}{2}}\left(\left(a+\dfrac{2}{T}\right)e^{+j\frac{\omega}{2}}+\left(a-\dfrac{2}{T}\right)e^{-j\frac{\omega}{2}}\right)} \\
&= \frac{2j\cos\dfrac{\omega}{2}}{2a\cos\dfrac{\omega}{2}+\dfrac{2}{T}2j\sin\dfrac{\omega}{2}} \\
&= \frac{1}{a+j\dfrac{2}{T}\dfrac{\sin\dfrac{\omega}{2}}{\cos\dfrac{\omega}{2}}} \\
&= \frac{1}{a+j\dfrac{2}{T}\tan\dfrac{\omega}{2}}.
\end{aligned} \tag{9.56}$$

So the mapping of the cutoff frequency Ω_c into the digital domain may be written as $\Omega_c \to (2/T)\tan(\omega/2)$. The conclusion is that we need to compensate for the nonlinear mapping in the frequency domain. We illustrate how this is done by considering a second-order system,

$$G(s) = \frac{1}{s^2 + 2\zeta s + 1}, \tag{9.57}$$

where ζ is a constant of the system (called the damping ratio).

As we have done previously, we design for a system with a normalized cutoff frequency of 1 rad/s and then scale the filter to the required cutoff frequency. This is easily accomplished by replacing s with $\frac{s}{\Omega_c}$. In the present case, this gives

$$G(s) = \frac{1}{\left(\frac{s}{\Omega_c}\right)^2 + \frac{2\zeta s}{\Omega_c} + 1}.$$

Transforming to a digital filter using the bilinear transform, we have

$$\begin{aligned} G(z) &= \frac{\Omega_c^2}{\left(\frac{2}{T}\right)^2\left(\frac{z-1}{z+1}\right)^2 + 2\zeta\Omega_c \frac{2}{T}\frac{z-1}{z+1} + \Omega_c^2} \\ &= \frac{\Omega_c^2 (z+1)^2}{\left(\frac{2}{T}\right)^2 (z-1)^2 + 2\zeta\Omega_c \frac{2}{T}(z-1)(z+1) + \Omega_c^2 (z+1)^2}. \end{aligned}$$

Letting $\alpha = 2/T$, we have

$$\begin{aligned} G(z) &= \frac{\Omega_c^2 (z^2 + 2z + 1)}{\alpha^2 (z^2 - 2z + 1) + 2\zeta\Omega_c\alpha(z^2 - 1) + \Omega_c^2 (z^2 + 2z + 1)} \\ &= \frac{\Omega_c^2 (z^2 + 2z + 1)}{z^2(\alpha^2 + 2\zeta\alpha\Omega_c + \Omega_c^2) + z(-2\alpha^2 + 2\Omega_c^2) + (\alpha^2 - 2\zeta\alpha\Omega_c + \Omega_c^2)}. \end{aligned}$$

Letting $\gamma = (\alpha^2 + 2\zeta\alpha\Omega_c + \Omega_c^2)$, to simplify,

$$G(z) = \frac{\Omega_c^2}{\gamma}\left(\frac{z^2 + 2z + 1}{z^2 + z\frac{2(\Omega_c^2 - \alpha^2)}{\gamma} + \frac{(\alpha^2 - 2\zeta\alpha\Omega_c + \Omega_c^2)}{\gamma}}\right). \tag{9.58}$$

This is the resulting discrete-time transfer function corresponding to the analog prototype. While not incorrect, it will suffer from distortion or warping, as was illustrated in Figure 9.18. The warping distortion gets worse the closer the frequency is to the Nyquist frequency. What are we to do about this? One possibility is to simply employ a higher sampling rate so that the effect of the warping due to aliasing is diminished. But it would be good to try to compensate for the nonlinear mapping of the BLT so that at least we can be sure the analog frequency response matches the digital frequency response at some critical frequency of interest (usually the cutoff frequency).

How do we do this? It depends on two pieces of information. First, we can scale a filter to any frequency Ω_c by dividing s by Ω_c (this is covered in detail in Section 9.7). Second, we want the responses to match at this "critical" frequency, and we know the mathematical mapping of this critical frequency when we go from analog to discrete—it is in fact $(2/T)\tan(\omega_c/2)$. So, we can combine the scaling and compensation (called "prewarping") using

$$\frac{s}{\Omega_c}=\frac{\frac{2}{T}\left(\frac{z-1}{z+1}\right)}{\frac{2}{T}\tan\frac{\omega_c}{2}}. \tag{9.59}$$

Clearly, $T/2$ cancels. To simplify the algebra, let $\beta = 1/(\tan\omega_c/2)$. Then, our modified transfer function becomes

$$\begin{aligned}
G(z) &= \frac{1}{\beta^2\left(\frac{z-1}{z+1}\right)^2+2\zeta\beta\left(\frac{z-1}{z+1}\right)+1} \\
&= \frac{(z+1)^2}{\beta^2(z-1)^2+2\zeta\beta(z-1)(z+1)+(z+1)^2} \\
&= \frac{(z+1)^2}{\beta^2(z^2-2z+1)+2\zeta\beta(z^2-1)+(z^2+2z+1)} \\
&= \frac{z^2+2z+1}{z^2(\beta^2+2\zeta\beta+1)+z(-2\beta^2+2)+(\beta^2-2\zeta\beta+1)} \\
&= \frac{1}{(\beta^2+2\zeta\beta+1)}\left(\frac{z^2+2z+1}{z^2+z\left(\frac{2(1-\beta^2)}{\beta^2+2\zeta\beta+1}\right)+\left(\frac{\beta^2-2\zeta\beta+1}{\beta^2+2\zeta\beta+1}\right)}\right).
\end{aligned} \tag{9.60}$$

Let $\gamma = (\beta^2 + 2\zeta\beta + 1)$,

$$G(z)=\frac{1}{\gamma}\left(\frac{z^2+2z+1}{z^2+z\left(\frac{2(1-\beta^2)}{\gamma}\right)+\left(\frac{\beta^2-2\zeta\beta+1}{\gamma}\right)}\right). \tag{9.61}$$

To illustrate the difference the prewarping may produce in practice, consider a first-order filter with a required cutoff of 2.2 kHz. Figure 9.19 shows the prototype analog

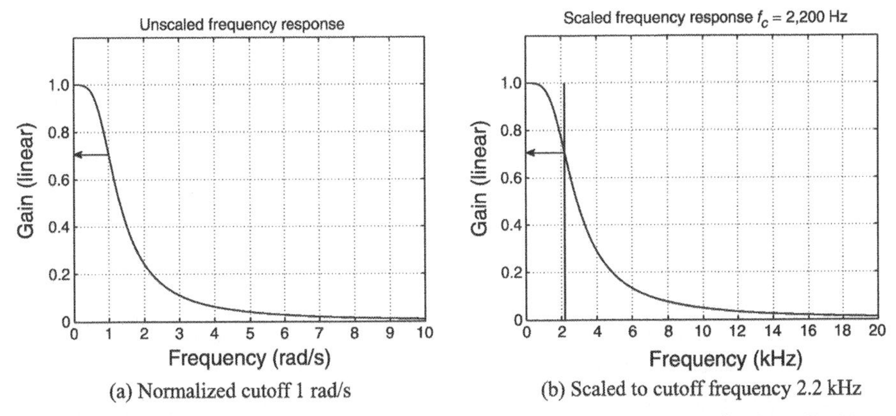

(a) Normalized cutoff 1 rad/s (b) Scaled to cutoff frequency 2.2 kHz

FIGURE 9.19 The bilinear transform (BLT) filter for a normalized cutoff of 1 rad/s (a). This is also called the prototype filter. The response scaled for the desired cutoff of 2,200 Hz is shown in (b).

filter, with a cutoff of 1 rad/s, which is then scaled for the required cutoff frequency (note that the gain at the cutoff is $1/\sqrt{2}$).

Next, we must convert the design into a digital filter and thus must choose the sample frequency. This is where the effect of the nonlinear mapping of the BLT becomes apparent. In order to choose the sampling rate, the cutoff of 2.2 kHz would correspond to a Nyquist rate of 2×2.2 kHz, so we choose a rate above that. The next highest rate commonly available in practical systems is 8 kHz. So, with $f_s = 8{,}000$ Hz, the Nyquist or folding frequency is $f_s/2 = 4{,}000$ Hz (corresponding to π radians per sample). Thus, the cutoff when normalized to the sample rate becomes $2{,}200/4{,}000 \times \pi = 0.55\pi$.

Figure 9.20 shows the calculated response (solid line) for the direct BLT. It is apparent that, as expected, the magnitude response is reduced because, as explained, the entire frequency range is "compressed," resulting in the analog gain curve being mapped into a finite discrete frequency range. The inevitable consequence is that the gain is reduced at all points, and at the cutoff frequency in this example, it is reduced to approximately 0.5.

However, if we prewarp the critical frequency from the analog to digital domain and incorporate that into the BLT, the result is as shown by the dashed line in Figure 9.20. Clearly, the cutoff frequency gain is correct, but this necessarily results in a slight modification of the overall frequency response.

Figure 9.20 also shows the result of using a higher sampling rate. We have chosen 20 kHz, and show both the prewarped and non-prewarped frequency response. The non-prewarped or direct BLT is very close to the desired gain at the critical cutoff frequency, and in this case, the difference between the two is insignificant. However, this has come at the expense of more than doubling the sampling rate.

9.6.5 Automating the Bilinear Transform

The algebra required for the bilinear transform is relatively straightforward, though tedious. We simply need to substitute

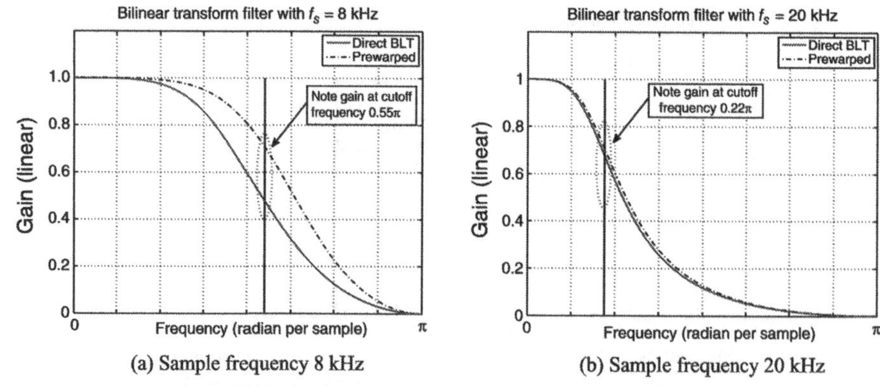

(a) Sample frequency 8 kHz

(b) Sample frequency 20 kHz

FIGURE 9.20 Comparison of the bilinear transform, with and without prewarping (a). Note that for the prewarped version, the gain at the cutoff frequency equals the desired gain, whereas the non-prewarped version has a somewhat lower gain at the cutoff frequency. The bilinear transformation (BLT) (and prewarped BLT), with a higher sample rate (20 kHz vs. 8 kHz, filter cutoff 2.2 kHz) is shown in (b). The higher sample rate clearly moves the response of the filter without prewarping much closer to the desired response.

$$s \to \left(\frac{2}{T}\right)\left(\frac{z-1}{z+1}\right)$$

in the continuous-domain transfer function. For first-order systems, this is simple; for second-order systems, it becomes more involved. For higher-order systems, the chances of mistakes in the algebraic manipulation become much higher. Thus, it is desirable to develop an automated procedure to handle the bilinear transform for any arbitrary filter transfer function $G(s)$.

We can do this by noting that any transfer function is the ratio of two polynomials, with a constant gain. For example, the transfer function

$$G(s) = \frac{4}{s^2 + 2s + 3}$$

could be decomposed into a numerator polynomial $G_n(s) = 1s^0$, a denominator polynomial $G_d(s) = 1s^2 + 2s^1 + 3s^0$ and gain $K = 4$. In the general case, we could write

$$G(s) = K\frac{\beta_0 s^M + \beta_1 s^{M-1} + \cdots + \beta_M s^0}{\alpha_0 s^N + \alpha_1 s^{N-1} + \cdots + \alpha_N s^0}. \tag{9.62}$$

The order of the numerator is M and that of the denominator is N. Both $G_n(s)$ and $G_d(s)$ could be factored into corresponding roots, which may be either real or complex. Thus, we have the product of roots

$$G(s) = K\frac{\Pi_M (s - q_m)}{\Pi_N (s - p_n)}, \tag{9.63}$$

where the q_m are the (possibly complex) zeros, and p_n are the (possibly complex) poles.

Consider first the case of the poles. We have terms such as $1/(s-p)$ with a pole at $s=p$, which are replaced in the bilinear transform with

$$\frac{1}{\left(\frac{2}{T}\right)\left(\frac{z-1}{z+1}\right)-p}.$$

We can rearrange this to get the pole/zero form as follows:

$$\begin{aligned}
\frac{1}{s-p} &= \frac{1}{\left(\frac{2}{T}\right)\left(\frac{z-1}{z+1}\right)-p} \\
&= \frac{(z+1)}{\left(\frac{2}{T}\right)(z-1)-p(z+1)} \\
&= \frac{(z+1)}{z\left(\frac{2}{T}-p\right)-\left(p+\frac{2}{T}\right)} \\
&= \frac{(z+1)}{\left(\frac{2}{T}-p\right)\left(z-\frac{\frac{2}{T}-p}{\frac{2}{T}-p}\right)} \\
&= \frac{(z+1)}{\left(\frac{2}{T}-p\right)\left(z+\frac{p+\frac{2}{T}}{p-\frac{2}{T}}\right)}.
\end{aligned} \tag{9.64}$$

Thus, the resulting z domain transfer function will have the following factors, and every pole in the s domain will produce these three terms in the z domain:

$$\begin{aligned}
&\text{Numerator} && (z+1) \\
&\text{Denominator} && \left(z+\frac{p+\frac{2}{T}}{p-\frac{2}{T}}\right) \\
&\text{Gain} && \frac{1}{\left(\frac{2}{T}-p\right)}.
\end{aligned}$$

Consider now the case of the zeros. We have terms such as $(s-q)$ with a zero at $s=q$, which are replaced in the bilinear transform with

$$\left(\frac{2}{T}\frac{z-1}{z+1}-q\right).$$

We can rearrange this into the pole/zero form as follows:

$$\begin{aligned}
s-q &= \frac{\left(\frac{2}{T}\right)\left(\frac{z-1}{z+1}\right)-q}{1} \\
&= \frac{\left(\frac{2}{T}\right)(z-1)-q(z+1)}{(z+1)} \\
&= \frac{z\left(\frac{2}{T}-q\right)-\left(q+\frac{2}{T}\right)}{(z+1)} \\
&= \frac{\left(\frac{2}{T}-q\right)\left(z+\frac{q+\frac{2}{T}}{q-\frac{2}{T}}\right)}{(z+1)}.
\end{aligned} \tag{9.65}$$

Thus, the resulting z domain transfer function will have the following factors:

$$\begin{array}{ll}
\text{Numerator} & \left(z+\frac{q+\frac{2}{T}}{q-\frac{2}{T}}\right) \\
\text{Denominator} & (z+1) \\
\text{Gain} & \frac{1}{\left(\frac{2}{T}-q\right)}.
\end{array}$$

To convert the above to MATLAB code, we proceed as follows. First, we must find the zeros, which are the roots of the numerator coefficients in vector bs, the number of zeros, and the corresponding poles. To determine the poles and zeroes, we recognize as per the previous dealings with polynomials that in fact, polynomial multiplication is the same as convolution. Thus, we need to convolve the existing coefficients with the newly determined coefficients.

The multiplication by

$$\left(z+\left(q+\frac{2}{T}\right)\Big/\left(q-\frac{2}{T}\right)\right)$$

becomes a convolution with vector

$$\mathbf{v}=\left[1 \quad \left(q+\frac{2}{T}\right)\Big/\left(q-\frac{2}{T}\right)\right]^T.$$

These are then the coefficients of z^1 and z^0, respectively. This could be implemented for all poles as follows:

```
bsroots = roots(bs);
M = length(bsroots);
asroots = roots(as);
N = length(asroots);
zgain = 1;
bz = [1];
az = [1];
for n = 1:N
    p = asroots(n);
    zgain = zgain/(2/T - p);
    eta = (p + 2/T)/(p - 2/T);
    bz = conv(bz, [1 1]);
    az = conv(az, [1 eta]);
end
```

A similar construction is used for the zeros. Note the convolution with the vector $\mathbf{v}^T = (1\ 1)$, which arises from the term $(z + 1)$. If we repeat this for the zeros, we end up with a problem—we have multiple poles and zeros at $z = -1$, that is, the factor $(z + 1)$ on both the top and bottom lines. Mathematically, this is not incorrect; it is simply that our code as developed thus far cannot "cancel" these terms as we would algebraically:

$$G(z) = \frac{(z - q_M)(z - q_{M-1})\cdots(z - q_0)(z + 1)\overbrace{(z + 1)\cdots}^{\text{terms cancel}}}{(z - p_N)(z - p_{N-1})\cdots(z - p_0)(z + 1)(z + 1)\cdots}. \tag{9.66}$$

The solution to this problem is to recognize that there will be exactly $(N - M)$ factors of $(z + 1)$ in the numerator, assuming $M < N$ (i.e., in the s domain, more poles than zeros). Thus, for the zeros, we remove the convolution with $\mathbf{v}^T$ to leave

```
for m = 1:M
    q = bsroots(m);
    zgain = zgain*(2/T - q);
    gamm = -(q + 2/T)/(2/T - q);
    bz = conv(bz, [1 gamm]);
end
```

Similarly, for the poles, we could use

```
for n = 1:N
    p = asroots(n);
    zgain = zgain/(2/T - p);
    eta = (p + 2/T)/(p - 2/T);
    az = conv(az, [1 eta]);
end
```

and finally factor in $(z+1)^{N-M}$ as follows

```
nz = N - M;
for n = 1:nz
    bz = conv(bz, [1 1]);
end
```

We shall now develop algebraically some test cases for our bilinear transform method. The first is an obvious and simple first-order system described by

$$G(s)=\frac{1}{s+1},$$

which becomes

$$G(s)=\frac{(z+1)}{\left(\frac{2}{T}+1\right)\left(z+\left[\frac{1-\frac{2}{T}}{1+\frac{2}{T}}\right]\right)}.$$

This may be tested against the algorithmic development as described. To extend the testing to a higher-order system, consider a simple second-order system described by

$$G(s)=\frac{1}{s^2+s+1},$$

which becomes

$$G(s)=\frac{(z^2+2z+1)}{\left(\left(\frac{2}{T}\right)^2+\left(\frac{2}{T}\right)+1\right)z^2+\left(2-2\left(\frac{2}{T}\right)^2\right)z+\left(\left(\frac{2}{T}\right)^2-\left(\frac{2}{T}\right)+1\right)}.$$

Finally, we extend to a second-order system with a zero. Consider

$$G(s)=\frac{s+1}{s^2+s+1},$$

which, when transformed, becomes

$$G(s)=\frac{\left(\frac{2}{T}+1\right)z^2+2z+\left(1-\frac{2}{T}\right)}{\left(\left(\frac{2}{T}\right)^2+\left(\frac{2}{T}\right)+1\right)z^2+\left(2-2\left(\frac{2}{T}\right)^2\right)z+\left(\left(\frac{2}{T}\right)^2-\left(\frac{2}{T}\right)+1\right)}.$$

Each of these algebraic test cases can be used to check the numerical/algorithmic solution approach as described.

9.7 SCALING AND TRANSFORMATION OF CONTINUOUS FILTERS

A natural question arising at this point is how to design for any cutoff frequency and how to design different responses. The former question is relatively easy to answer; the latter is a little more difficult but follows along similar lines. In this section, we develop a method to cater for any arbitrary filter cutoff frequency, followed by an investigation of methods to convert the low-pass response "prototype" filter into high-pass, bandpass, or bandstop filter response shapes.

Scaling a low-pass filter to a cutoff frequency Ω_c is accomplished relatively easily, by the substitution of

$$s \rightarrow \frac{s}{\Omega_c}. \tag{9.67}$$

This seems intuitively reasonable since to determine the frequency response, we substitute $s = j\Omega$, and the resulting division by Ω_c in the scaling operation effectively gives a ratio of Ω/Ω_c. When $\Omega = \Omega_c$, the ratio is unity, and thus we effectively have a scaled or normalized response.

Along similar lines, transforming a low-pass filter to a high-pass filter requires the substitution

$$s \rightarrow \frac{1}{s}. \tag{9.68}$$

Combining these two results, it is not difficult to see that converting a normalized low-pass filter (i.e., one with cutoff 1 rad/s) into a high-pass filter with cutoff frequency Ω_c requires the substitution

$$s \rightarrow \frac{\Omega_c}{s}. \tag{9.69}$$

Converting a low-pass filter to a bandpass filter is somewhat less intuitive. We need to know the range of frequencies to pass, and these are specified by the passband frequency range Ω_l to Ω_u. To develop the method, we define the filter passband (or bandwidth) Ω_b as the difference between upper and lower frequencies:

$$\Omega_b = \Omega_u - \Omega_l. \tag{9.70}$$

We then need to define a center frequency Ω_o, which is calculated as the *geometric* mean of the lower and upper cutoffs:

$$\Omega_o^2 = \Omega_l \Omega_u. \tag{9.71}$$

Finally, the substitution into the original low-pass filter with normalized cutoff becomes

$$s \rightarrow \frac{1}{\Omega_b} \frac{s^2 + \Omega_o^2}{s} \tag{9.72}$$

with Ω_b and Ω_o as above. Note that this increases the order of the filter: For every s, we have an s^2 term after substitution. This aspect may be understood by considering that a bandpass filter is effectively a high-pass filter cascaded with a low-pass filter.

To effect a bandstop response, the related transformation

$$s \to \Omega_b \frac{s}{s^2 + \Omega_o^2} \tag{9.73}$$

is applied.

The following sections elaborate on each of these methods in turn by giving an algebraic solution and then a corresponding algorithmic solution.

9.7.1 Scaling the Filter Response

To investigate the scaling of the normalized low-pass response into a low-pass response with an arbitrary cutoff, consider the transfer function

$$G(s) = \frac{6s+3}{s^2+2s+4}.$$

The order of the numerator is $M = 1$ and that of the denominator is $N = 2$. Normally, this would be a filter with a cutoff of 1 rad/s, for the Butterworth and Chebyshev designs presented earlier. We scale this low-pass filter (keeping a low-pass filter response) to a cutoff frequency, Ω_c, using

$$s \to \frac{s}{\Omega_c}$$

to obtain

$$\begin{aligned} G(s) &= \frac{6\left(\frac{s}{\Omega_c}\right)+3}{\left(\frac{s}{\Omega_c}\right)^2+2\left(\frac{s}{\Omega_c}\right)+4} \\ &= \frac{\Omega_c^2}{\Omega_c^2} \cdot \frac{6\left(\frac{s}{\Omega_c}\right)+3}{\left(\frac{s}{\Omega_c}\right)^2+2\left(\frac{s}{\Omega_c}\right)+4} \\ &= \frac{6\Omega_c s+3\Omega_c^2}{s^2+2\Omega_c s+4\Omega_c^2}. \end{aligned}$$

We can generalize this as illustrated in Figure 9.21. Start with the low-pass function

$$G(s) = \frac{b_0 s^M + b_1 s^{M-1} + \cdots + b_M s^0}{a_0 s^N + a_1 s^{N-1} + \cdots + a_N s^0}. \tag{9.74}$$

Then, we scale using $s \to s/\Omega_c$ to give

$$G(s) = \frac{b_0\left(\frac{s}{\Omega_c}\right)^M + b_1\left(\frac{s}{\Omega_c}\right)^{M-1} + \cdots + b_M\left(\frac{s}{\Omega_c}\right)^0}{a_0\left(\frac{s}{\Omega_c}\right)^N + a_1\left(\frac{s}{\Omega_c}\right)^{N-1} + \cdots + a_N\left(\frac{s}{\Omega_c}\right)^0}$$

$$= \frac{\Omega_c^N}{\Omega_c^N} \cdot \frac{b_0\left(\frac{s}{\Omega_c}\right)^M + b_1\left(\frac{s}{\Omega_c}\right)^{M-1} + \cdots + b_M\left(\frac{s}{\Omega_c}\right)^0}{a_0\left(\frac{s}{\Omega_c}\right)^N + a_1\left(\frac{s}{\Omega_c}\right)^{N-1} + \cdots + a_N\left(\frac{s}{\Omega_c}\right)^0} \tag{9.75}$$

$$= \frac{b_0\Omega_c^{N-M}s^M + b_1\Omega_c^{N-M+1}s^{M-1} + \cdots + b_M\Omega_c^N s^0}{a_0\Omega_c^0 s^N + a_1\Omega_c^1 s^{N-1} + \cdots + a_N\Omega_c^N s^0}.$$

The highest power of variable s in the numerator M will remain unchanged. Similarly, the order of the denominator will remain N.

To develop an algorithm for computing the coefficients in the polynomial ratio, we proceed as follows. We need to input the coefficients b_M and a_N, and the scaling factor Ω_c.

The orders of numerator M and denominator N are then initialized as follows. We also normalize by coefficient a_0:

```
% powers of s
M = length(bs) - 1;
N = length(as) - 1;
% coefficients in scaled transfer function
k = as(1);
bss = bs/k;
ass = as/k;
```

If we let p be the power of Ω_c, which starts at $N - M$, we can calculate the numerator using the following loop. Remember that for an order M, the number of coefficients will be $M + 1$.

```
p = N - M;
for m = 1:M + 1
    sc = OmegaC^p;
    bss(m) = bss(m)*sc;
    p = p + 1;
end
```

For the denominator, we can use a similar approach, noting that p starts at zero:

```
p = 0;
for n = 1:N + 1
    sc = OmegaC^p;
    ass(n) = ass(n)*sc;
    p = p + 1;
end
```

The polynomials bss and ass now define the scaled transfer function, with the scaling of the cutoff frequency defined by Ω_c. The low-pass response at 1 rad/s is translated in frequency to Ω_c.

9.7.2 Converting Low-Pass to High-Pass

Suppose we have a transfer function,

$$G(s)=\frac{6s+3}{s^2+2s+4}.$$

The order of the numerator is $M = 1$ and that of the denominator is $N = 2$. Normally, this would be a filter with a cutoff of 1 rad/s, for the Butterworth and Chebyshev designs presented earlier (Fig. 9.21).

We can convert this low-pass filter to a high-pass filter with cutoff frequency Ω_c using the substitution defined previously,

$$s \to \frac{\Omega_c}{s},$$

to obtain

$$\begin{aligned}G(s)&=\frac{6\left(\frac{\Omega_c}{s}\right)+3}{\left(\frac{\Omega_c}{s}\right)^2+2\left(\frac{\Omega_c}{s}\right)+4}\\&=\frac{s^2}{s^2}\cdot\frac{6\left(\frac{\Omega_c}{s}\right)+3}{\left(\frac{\Omega_c}{s}\right)^2+2\left(\frac{\Omega_c}{s}\right)+4}\\&=\frac{6s^1\Omega_c^1+3s^2\Omega_c^0}{1\Omega_c^2s^0+2\Omega_c^1s^1+4\Omega_c^0s^2}.\end{aligned}$$

We can generalize this as illustrated in Figure 9.22. Start with the low-pass function

$$G(s)=\frac{b_0s^M+b_1s^{M-1}+\cdots b_Ms^0}{a_0s^N+a_1s^{N-1}+\cdots a_Ns^0}. \tag{9.76}$$

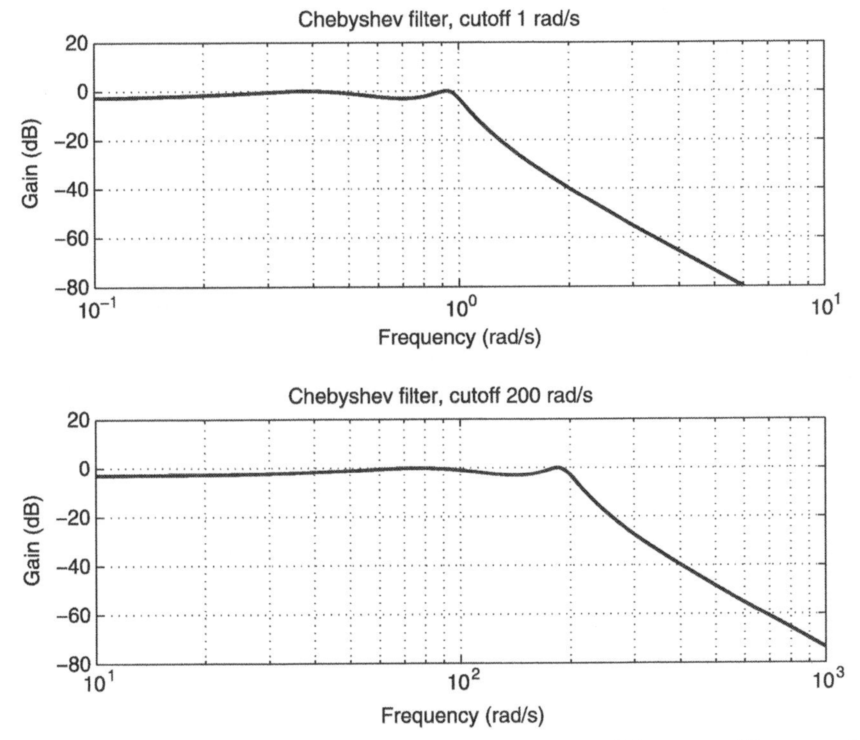

FIGURE 9.21 Scaling a low-pass filter for an arbitrary cutoff frequency. The upper plot shows the normalized response with a cutoff of 1 rad/s. The lower plot shows the response scaled using the method described in the text to a cutoff of 200 rad/s. The response *shape* remains unchanged; only the cutoff frequency is altered.

Then, we scale using $s \to \frac{\Omega_c}{s}$ to give

$$\begin{aligned} G(s) &= \frac{b_0\left(\frac{\Omega_c}{s}\right)^M + b_1\left(\frac{\Omega_c}{s}\right)^{M-1} + \cdots + b_M\left(\frac{\Omega_c}{s}\right)^0}{a_0\left(\frac{\Omega_c}{s}\right)^N + a_1\left(\frac{\Omega_c}{s}\right)^{N-1} + \cdots + a_N\left(\frac{\Omega_c}{s}\right)^0} \\ &= \frac{s^N}{s^N}\cdot\frac{b_0\left(\frac{\Omega_c}{s}\right)^M + b_1\left(\frac{\Omega_c}{s}\right)^{M-1} + \cdots + b_M\left(\frac{\Omega_c}{s}\right)^0}{a_0\left(\frac{\Omega_c}{s}\right)^N + a_1\left(\frac{\Omega_c}{s}\right)^{N-1} + \cdots + a_N\left(\frac{\Omega_c}{s}\right)^0} \\ &= \frac{b_0\Omega_c^M s^{N-M} + b_1\Omega_c^{M-1}s^{N-M+1} + \cdots + b_M\Omega_c^0 s^N}{a_0\Omega_c^N s^0 + a_1\Omega_c^{N-1}s^1 + \cdots + a_N\Omega_c^0 s^N}. \end{aligned} \tag{9.77}$$

The highest power of variable s in the numerator M will remain unchanged. Similarly, the order of the denominator will remain N.

To develop an algorithm for computing the coefficients in the polynomial ratio, we proceed as follows. We need to input the coefficients b_m and a_n, and the scaling factor Ω_c. The order of numerator and denominator are now both N (since $N > M$, we assume more poles than zeros in the low-pass prototype).

```
M = length(bs)-1;
N = length(as)-1;
Ms = N;
Ns = N;
bshp = zeros(Ms + 1, 1);
ashp = zeros(Ns + 1, 1);
```

It is important to remember that polynomial coefficients are stored in decreasing powers from left to right. For example, if we have $6s^2 + 5s + 4$, we would store the coefficients in order of $s^2 s^1 s^0$ as the array c = [6 5 4]. So, c(1) = 6 is the coefficient of s^2, and so forth. This appears when printed in the "natural" order (i.e., 6, 5, 4), but the indexing then becomes a little tricky, since in this example powers 2, 1, 0 are at index 1, 2, 3 into array c.

If we let q be the power of s, which starts at $N - M$, we can calculate the numerator using the following loop (remember that for an order M, the number of coefficients will be $M + 1$).

```
q = N - M;
p = M;
for m = 1:M+1
    bshp(Ms+1-q) = bs(m)*(OmegaC^p);
    q = q + 1;
    p = p - 1;
end
```

For the denominator, we can use a similar approach, noting that q starts at zero since we start from s^0 at the leftmost term.

```
q = 0;
p = N;
for n = 1:N+1
    ashp(Ns+1-q) = as(n)*(OmegaC^p);
    q = q + 1;
    p = p - 1;
end
```

9.7.3 Converting Low-Pass to Band-Pass

In order to convert a low-pass filter into a bandpass filter with passband frequency Ω_l to Ω_u, we use the substitution defined previously:

$$s \to \frac{s^2 + \Omega_o^2}{s\Omega_b}, \tag{9.78}$$

where the center frequency Ω_o is defined by

$$\Omega_o^2 = \Omega_l \Omega_u, \tag{9.79}$$

and the bandwidth Ω_b is defined by

$$\Omega_b = \Omega_u - \Omega_l. \tag{9.80}$$

To give an example, suppose we have a transfer function,

$$G(s) = \frac{6s + 3}{s^2 + 2s + 4}.$$

The order of the numerator is $M = 1$ and that of the denominator is $N = 2$. Normally, this would be a filter with a cutoff of 1 rad/s for the Butterworth and Chebyshev designs presented earlier (Fig. 9.22).

We convert this low-pass filter to a bandpass filter with center frequency Ω_o and bandwidth Ω_b using

$$s \to \frac{s^2 + \Omega_o^2}{s\Omega_b}, \tag{9.81}$$

where Ω_o and Ω_b are defined as above. Note that now, Ω_o is not the "cutoff" frequency but rather the geometric mean of the lower and upper cutoff frequencies. The present example then becomes

$$\begin{aligned} G(s) &= \frac{6\left(\dfrac{s^2 + \Omega_c^2}{s\Omega_b}\right) + 3}{\left(\dfrac{s^2 + \Omega_c^2}{s\Omega_b}\right)^2 + 2\left(\dfrac{s^2 + \Omega_c^2}{s\Omega_b}\right) + 4} \\ &= \left(\frac{s\Omega_b}{s\Omega_b}\right)^2 \cdot \frac{6\left(\dfrac{s^2 + \Omega_c^2}{s\Omega_b}\right) + 3}{\left(\dfrac{s^2 + \Omega_c^2}{s\Omega_b}\right)^2 + 2\left(\dfrac{s^2 + \Omega_c^2}{s\Omega_b}\right) + 4} \\ &= \frac{6\left(s^2 + \Omega_c^2\right)^1 \left(s\Omega_b\right)^1 + 3\left(s^2 + \Omega_c^2\right)^0 \left(s\Omega_b\right)^2}{\left(s^2 + \Omega_c^2\right)^2 \left(s\Omega_b\right)^0 + 2\left(s^2 + \Omega_c^2\right)^1 \left(s\Omega_b\right)^1 + 4\left(s^2 + \Omega_c^2\right)^0 \left(s\Omega_b\right)^2}. \end{aligned} \tag{9.82}$$

We can generalize this approach as follows. Start with the low-pass function

$$G(s) = \frac{b_0 s^M + b_1 s^{M-1} + \cdots b_M s^0}{a_0 s^N + a_1 s^{N-1} + \cdots a_N s^0}. \tag{9.83}$$

Then, scale using

$$s \to \frac{s^2 + \Omega_o^2}{s\Omega_b}. \tag{9.84}$$

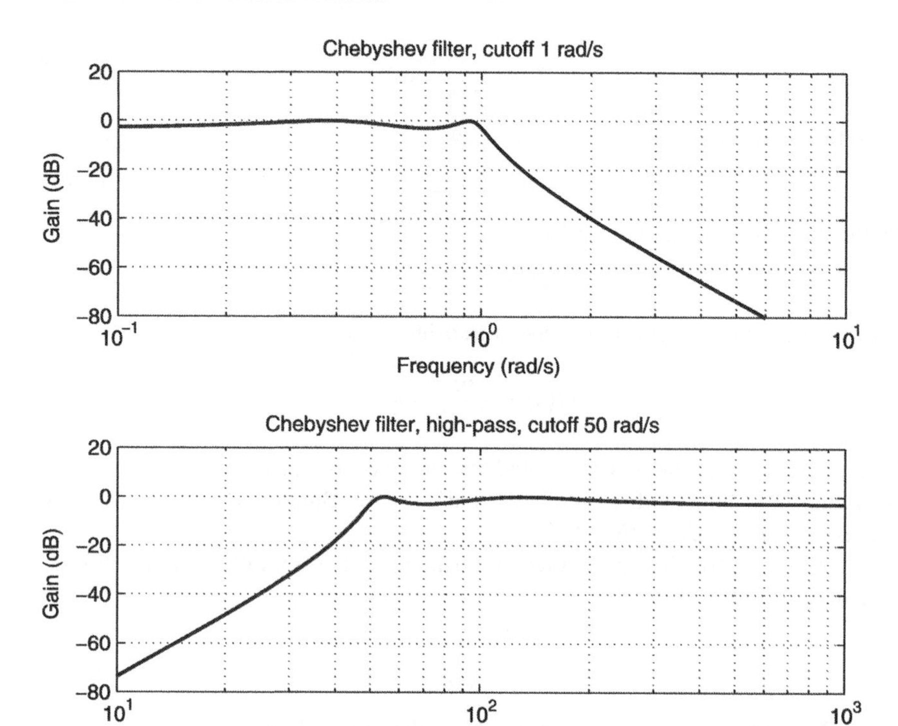

FIGURE 9.22 Converting a low-pass filter into a high-pass filter. The upper plot shows the normalized low-pass response with a cutoff of 1 rad/s. The lower plot shows the response scaled and mapped into a high-pass response using the method described in the text to a cutoff of 50 rad/s. The response shape remains unchanged, though of course "flipped" with respect to frequency.

Then, we have

$$
\begin{aligned}
G(s) &= \frac{b_0\left(\dfrac{s^2+\Omega_c^2}{s\Omega_b}\right)^M + b_1\left(\dfrac{s^2+\Omega_c^2}{s\Omega_b}\right)^{M-1} + \cdots + b_M\left(\dfrac{s^2+\Omega_c^2}{s\Omega_b}\right)^0}{a_0\left(\dfrac{s^2+\Omega_c^2}{s\Omega_b}\right)^N + a_1\left(\dfrac{s^2+\Omega_c^2}{s\Omega_b}\right)^{N-1} + \cdots + a_N\left(\dfrac{s^2+\Omega_c^2}{s\Omega_b}\right)^0} \\
&= \left(\frac{s\Omega_b}{s\Omega_b}\right)^N \cdot \frac{b_0\left(\dfrac{s^2+\Omega_c^2}{s\Omega_b}\right)^M + b_1\left(\dfrac{s^2+\Omega_c^2}{s\Omega_b}\right)^{M-1} + \cdots + b_M\left(\dfrac{s^2+\Omega_c^2}{s\Omega_b}\right)^0}{a_0\left(\dfrac{s^2+\Omega_c^2}{s\Omega_b}\right)^N + a_1\left(\dfrac{s^2+\Omega_c^2}{s\Omega_b}\right)^{N-1} + \cdots + a_N\left(\dfrac{s^2+\Omega_c^2}{s\Omega_b}\right)^0} \\
&= \frac{\begin{array}{c} b_0\left(s^2+\Omega_c^2\right)^M (s\Omega_b)^{N-M} + b_1\left(s^2+\Omega_c^2\right)^{M-1}(s\Omega_b)^{N-M+1} + \cdots + \\ b_M\left(s^2+\Omega_c^2\right)^0 (s\Omega_b)^N \end{array}}{a_0\left(s^2+\Omega_c^2\right)^N (s\Omega_b)^0 + a_1\left(s^2+\Omega_c^2\right)^{N-1}(s\Omega_b)^1 + \cdots + a_N\left(s^2+\Omega_c^2\right)^0 (s\Omega_b)^N}.
\end{aligned}
\tag{9.85}
$$

The order of the denominator will be $2N$, whereas the numerator will be $2M + (N - M)$. To see this, note that we will have a $(s^2)^N$ term in the denominator, and the numerator will have a $(s^2)^M$ term multiplied by s^{N-M}. Thus, adding the indices, we have the order $2M + (N - M) = M + N$. To develop an algorithm for computing the coefficients in the polynomial ratio, we proceed as follows. We need to input the coefficients b_m and a_n, and the bandwidth defined by (Ω_u, Ω_l). According to the reasoning above, we have

```
M = length(bs)-1;
N = length(as)-1;
Ms = M*2 + (N-M);
Ns = N*2;
OmegaB = OmegaU - OmegaL;
OmegaC = sqrt(OmegaL*OmegaU);
bsbp = zeros(Ms+1, 1);
asbp = zeros(Ns+1, 1);
```

The added complication we have now is that we have polynomials multiplied out. In the numerical example, we had terms such as $(s^2 + \Omega_c^2)(s\Omega_b)$. The MATLAB operator **conv()** may be used to multiply out a polynomial. If we express the first factor as $1s^2 + 0s^1 + \Omega_c^2 s^0$ and the second as $\Omega_b s^1 + 0s^0$, we would calculate the product as the convolution of the vectors $[1\, 0\, \Omega_c^2]$ and $[\Omega_b, 0]$.

Where we have a factor such as $(s^2 + \Omega_c^2)^N$, we perform the convolution N times. So, we let p be the power of $(s^2 + \Omega_c^2)$ and q be the power of $s\Omega_b$, and the denominator may then be calculated as follows.

```
p = N;  % power of (s^2+OmegaC^2)
q = 0;  % power of s OmegaB
for n = 1:N+1
    % calc (s^2 + OmegaC^2)^p;
    tv = [1 0 OmegaC^2];
    rp = [1]*as(n);
    for k = 1:p
        rp = conv(rp, tv);
    end
    % calc(s Omega)^q;
    tv = [OmegaB 0];
    rq = [1];
    for k = 1:q
        rq = conv(rq, tv);
    end
```

```
    % (s^2 + OmegaC^2)^p (s OmegaB)^q
    r = conv(rp, rq);
    is = Ns+1 - length(r)+1;
    ie = Ns+1;
    asbp(is:ie) = asbp(is:ie) + r(:);
    p = p - 1;
    q = q + 1;
end
```

We have used `tv` as a temporary vector in the above. Note also the alignment of the resulting polynomials with the final output coefficient vector, performed by the start/end indexing `ie : ie`. For the numerator, we use a similar approach since we start from s^0 at the leftmost term.

```
p = M;        % power of ( s^2+OmegaC^2)
q = N - M;  % power of s OmegaB
for m = 1:M+1
    % calculate (s^2 + OmegaC^2)^p;
    tv = [1 0 OmegaC^2];
    rp = [1]*bs(m);
    for k = 1:p
        rp = conv(rp, tv);
    end
    % calculate the product (s Omega)^q;
    tv = [OmegaB 0];
    rq = [1];
    for k = 1:q
        rq = conv(rq, tv);
    end
    % (s^2+OmegaC^2)^p (s OmegaB)^q
    r = conv(rp, rq);
    is = Ms+1 - length(r)+1;
    ie = Ms+1;
    bsbp(is:ie) = bsbp(is:ie) + r(:);
    p = p - 1;
    q = q + 1;
end
```

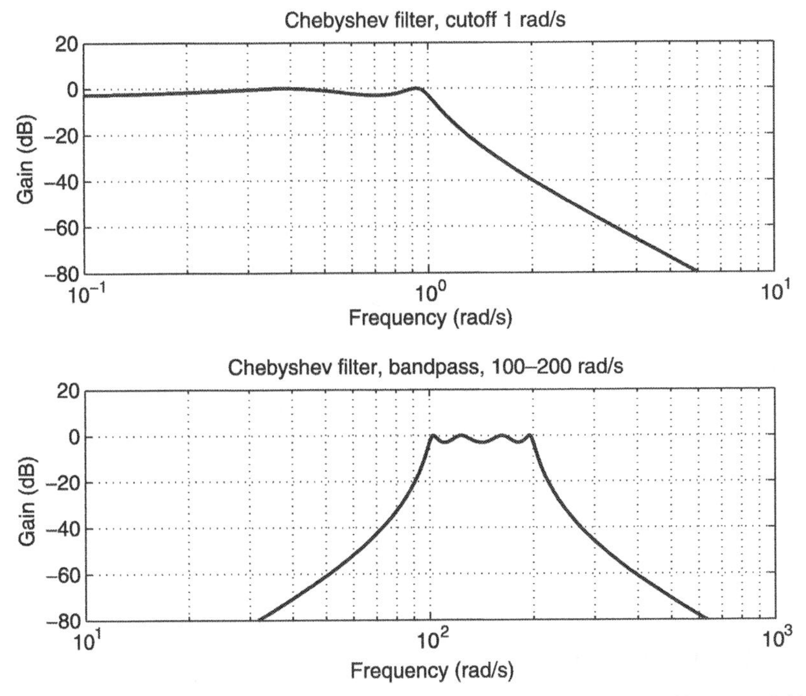

FIGURE 9.23 Converting a low-pass filter into a bandpass filter. This type of filter is effectively a high-pass and a low-pass filter in cascade, with the high-pass cutoff being *lower* than the low-pass cutoff. The example shows a cutoff range of 100–200 rad/s. It is clear that the low-pass response shape is mapped into the lower and upper components of the bandpass response.

Figure 9.23 shows the analog low-pass prototype, and the result of the conversion to a bandpass filter. Once this filter response is available, one of the methods discussed earlier (BLT or impulse invariant design) may be used to convert the filter into a digital or discrete-time design.

9.8 SUMMARY OF DIGITAL FILTER DESIGN VIA ANALOG APPROXIMATION

There are many steps to be undertaken in designing a suitable filter for a given application. This almost always requires a computer-based approach, combined with a sound understanding of the underlying theoretical and mathematical principles. For anything other than very low-order filters, software is utilized to speed up the design process and to reduce the likelihood of errors. Usually, filter design involves an iterative approach—a design is created, tested theoretically, and, if need be, further refined. There are a great many approaches to design, and we have only attempted to summarize some of the common approaches here. The reader wishing to develop further specialist knowledge in this field is referred to the many technical books on the topic, as well as the academic literature in the bibliography.

Recursive Filter Design Steps

1. Specify the response shape and determine the f_s.
2. Determine the appropriate filter type (Butterworth, Chebyshev, other).
3. Select the filter order and other parameters as appropriate (e.g., allowable ripple).
4. Design the low-pass prototype with normalized frequency response.
5. Scale to the required frequency and convert the filter type (if not low-pass).
6. Convert to a discrete-time transfer function using either the BLT method or the impulse invariant method.
7. Plot the sampled frequency response and iterate the above steps as necessary.
8. Convert $Y(z)$ into a difference equation; find $y(n)$ in terms of $x(n)$, $x(n-1)$, $y(n-1)$, and so on.
9. Code a loop to sample the input $x(n)$, calculate $y(n)$, output $y(n)$, and buffer the samples.

9.9 CHAPTER SUMMARY

The following are the key elements covered in this chapter:

- A review of *analog filter theory* including time and frequency response methods
- The role of *IIR* filters
- The theory behind two commonly employed types of analog filter: the *Butterworth* filter and the *Chebyshev* filter. The former is maximally flat in the passband; the latter has a sharper transition but has a nonflat passband response.
- The transformation and scaling of filters from the basic low-pass design to *high-pass* and *bandpass* responses
- How to *transform analog filters* into discrete-time or digital filters. Two approaches were considered: the *impulse invariant method*, which matches the impulse response in the time domain, and the *bilinear transform*, which is a direct substitution for the continuous-to-discrete approximation.

PROBLEMS

9.1. Using the approach outlined in Section 9.3.2, plot the impulse response of

$$\frac{1}{s^2 + \frac{1}{2}s + 1}$$

and compare to Figure 9.3.

9.2. Using the approach outlined in Section 9.3.3, plot the frequency response of

$$\frac{1}{s^2+\frac{1}{2}s+1}$$

and compare to Figure 9.4. Note that you will need to choose appropriate values for some of the parameters. Remember to plot decibels on the gain axis, and use a logarithmic scale for the frequency axis using `set(gca, 'xscale', 'log');`.

9.3. Consider the Butterworth filter defined in Section 9.4.1.

(a) Show that as $N \to \infty$, the filter approaches the ideal or brick wall response. Do this by finding the gain of the filter in Equation 9.13 as $\Omega \to 0$ and as $\Omega \to \infty$, and then by determining the gain at the cutoff, where $\Omega = \Omega_c$. From these results, sketch the filter gain characteristic.

(b) Apply the normalized low-pass to high-pass conversion as explained in Section 9.7. Show that the filter approaches the ideal as $N \to \infty$; that is, find the limit as $\Omega \to 0$ and as $\Omega \to \infty$, then determine the gain at the cutoff frequency where $\Omega = \Omega_c$; and finally, sketch the gain characteristic.

(c) Apply the normalized low-pass to bandpass transformation and find the gain for low frequencies, high frequencies, at the cutoff frequencies ($\Omega = \Omega_l$ and $\Omega = \Omega_u$), and within the passband at $\Omega = \Omega_c$. Sketch the filter gain characteristic.

(d) Repeat the previous question for the bandstop transformation.

9.4. Chebyshev filters (Section 9.4.2) are based on the theory and mathematics of the Chebyshev polynomial. The Nth order Chebyshev polynomial is denoted $T_N(x)$. $T_N(x)$ may be defined in terms of cosine functions as $T_N(x) = \cos(N\cos^{-1}x)$.

(a) Show that $T_0(x) = 1$ and $T_1(x) = x$.

(b) Show that the recursive definition $T_N(x) = 2xT_{N-1}(x) - T_{N-2}(x)$ for $N > 0$ holds.

(c) Explain how the recursive definition of $T_N(x)$ applies to the graphs of the Chebyshev polynomials as shown in Figure 9.8.

9.5. Plot the continuous-time response for a fourth-order Butterworth low-pass filter with cutoff frequency 25 kHz. Repeat this for a Chebyshev filter. How do the responses compare?

9.6. For the continuous-time system $G(s) = 1/(s^2 + 2s + 1)$,

(a) plot the frequency response over a reasonable range of frequencies and explain your choice of frequencies by referring to the transfer function;

(b) scale to a frequency of 10 Hz and again plot the response;

(c) use the direct bilinear transform method to determine the corresponding z transform; use a sample frequency of 100 Hz; and

(d) use the bilinear transform method with prewarping, and determine the corresponding z transform; use the same sampling frequency and compare to the previous result and to the analog filter response.

9.7. Complete the first-order example using both the original and corrected impulse invariant methods. Compare the resulting difference equation for each filter. Using the pole–zero approach, sketch their respective frequency responses.

9.8. Complete the second-order example using both the original and corrected impulse invariant methods.

9.9. A discrete filter with a low-pass response, cutoff 100 Hz, and a sample rate of 10 kHz is required.

- **(a)** Design the continuous-time Butterworth filter for these parameters using a normalized 1 rad per sample cutoff. Calculate the transfer function using algebra, and check your answer using the MATLAB code outlined in the text.
- **(b)** Scale the Butterworth filter to the desired cutoff and plot the frequency response.
- **(c)** Convert the normalized Butterworth filter to a discrete-time one using the bilinear transform.
- **(d)** Plot the frequency response and compare to the design specification.

BIBLIOGRAPHY

SELECTED WEB REFERENCES

Frigo, M., and S. G. Johnson. n.d.. FFTW—fastest Fourier transform in the West. http://www.fftw.org.

Goldberger, A. L., L. A. N. Amaral, L. Glass, J. M. Hausdorff, P. C. Ivanov, R. G. Mark, J. E. Mietus, G. B. Moody, C.-K. Peng, and H. E. Stanley. 2000. PhysioBank, PhysioToolkit, and PhysioNet: components of a new research resource for complex physiologic signals. *Circulation* 101(23):e215–e220. http://circ.ahajournals.org/cgi/content/full/101/23/e215.

IEEE 754 Group. 2004. IEEE 754: standard for binary floating-point arithmetic. http://grouper.ieee.org/groups/754.

IEEE/NSF. n.d.. Signal processing information base. http://spib.rice.edu/spib.html.

Kak, A. C., and M. Slaney. 1988. *Principles of Computerized Tomographic Imaging*. New York: IEEE Press. http://www.slaney.org/pct/index.html.

SIDC-team. 1749–2010. Monthly report on the international sunspot number. http://www.sidc.be/sunspot-data/.

SELECTED ACADEMIC PAPERS

Ahmed, N., T. Natarajan, and K. R. Rao. 1974. Discrete cosine transform. *IEEE Transactions on Computers* C23(1):90–93.

Black, H. S. 1934. Stabilized feed-back amplifiers. *American Institute of Electrical Engineers* 53:114–120.

Butterworth, S. 1930. On the theory of filter amplifiers. *Experimental Wireless and the Wireless Engineer* 7:536–541.

Cooley, J. W., P. A. Lewis, and P. D. Welch. 1967a. Application of the fast Fourier transform to computation of Fourier integrals, Fourier series, and convolution integrals. *IEEE Transactions on Audio and Electroacoustics* AU-15(2):79.

Cooley, J. W., P. A. Lewis, and P. D. Welch. 1967b. Historical notes on the fast Fourier transform. *IEEE Transactions on Audio and Electroacoustics* AU-15(2):76.

Fernandes, C. W., M. D. Bellar, and M. M. Werneck. 2010. Cross-correlation-based optical flowmeter. *IEEE Transactions on Instrumentation and Measurement* 59(4):840–846.

Frigo, M., and S. G. Johnson. 1998. FFTW: an adaptive software architecture for the FFT. In *Proceedings of the International Conference on Acoustics, Speech, and Signal Processing*. 1381–1384.

Digital Signal Processing Using MATLAB for Students and Researchers, First Edition. John W. Leis.

Frigo, M., and S. G. Johnson. 2005. The design and implementation of FFTW3. *Proceedings of the IEEE* 93(2):216–231.

Goldberg, D. 1991. What every computer scientist should know about floating-point arithmetic. *ACM Computing Surveys* 23(1).

Harris, F. J. 1978. On the use of Windows for harmonic analysis with the discrete Fourier transform. *Proceedings of the IEEE* 66(1):51–83.

Hayes, A. M., and G. Musgrave. 1973. Correlator design for flow measurement. *The Radio and Electronic Engineer* 43(6):363–368.

Heideman, M. T., D. H. Johnson, and C. S. Burrus. 1984. Gauss and the history of the fast Fourier transform. *IEEE ASSP Magazine* 1(4):14–19.

Helms, H. D. 1967. Fast Fourier transform method of computing difference equations and simulating filters. *IEEE Transactions on Audio and Electroacoustics* AU-15(2):85.

Jackson, L. B. 2000. A correction to impulse invariance. *IEEE Signal Processing Letters* 7(10):273–275.

Lüke, H. D. 1999. The origins of the sampling theorem. *IEEE Communications Magazine* 37(4):106–108.

McClellan, J. H., and T. W. Parks. 1973. A unified approach to the design of optimum FIR linear-phase digital filters. *IEEE Transactions on Circuit Theory* CT-20(6):697–701.

Mecklenbraüker, W. F. G. 2000. Remarks on and correction to the impulse invariant method for the design of IIR digital filters. *Signal Processing* 80:1687–1690.

Meijering, E. 2002. A chronology of interpolation: from ancient astronomy to modern signal and image processing. *Proceedings of the IEEE* 90(3):319–342.

Narashima, M. J., and A. M. Peterson. 1978. On the computation of the discrete cosine transform. *IEEE Signal Processing Magazine* COM-26(6):934–936.

Nyquist, H. 1928. Certain topics in telegraph transmission theory. *American Institute of Electrical Engineers* 47:617–644. Reprinted Proc IEE, 90(2), Feb 2002.

Nyquist, H. 2002. Certain topics in telegraph transmission theory. *Proceedings of the IEEE* 90(2):280–305. Reprint of classic paper.

Pitas, I., and A. N. Venetsanopoulos. 1992. Order statistics in digital image processing. *Proceedings of the IEEE* 80:1893–1921.

Radon, J. 1986. On the determination of functions from their integral values along certain manifolds. *IEEE Transactions on Medical Imaging* MI-5(4):170–176. Translated from the German text of 1917 by P. C. Parks.

Shannon, C. E. 1948. A mathematical theory of communication. *Bell System Technical Journal* 27:379–423, 623–656. http://cm.bell-labs.com/cm/ms/what/shannonday/paper.html.

Shepp, L. A., and B. F. Logan. 1974. Reconstructing interior head tissue from X-ray transmissions. *IEEE Transactions on Nuclear Science* 21(1):228–236.

Stockham, T. G. 1966. High-speed convolution and correlation. *AFIPS Joint Computer Conferences* AU-15:229–233.

Stylianou, Y. 2000. A simple and fast way of generating a harmonic signal. *IEEE Signal Processing Letters* 7(5):111–113.

Tustin, A. 1947. A method of analysing the behaviour of linear systems in terms of time series. *Journal of the Institution of Electrical Engineers* 94:130–142.

Widrow, B. 2005. Thinking about thinking: the discovery of the LMS algorithm. *IEEE Signal Processing Magazine* 22(1):100–106.

Wilson, D. R., D. R. Corrall, and R. F. Mathias. 1973. The design and application of digital filters. *IEEE Transactions on Industrial Electronics and Control Instrumentation* IECI-20(2):68–74.

SELECTED TEXTBOOKS

Acharya, T., and P.-S. Tsai. 2005. *JPEG2000 Standard for Image Compression*. Hoboken, NJ: Wiley.

Elali, T. S., ed. 2004. *Discrete Systems and Digital Signal Processing with MATLAB*. Boca Raton, FL: CRC.

Hamming, R. W. 1983. *Digital Filters*. Englewood Cliffs, NJ: Prentice-Hall.

Harris, R. W., and T. J. Ledwidge. 1974. *Introduction to Noise Analysis*. London: Pion.

Haykin, S. n.d.. *Adaptive Filter Theory*. Englewood Cliffs, NJ: Prentice-Hall.

Ifeachor, E. C., and B. W. Jervis. 1993. *Digital Signal Processing: A Practical Approach*. Boston: Addison-Wesley.

Ingle, V. K., and J. G. Proakis. 2003. *Digital Signal Processing Using MATLAB*. Stamford, CT: Thomson Brooks Cole.

Jain, A. K. 1989. *Fundamentals of Digital Image Processing*. Englewood Cliffs, NJ: Prentice-Hall.

Jayant, N. S., and P. Noll. 1984. *Digital Coding of Waveforms*. Englewood Cliffs, NJ: Prentice-Hall.

Kreyszig, E. n.d.. *Advanced Engineering Mathematics*. Boston: Addison-Wesley.

Kronenburger, J., and J. Sebeson. 2008. *Analog and Digital Signal Processing: an Integrated Computational Approach with MATLAB*. Clifton Park, NY: Thomson Delmar Learning.

Kumar, B. P. 2005. *Digital Signal Processing Laboratory*. Oxford: Taylor and Francis.

Lathi, L. P. 1998. *Signal Processing and Linear Systems*. Carmichael: Berkeley Cambridge Press.

Leis, J. W. 2002. *Digital Signal Processing: A MATLAB-Based Tutorial Approach*. Baldock, UK: Research Studies Press.

Lockhart, G. B., and B. M. G. Cheetham. 1989. *Basic Digital Signal Processing*. London: Butterworths.

Lyons, R. G. 1997. *Understanding Digital Signal Processing*. Boston: Addison Wesley.

Mitra, S. K. 2001. *Digital Signal Processing: a Computer-Based Approach*. New York: McGraw-Hill.

Oppenheim, A. V., and R. W. Schafer. 1989. *Discrete-Time Signal Processing*. Englewood Cliffs, NJ: Prentice-Hall.

Orfanidis, S. J. 1996. *Introduction to Signal Processing*. Englewood Cliffs, NJ: Prentice-Hall.

Papoulis, A. 1980. *Circuits and Systems: A Modern Approach*. New York: Holt-Saunders.

Pennebaker, W. B., and J. L. Mitchell. 1992. *JPEG: Still Image Data Compression Standard*. New York: Van Nostrand Reinhold.

Porat, B. 1997. *A Course in Digital Signal Processing*. New York: John Wiley & Sons.

Proakis, J. G., and D. G. Manolakis. n.d.. *Digital Signal Processing: Principles, Algorithms, and Applications*. Englewood Cliffs, NJ: Prentice-Hall.

Rosenfeld, A., and A. C. Kak. 1982. *Digital Picture Processing*. Burlington, VT: Academic Press/Elsevier.

Terrell, T. J., and L.-K. Shark. 1996. *Digital Signal Processing: A Student Guide*. New York: Macmillan/McGraw-Hill.

Widrow, B., and S. D. Stearns. 1995. *Adaptive Signal Processing*. Englewood Cliffs, NJ: Prentice-Hall.

INDEX

Digital Signal Processing Using MATLAB for Students and Researchers, First Edition. John W. Leis.

9 780470 880913

Printed and bound by CPI Group (UK) Ltd, Croydon, CR0 4YY

09/06/2025

14685904-0002